REPRESENTATIONS OF DISCRETE FUNCTIONS

REPRESENTATIONS OF DISCRETE FUNCTIONS

EDITED BY

Tsutomu SASAO
Kyushu Institute of Technology
Iizuka, Japan

Masahiro FUJITA
Fujitsu Laboratories of America Inc.
Santa Clara, California, USA

KLUWER ACADEMIC PUBLISHERS
Boston/London/Dordrecht

Distributors for North America:
Kluwer Academic Publishers
101 Philip Drive
Assinippi Park
Norwell, Massachusetts 02061 USA

Distributors for all other countries:
Kluwer Academic Publishers Group
Distribution Centre
Post Office Box 322
3300 AH Dordrecht, THE NETHERLANDS

Library of Congress Cataloging-in-Publication Data

A C.I.P. Catalogue record for this book is available
from the Library of Congress.

Printed on acid-free paper.

Printed in the United States of America

CONTENTS

CONTRIBUTORS

Bernd BECKER
Institute of Computer Science
Albert-Ludwigs-University
Freiburg, Germany

Beate BOLLIG
Department of Computer Science
University of Dortmund
Dortmund, Germany

Jörg BULLMANN
Wilhelm-Schickard-Institut für Informatik
Eberhard-Karls-Universität Tübingen
Tübingen, Germany

Edmund M. CLARKE
School of Computer Science
Carnegie Mellon University
Pittsburgh, PA 15213, USA

Rolf DRECHSLER
Institute of Computer Science
Albert-Ludwigs-University
Freiburg, Germany

Masahiro FUJITA
Fujitsu Labs of America Inc.
Santa Clara, CA, U.S.A.

Thilo HARICH
Computer Science Departement
J.W. Goethe-University
Frankfurt, Germany

Fumitaka IZUHARA
Department of Computer Science and Electronics
Kyushu Institute of Technology
Iizuka, Japan

Jawahar JAIN
Fujitsu Laboratories of America
Santa Clara, CA, U.S.A.

Udo KEBSCHULL
Forschungszentrum Informatik
Karlsruhe, Germany

Yung-Te LAI
Hitachi Labs
179 East Tasman Dr.
San Jose, CA, U.S.A.

Martin LÖBBING
Department of Computer Science
University of Dortmund
Dortmund, Germany

Shin-ichi MINATO
NTT LSI Laboratories
Kanagawa, Japan

Claudio MORAGA
Department of Computer Science
Dortmund University
Germany

Massoud PEDRAM
Department of Electrical Engineering - Systems
University of Southern California
Los Angeles, CA, U.S.A.

Tsutomu SASAO
Department of Computer Science and Electronics
Kyushu Institute of Technology
Iizuka, Japan

Martin SAUERHOFF
Department of Computer Science
University of Dortmund
Dortmund, Germany

Radomir S. STANKOVIĆ
Department of Computer Science
Faculty of Electronics
18 000 Niš, Yugoslavia

Sarma B. K. VRUDHULA
Department of Electrical and Computer Engineering
University of Arizona
Tuscon, Arizona, U.S.A.

Ingo WEGENER
Department of Computer Science
University of Dortmund
Dortmund, Germany

Ralph WERCHNER
Mathematics Departement
J.W. Goethe-University
Frankfurt, Germany

Xudong ZHAO
School of Computer Science
Carnegie Mellon University
Pittsburgh, PA 15213, U.S.A.

PREFACE

In August 27-29, 1995, the IFIP WG 10.5 Workshop on Applications of the Reed-Muller Expansions in Circuit Design (Reed-Muller '95) was held in Makuhari, Chiba, Japan. Many papers focused on the representation of discrete functions. They were so instructive that we collected the most outstanding papers in a book form.

This book consists of 13 chapters. To enhance their readability, all chapters here were reviewed again and were revised. In the original papers in the proceedings, authors used their own terminology. Thus, the same notion was represented by different terminology. In the book version, however, we have insisted on a unified terminology. We include three chapters showing standard terminology. Additional examples and illustrations designed to increase the readers' understandings were also incorporated in the book version. To make the book more comprehensive, we also included two relevant chapters, which were presented at the International Workshops on Logic Synthesis. Exercises prepared by editors will make this book appropriate for a textbook of a course.

This book deals with the following topics: Binary decision diagrams (BDDs), multi-terminal binary decision diagrams (MTBDDs), edge-valued binary decision diagrams (EVBDDs), functional decision diagrams (FDDs), Kronecker decision diagrams (KDDs), binary moment diagrams (BMDs), spectral transform decision diagrams (STDDs), ternary decision diagrams (TDDs), spectral transformation of logic functions, other transformations of logic functions, EXOR-based two-level expressions, FPRM minimization with TDDs and MTBDDs, complexity theories on FDDs, multi-level logic synthesis, and complexity of three-level logic networks.

In this book, a **discrete function** denotes a mapping $S \to L$, where S and L are finite non-empty sets. A **switching function** denotes a discrete function, where $S = \{0,1\}^n$ and $L = \{0,1\}$.

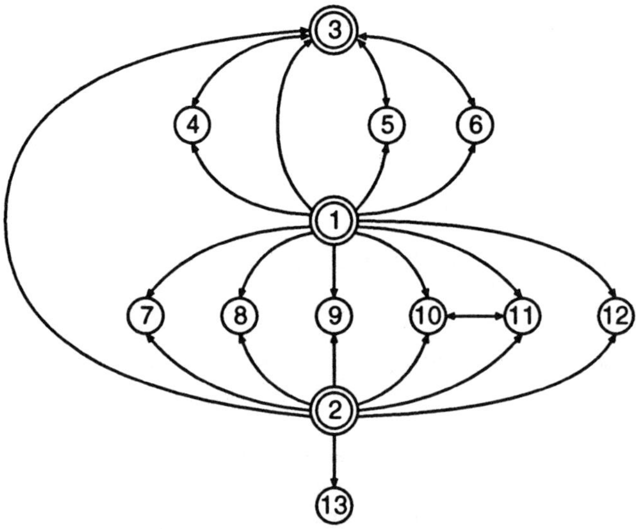

Figure 1 Relation of chapters.

Overview of the Book

This book consists of 13 chapters. Fig. 1 shows the relation of these chapters. The first three chapters are introductions. They contain many illustrations to clarify concepts presented in the text. It is recommended to read these three chapter first. After this, the other chapters will be easy to read.

Chapter 1 introduces BDDs, and their variants: it surveys the techniques for BDDs, and shows the relations between different types of decision diagrams.

Chapter 2 introduces EXOR-based representations: it compares complexities of various representations.

Chapter 3 shows relation of various decision diagrams: One decision diagram represents a function f, and at the same time, the spectrum S_f. For example, a BDD represents f, and at the same time, the Reed-Muller spectrum of f. It also shows methods to represent arithmetic functions by using decision diagrams. This chapter is related to Chapters 4, 5, and 6.

Chapter 4 shows applications of MTBDDs. It shows an MTBDD-based computation of spectral transformations, which is a revolutionary idea. With this technique, we can compute the spectra of a logic function with more than 200 variables. This chapter also proposes HDDs (hybrid decision diagrams).

Chapter 5 introduces EVBDDs, which are useful to represent both Boolean and arithmetic operations. This chapter shows EVBDD algorithms for binary operations. An extension of EVBDDs is also described.

Chapter 6 presents arithmetic transformations, which are useful for verification of large networks.

Chapter 7 considers OKFDDs (KDDs). OKFDDs are generalization of BDDs and FDDs, and can represent logic functions more compactly than BDDs and FDDs.

Chapter 8 shows a method for exact minimization of FPRMs. With this method, we could obtain an exact minimum FPRM with 94 inputs. MTBDDs, and MTTDDs are the key data structure in this method.

Chapter 9 shows a multi-level logic synthesis method. It considers both AND-OR and EXOR-based designs.

The next two chapters deal with complexity issues of FDDs.

Chapter 10 considers the satisfiability problems. It shows that the counting the satisfying assignments in an OFDD cannot be solved in polynomial time unless $NP = P$.

Chapter 11 considers the complexity of changing the order of input variables. It also shows that the replacement of variables with constants can cause an exponential blow up of the FDD size.

Chapter 12 presents ternary decision diagrams (TDDs). A TDD represents a logic expression rather than a logical function. A BDD is a special case of a TDD, where only two edges representing f_0 and f_1 are used. There is a $2n$ variable function whose size of the TDD is $2n^2 + 2$, while the size of the BDD is 2^{n+1}. TDD can be a mathematical model for analyzing the complexity of Boolean functions.

Chapter 13 deals the complexity of OR-AND-OR three-level logic networks. Any logic functions with $n = 2r$ variables are represented with at most 2^{r+1} gates.

Unlike the other chapters, preliminary versions of last two chapters were presented at the International Workshop on Logic Synthesis in May 1993, and 1989, respectively.

Acknowledgments

We would like to express our appreciation to all the contributors for their participation and cooperation in this project. They patiently revised the manuscript many times to improve the readability. Many people spent their precious time for review under extremely tight time schedule: Their names appear at the end of this book. We also thank all the attendee for the IFIP WG 10.5 Workshop on Applications of the Reed-Muller Expansions in Circuit Design. Conversation at the workshop inspired improvements in the enclosed chapters. We acknowledge with gratitude the support from the Fujitsu Systems Laboratory, The Telecommunication Advancement Foundation (TAF), and the Chiba Convention Bureau. Prof. Kazuhiko Iwasaki did a very good job as the local arrangement chairman.

Students of Sasao's Laboratory made important suggestions to improve the readability of the chapters during seminars using the preliminary version of the book.

Finally, we would like to thank Mr. Debatosh Debnath and Mr. Munehiro Matsuura for editing the LaTeX files to produce the camera-ready copy. Without their help, the book would never have materialized.

Tsutomu Sasao, Iizuka, Japan.

Masahiro Fujita, Santa Clara, California, U.S.A.

1

GRAPH-BASED REPRESENTATIONS OF DISCRETE FUNCTIONS

Shin-ichi MINATO

NTT LSI Laboratories
Kanagawa, Japan

Abstract — BDDs are now commonly used for representing Boolean functions because of their efficiency in terms of time and space. There are many cases in which conventional algorithms can be significantly improved by using BDDs. Recently, several variants of BDDs have been developed to represent other kinds of discrete functions, such as multi-valued functions, cube sets, or arithmetic formulas. These techniques are useful not only for VLSI CAD but also for various areas in Computer Science.
In this chapter, we survey the techniques of BDD and its variants. We explain the basic method of BDD manipulation, and show the relationships between the different types of BDDs.

1.1 INTRODUCTION

Manipulation of discrete functions is fundamental to Computer Science. Many problems in digital system design and testing can be expressed as a sequence of operations on discrete functions. With the recent advance in VLSI technology, the problems have grown well beyond the scope of manual design, and VLSI CAD systems have become widely used. The performance of these systems greatly depends on the efficiency of the data structure for representing discrete functions.

Recently, **BDDs (Binary Decision Diagrams)** have attracted much attention because they enable us to manipulate Boolean functions efficiently in terms of time and space. There are many cases in which conventional algorithms can be significantly improved by using BDDs.

As our understanding of BDDs has deepened, the range of applications has broadened. Besides Boolean functions, we sometimes deal with other kinds of discrete functions, such as multi-valued functions, cube sets, or arithmetic formulas. Recently, several variants of BDDs have been devised to represent such data models. These techniques are useful not only for VLSI CAD but also for various areas in Computer Science.

In this chapter, we survey the techniques of BDD and its variants. We explain the basic method of BDD manipulation, and show the relationships between the different types of BDD. For detailed information on each method, refer to the original papers.

The rest of this chapter is organized as follows. In Section 1.2, we describe the basic concept of BDDs and the implementation techniques of the BDD manipulator. We then discuss the method of using **attribute edges** and variable ordering to reduce the size of BDDs. In Section 1.3, we present several variants of BDDs to represent multi-valued functions, that include MTBDDs, BDD vectors, EVBDDs and BMDs. Section 1.4 discusses the method of representing cube sets or combinatorial expressions. ZBDDs, FDDs, and KFDDs are described here. Section 1.5 is summary of this chapter.

1.2 BDDS

This section introduces the basic concept of BDDs, which is now commonly used for Boolean function representation. We discuss the techniques to manipulate BDDs on computers.

1.2.1 Data Structure

BDDs are graph representations of Boolean functions, as shown in Fig. 1.2.1(a). The basic concept was introduced by Akers [1], and an efficient manipulation method was developed by Bryant [2].

A *BDD* is a directed acyclic graph with two terminal nodes, which we call the **0-terminal** and **1-terminal** node. Every non-terminal node has an index to identify an input variable of the Boolean function, and has two outgoing edges, called the **0-edge** and **1-edge**.

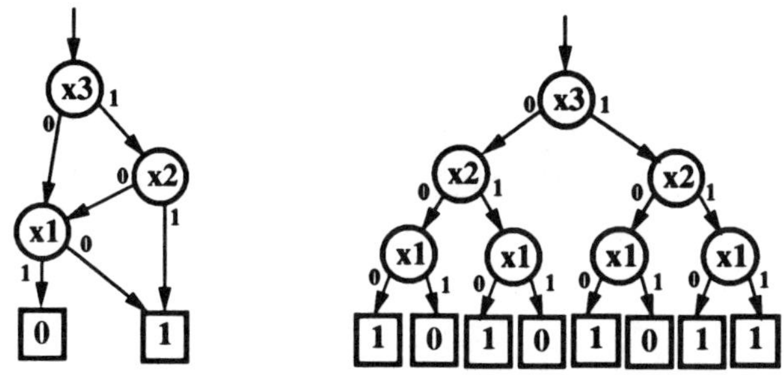

(a) A BDD for $(x3 \cdot x2) \vee \overline{x1}$.

(b) A binary decision tree for $(x3 \cdot x2) \vee \overline{x1}$.

Figure 1.2.1 A BDD and a binary decision tree.

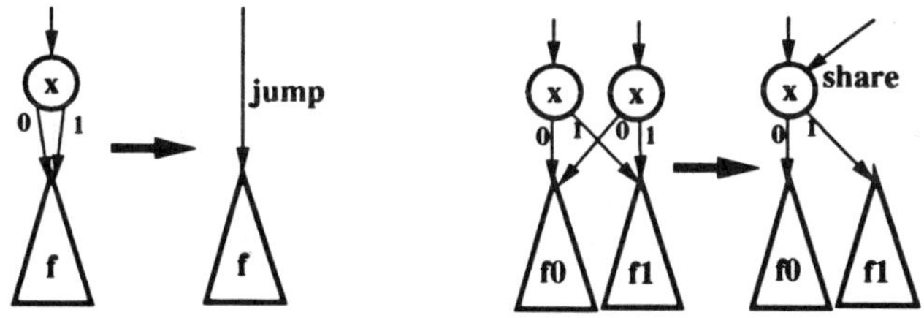

(a) Node elimination.

(b) Node sharing.

Figure 1.2.2 Reduction rules of BDDs.

An **Ordered BDD (OBDD)** is a BDD such that the input variables appear in a fixed order on all the paths of the graph, and that no variable appears more than once in a path. In this chapter, we use natural numbers $1, 2, \ldots$ for the indices of the input variables, and every non-terminal node has a greater index than its descendant nodes.

A compact OBDD is derived by reducing a binary tree graph, as shown in Fig. 1.2.1(b). In the binary tree, 0- and 1-terminal nodes represent logic values (0/1), and each node represents the **Shannon's expansion** of the Boolean function:

$$f = (\overline{x_i} \cdot f_0) \vee (x_i \cdot f_1),$$

where i is the index of the node. f_0 and f_1 are the functions of the nodes pointed to by 0- and 1-edges, respectively.

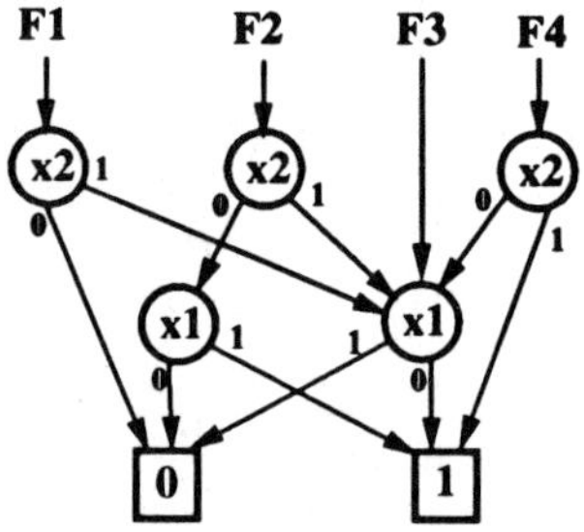

$(F1 = x2 \cdot \overline{x1},\ F2 = x2 \oplus x1,\ F3 = \overline{x1},\ F4 = x2 \vee \overline{x1}.)$

Figure 1.2.3 A shared BDD.

The following reduction rules give a *Reduced Ordered BDD (ROBDD).*

1. Eliminate all the redundant nodes whose two edges point to the same node. (Fig. 1.2.2(a))
2. Share all the equivalent sub-graphs.(Fig. 1.2.2(b))

ROBDDs give canonical forms for Boolean functions when the variable order is fixed. This property is very important to practical applications, as we can easily check the equivalence of two Boolean function by only checking isomorphism of their ROBDDs. Most works relating to BDDs use ROBDD techniques. In this chapter, we refer to ROBDDs as BDDs for the sake of simplification.

It is known that a BDD for an n-input function requires an exponential amount of memory in the worst case [3]. However, the size of the BDD varies with the kind of function, unlike truth tables which always require 2^n bit of memory. There is a class of Boolean functions that can be represented by BDDs with polynomial size, and many practical functions fall into this class [4]. This is an attractive feature of BDDs.

A set of BDDs representing multiple functions can be united into a graph which consists of BDDs sharing sub-graphs with each other, as shown in Fig. 1.2.3. This idea saves time and space to duplicate BDDs. When the isomorphic sub-graphs are completely shared, two equivalent nodes never coexist. We call such graphs **Shared BDDs (SBDDs)** [5], or **multi-rooted BDDs**. In the shared

(address)	(index)	(0-edge)	(1-edge)	
N_0	-	-	-	$\longleftarrow 0$
N_1	-	-	-	$\longleftarrow 1$
N_2	x_1	N_0	N_1	
N_3	x_1	N_1	N_0	$\longleftarrow F_3(=\overline{x_1})$
N_4	x_2	N_0	N_3	$\longleftarrow F_1(=x_2 \cdot \overline{x_1})$
N_5	x_2	N_2	N_3	$\longleftarrow F_2(=x_2 \oplus x_1)$
N_6	x_2	N_3	N_1	$\longleftarrow F_4(=x_2 \vee \overline{x_1})$

Figure 1.2.4 BDD representation using a table.

BDD environment, equivalence checking can be performed immediately by just looking at the root nodes.

Bryant's original paper [2] presents algorithms for manipulating non-shared (separated) BDDs. However, shared BDDs are now widely used and those algorithms are more concise than the original ones. In the remainder of this chapter, we assume the shared BDD environment.

In a typical implementation of a BDD manipulator, all the nodes are stored in a single table in the main memory of the computer. Fig. 1.2.4 shows an example of the table representing BDDs shown in Fig. 1.2.3. Each node has three basic attributes; an index of the input variable and two pointers of the 0- and 1-edges. Some additional pointers and counters are attached to the node data for maintaining the table. 0- and 1-terminal nodes are first allocated in the table as special nodes. The other non-terminal nodes are gradually generated as the results of logic operations.

Before creating a new node, we check the reduction rules shown in Fig. 1.2.2. If the 0- and 1-edges have the same destination, or if an equivalent node already exists, then we do not create a new node but simply copy the pointer to the existing node. To find an equivalent node, we check a table which indices all the nodes. The hash table technique is very effective to accelerate this checking. In this way, the uniqueness of the nodes is maintained, and every Boolean function data can be identified by a 1-word address of the root node.

When generating BDDs for Boolean expressions, many intermediate BDDs are temporarily generated. It is important for memory efficiency to delete such unneeded BDDs. In order to determine the necessity of the nodes, a **reference**

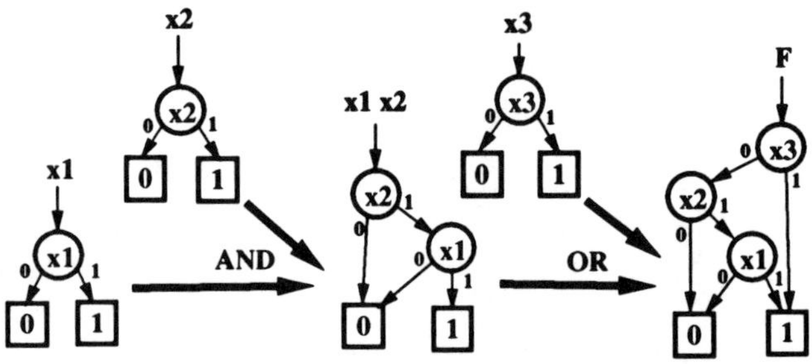

Figure 1.2.5 Generation of BDDs for $(x1 \cdot x2) \vee x3$.

counter is attached to each node, which shows the number of incoming edges to the node.

In a typical implementation, the BDD manipulator consumes 20 to 30 bytes of memory for each node. Today there are workstations with more than 100M byte of memory, and those facilitate us to generate BDDs containing as many as millions of nodes. However, the BDDs still grow large beyond the memory capacity in some practical applications.

1.2.2 Logic Operations

In many cases, BDDs are generated as the results of logic operations. Fig. 1.2.5 shows an example for $(x_1 \cdot x_2) \vee x_3$. First, trivial BDDs for x_1, x_2, x_3 are created. Then applying the AND operation between x_1 and x_2, the BDD for $(x_1 \cdot x_2)$ is generated. The final BDD for the entire expression is obtained as the result of the OR operation between $(x_1 \cdot x_2)$ and x_3.

The basic logic operations are summarized as follows:

- Creating a BDD for a single variable function x_i.
- Generating $\overline{f}$ for a given BDD f.
- Generating a BDD as the result of a binary operation $(f \circ g)$, that includes $(f \cdot g), (f \vee g)$ and $(f \oplus g)$.
- Generating a BDD for $f_{(x_i=0)}$ or $f_{(x_i=1)}$.

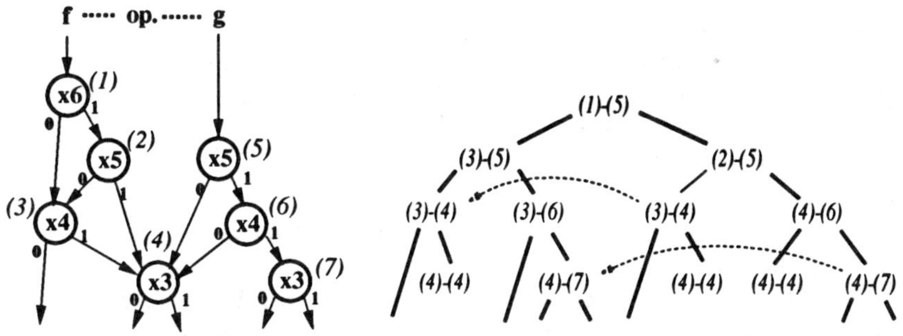

(a) An example. (b) Structure of procedure calls.

Figure 1.2.6 Procedure of binary operation.

- Equivalence checking of two BDDs.
- Finding the input assignments to satisfy $f = 1$.

(Binary operation)
The binary operation $(f \circ g)$ is the most important technique of BDD manipulation. It is computed based on the following expansion:

$$f \circ g = \overline{v} \cdot (f_{(v=0)} \circ g_{(v=0)}) \vee v \cdot (f_{(v=1)} \circ g_{(v=1)}),$$

where v is the highest ordered variable in f and g. It creates a new node with the variable v, having two sub-graphs generated by sub-operations $(f_{(v=0)} \circ g_{(v=0)})$ and $(f_{(v=1)} \circ g_{(v=1)})$. Repeating this expansion recursively for all the input variables, eventually trivial operations appear, and the result is obtained. (For instance, $f \cdot 1 = f$, $f \oplus f = 0$, etc.) As mentioned in previous section, we check the reduction rules before creating a new node.

Fig. 1.2.6(a) shows an example. When we perform the operation between the nodes (1) and (5), the procedure is broken down into the binary tree, as shown in Fig. 1.2.6(b). The number of sub-operations become exponential; however, they are sometimes redundant, for instance, (3)-(4) and (4)-(7) appears more than once. We can avoid the redundant executions for equivalent sub-operations by using a hash-based cache which record the results of recent sub-operations. In this technique, we can execute the operation in a time almost proportional to the size of the BDDs.

(Negation)
By using binary operation $(f \oplus 1)$, $\overline{f}$ can be computed in a linear time to the size of BDDs. However, it is improved to a constant time by using the **negative edge**, which is discussed in the following section. This technique is now commonly used in many implementation.

(Restriction)
The operation $f_{(v=0)}$ or $f_{(v=1)}$ is called **restriction**, or *cofactoring.* If v is the variable of the root node, the sub-graph pointed by 0- or 1-edge is immediately returned. Otherwise, we have to expand the BDDs until v appears in the root node, and then re-combine them into a BDD. The computation time is linear to the number of nodes which have an index greater than v.

(Satisfiable Assignment)
When a BDD for f is given, it is easy to find an input assignment to satisfy $f = 1$. A path from the root node to the 1-terminal node, which we call **1-path**, gives a solution to $f = 1$. BDDs have a good property in that every non-terminal node has at least one 1-path. By traversing the BDD from the root node, we can easily find a certain 1-path in a time proportional to the number of inputs, independent of the BDD size.

It is an interesting problem to find a 1-path having the minimum cost [6], which is defined as $\sum_{i=1}^{n}(C_i \times x_i)$, where C_i is a non-negative cost for input variable $x_i(\in \{0,1\})$. Many NP complete problems can be described in this format. A minimum cost 1-path is found by backtrack search of the BDD. It can be executed in a liner time to the BDD size, by using the cache technique to store and refer the minimum cost on each sub-graph. This method is applied for solving some kinds of combinatorial problems.

We can also count the number of solutions to $f = 1$, within a time linear to the BDD size. This technique is useful for computing signatures of Boolean functions.

1.2.3 Attributed Edges

Attributed edge is a technique to reduce computation time and memory requirement of BDDs by using edges with an attribute representing a certain operation, as shown in Fig. 1.2.7. Several kinds of attributed edges have been proposed [5]. Particularly, the **negative edge** is very effective and now widely implemented in BDD manipulators.

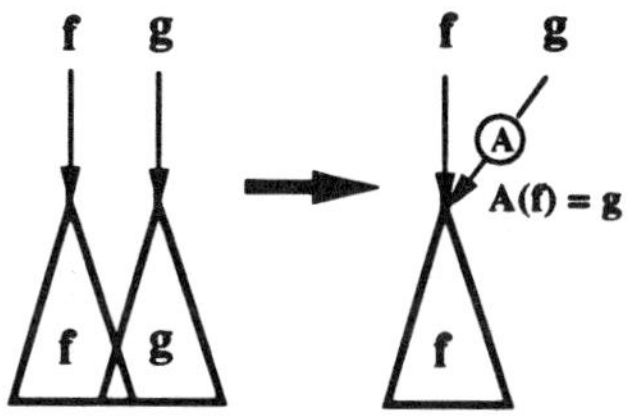

Figure 1.2.7 Attributed edges.

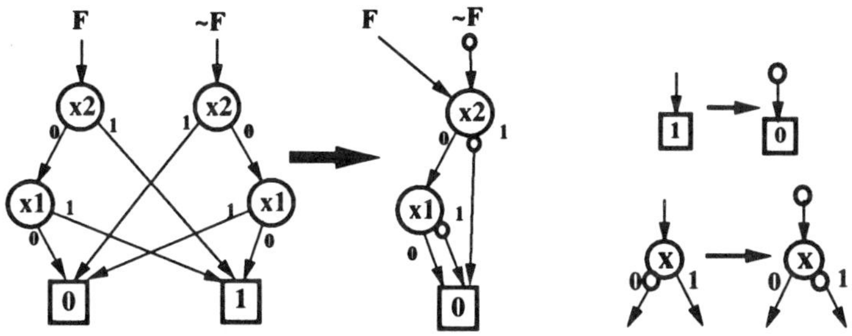

(a) Effect of negative edges (b) Rule of usage

Figure 1.2.8 Negative edges.

The negative edge is an attribute which indicates to complement the function of the sub-graph pointed to by the edge, as shown in Fig. 1.2.8(a). This idea is introduced as Akers's **inverter** [1] and Madre and Billon's **typed edge** [7]. The use of negative edges brings outstanding merits as follows.

- Reduces the BDD size by up to a half.
- Negation can be performed in constant time.
- Accelerate logic operations by applying the rules, such as $f \cdot \overline{f} = 0$, $f \vee \overline{f} = 1$, $f \oplus \overline{f} = 1$, etc.

Use of the negative edges may break the uniqueness of BDDs. To preserve the uniqueness, we have to adhere to two rules:

(1) Using a 0-terminal node only.
(2) Not using a negative edge as the 0-edges of any node. If necessary, the negative edges can be carried over as shown in Fig. 1.2.8(b).

Besides negative edges, some other kinds of attributed edges are proposed [5]. For instance, **input inverter** indicates to exchange the 0- and 1-edges at the pointing node. **Variable shifter** is another attributed edge to share subgraphs which are isomorphic except a difference of their index numbers of input variables.

Here we show a general method to define an attributed edge. Let S be the set of Boolean functions to be represented.

1. Divide S into the two subset S_0 and S_1
2. Define a mapping $\mathcal{F} : (S \to S)$, such that for any $f_1 \in S_1$ there is a unique $f_0 \in S_0$ to satisfy $f_1 = \mathcal{F}(f_0)$.

$\mathcal{F}$ represents the operation of the attributed edge. By applying $\mathcal{F}$, every subgraph for a function in S_1 is mapped into S_0, and it promotes BDD reduction. Notice that the reversal mapping from S_0 to S_1 breaks the uniqueness of BDD representation, so that we have to keep the rule on the location of using attributed edges.

This method can be extended to the attributed edges having more than two partitions, as follows.

1. Partition S into a number of subsets $S_0, S_1, \ldots S_n$.
2. For any $k \geq 1$, define a mapping $\mathcal{F}_k : (S \to S)$, such that for any $f_k \in S_k$ there is a unique $f_0 \in S_0$ to satisfy $f_k = \mathcal{F}_k(f_0)$.

1.2.4 Variable Ordering for BDDs

BDDs are a canonical representation of Boolean functions under a fixed order of input variables; however, the permutation of variable order may yields different BDDs for the same function. The effect of variable ordering depends on the nature of functions to be handled, and sometimes the size of BDDs greatly varies. Variable ordering is an important problem in the BDD technique.

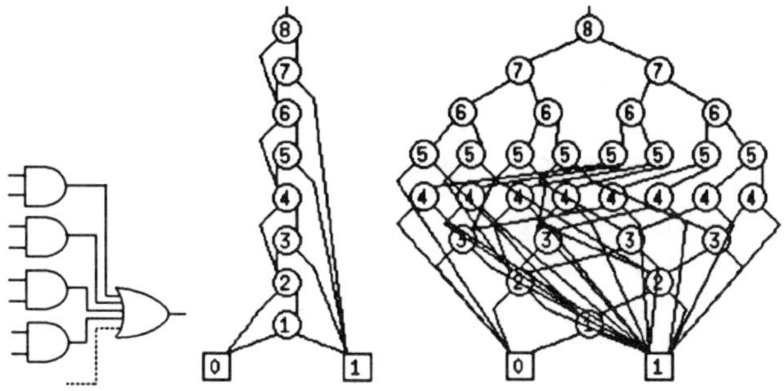

(a) Circuit. (b) Best order. (c) Worst order.

Figure 1.2.9 BDDs for 2-level AND-OR circuit.

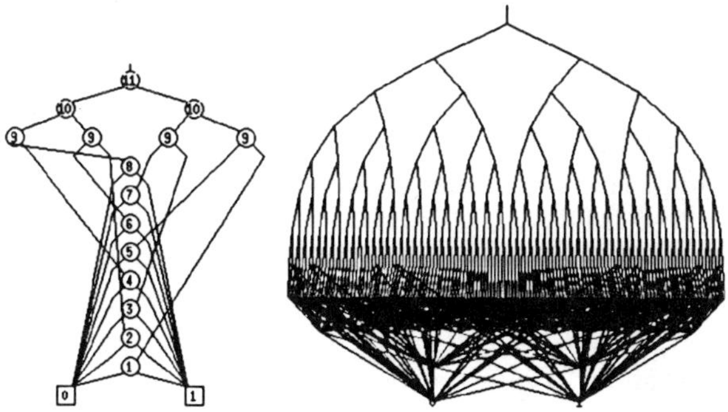

(a) Best order. (b) Worst order.

Figure 1.2.10 BDDs for 8-bit data selector.

Concerning the problem of finding the best order, it is known that the problem is in NP complete [8]. Existing algorithms are limited to run on the small size of BDDs with about 17 inputs [9]. It is difficult to find the best order for larger problems in practical time. However, a fairly good order is efficient for practical applications. There are some works on heuristic methods of variable ordering.

Empirically, the following properties are known on the variable ordering.

(Local computability)
The variable groups with close relation should be near. As shown in Fig. 1.2.9, the *AND-OR* 2-level circuit with $2n$ inputs takes $2n$ nodes in the best order $(x_1 \cdot x_2) \vee (x_3 \cdot x_4) \vee \ldots \vee (x_{2n-1} \cdot x_{2n})$, while it needs $(2^{n+1} - 2)$ nodes in the worst order $(x_1 \cdot x_{n+1}) \vee (x_2 \cdot x_{n+2}) \vee \ldots \vee (x_n \cdot x_{2n})$.

(Power to control)
The variables which greatly control the function should be at higher position. For example, the **8-bit data selector** can be represented by a linear size of BDD when the three control inputs are ordered high (Fig. 1.2.10(a)), while it becomes an exponential size using the reverse order (Fig. 1.2.10(b)).

Based on these empirical rules, Fujita et al. [10] and Malik et al. [11] presented methods which traverse given logic circuits in a depth first manner, and the input first arrived at gets highest position. Minato [5] devised another heuristic method based on weight assignment to given logic circuits. Butler et al. [12] uses testability measure for the heuristics, which reflect not only topological but logical information of the circuit. These methods find a (maybe) good order before generating BDDs. They are applied to the practical benchmark circuits and compute a good order in many cases.

On the other hand, there is another approach to improve the order for the given BDD by exchange of variables. A greedy local exchange method is shown by Fujita et al. [13]. Minato [14] presented another exchange-based method which measures the **width** of BDDs as a cost function. In many cases, these methods find a fairly good order using no additional information. A drawback of this approach is that it cannot start if we fail to make an initial BDD of reasonable size.

In Table 1.2.1, we show experimental results on the effect of variable ordering. The circuit "sel8" is an 8-bit data selector, and "enc8" is an 8-bit encoder. "add8" and "mult6" are an 8-bit adder and a 6-bit multiplier. The rests are chosen from benchmark circuits in MCNC'90. We compared four different orders: the original order, a random order, and two heuristic orders. The results show that the heuristic ordering methods are very effective except in a few cases which are insensitive to the order.

The techniques of variable ordering are intensively researched still now. One remarkable work is the **dynamic variable ordering**, presented by Rudell [15]. In this technique, the BDD package itself determines and maintains the variable order. Every time the BDDs grow to a certain size, the re-ordering process

Table 1.2.1 Effect of variable ordering.

Name	Circuit size			BDD size (with negative edges)			
	In.	Out.	Nets	Original	Random	Heur-1	Heur-2
sel8	12	2	29	16	88	23	19
enc8	9	4	31	28	29	28	27
add8	17	9	65	83	885	41	41
mult6	12	12	187	2183	2927	2471	2281
vg2	25	8	97	117	842	97	86
c432	36	7	203	3987	(>500k)	27302	1361
c499	41	32	275	115654	(>500k)	52369	40288
c880	60	26	464	(>500k)	(>500k)	23364	9114

Heur-1: heuristic order based on net list topology[5].
Heur-2: BDD reduction by exchanging variables[14].

is invoked automatically, just like garbage collection. This method is very effective in terms of the reduction of BDD size although it sometimes takes a long computation time.

1.3 REPRESENTATION OF MULTI-VALUED FUNCTIONS

The BDD techniques can be extended to represent multi-valued functions. This is useful for various areas in Computer Science. In this section, we show several methods to manipulate multi-valued functions.

1.3.1 MTBDDs

MTBDDs (Multi-Terminal BDDs) [16] are extended BDDs with multiple terminal nodes, each of which has an integer value. This method is a natural extension of BDDs to represent functions from binary-vector to integer: $\{0,1\}^n \rightarrow I$. Hachtel and Somenzi et al. reported several works on solving combinatorial problems [17]. They call this graph **ADDs (Algebraic Decision Diagrams)**. Jain et al. [18] presented a probabilistic verification method using this data structure, which they call **snDDs (Semi-Numeric Decision**

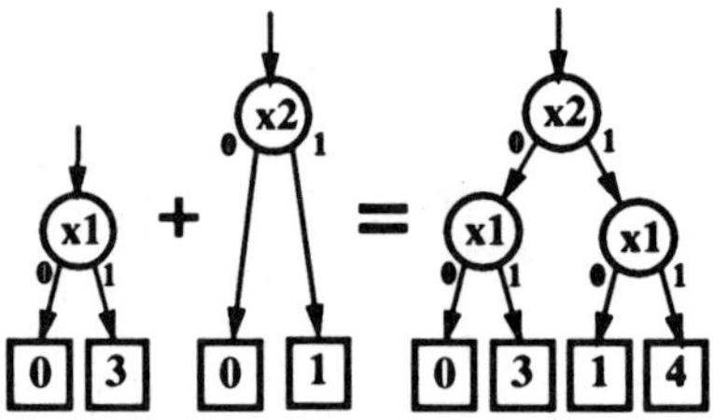

Figure 1.3.1 MTBDDs for $(3x_1 + x_2)$.

Diagrams). MTBDDs can also be utilized for manipulating certain matrix data structures.

As shown in Fig. 1.3.1, MTBDDs are handled by arithmetic operations, such as addition, subtraction, and multiplication. Like logic operations, they are implemented by recursive algorithms and executed in almost linear time to the graph size.

1.3.2 BDD Vectors

There is another way to represent multi-valued functions with a number of ordinary BDDs. By encoding the integer numbers into n-bit binary codes, a multi-valued function can be decomposed into a vector of BDDs which share the sub-graphs, as shown in Fig. 1.3.2. Minato presented an arithmetic Boolean expression manipulator [20] based on this representation.

Here we compare BDD vectors and MTBDDs. Assume the BDD shown in Fig. 1.3.3(a), which was obtained by combining the BDD vector shown in Fig. 1.3.2 with what we call **bit-selection variables.** If we change the variable order to move the bit-selection variables from higher to lower position, the BDD becomes as shown in Fig. 1.3.3(b). In this BDD, the sub-graphs with bit-selection variables correspond to the terminal nodes in the MTBDD. Namely, MTBDDs and BDD vectors can be transformed into each other by changing the position of the bit-selection variables.

B-to-I function

x1	x2	f ($2^2 2^1 2^0$)
0	0	0 (0 0 0)
0	1	1 (0 0 1)
1	0	3 (0 1 1)
1	1	4 (1 0 0)

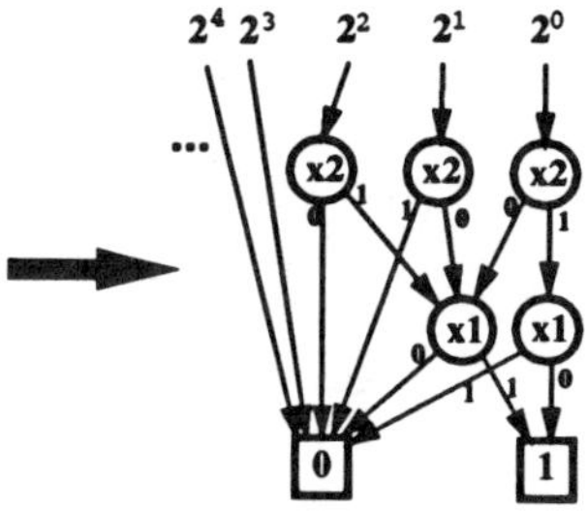

Figure 1.3.2 A BDD vector for $(3x_1 + x_2)$.

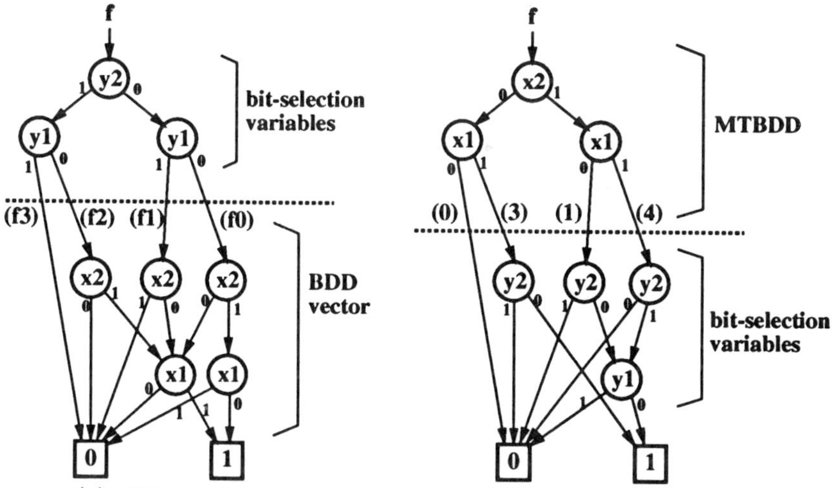

(a) BDD vector combined by bit-selection variables.

(b) After variable reordering.

Figure 1.3.3 Comparison of BDD vectors and MTBDDs.

1.3.3 EVBDDs

Lai et al. [21] proposed **EVBDDs (Edge-Valued BDDs)**, which are MTB-DDs with a kind of attributed edge. EVBDDs contain the attributed edges which indicate to add the value to the functions. As shown in Fig. 1.3.4, the

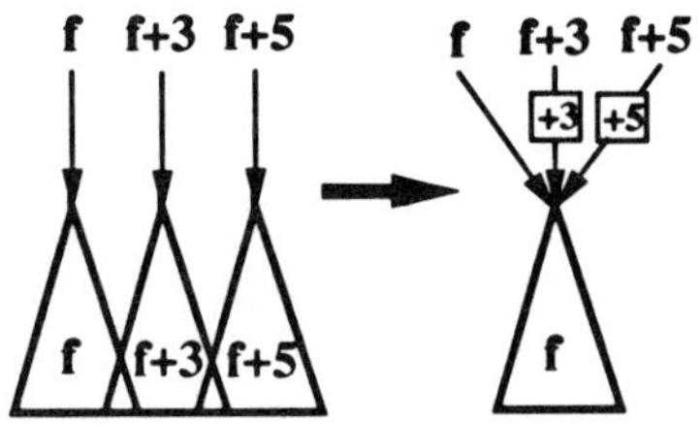

Figure 1.3.4 Reduction using EVBDDs.

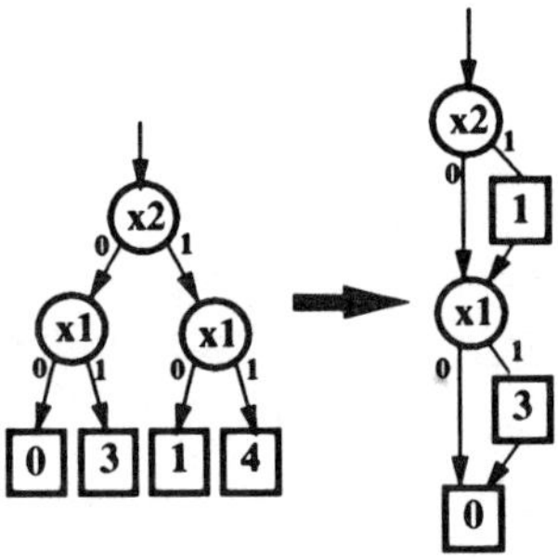

Figure 1.3.5 EVBDDs for $(3x_1 + x_2)$.

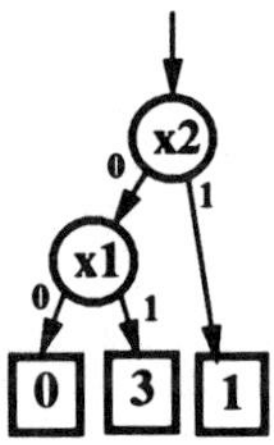

Figure 1.3.6 BMDs for $(3x_1 + x_2)$.

MTBDDs for $f, (f + 3)$, and $(f + 5)$ can be shared into one graph by using EVBDDs. To preserve the uniqueness of the representation, EVBDDs have 0-terminal node only, and do not use the attribute on 0-edges. This technique is especially effective for representing multi-valued functions of linear expressions. Fig. 1.3.5 shows an example of using EVBDD for the function $(3x_1 + x_2)$.

1.3.4 BMDs

BMDs (Binary Moment Diagrams), presented by Bryant [22], provide the representation of multi-valued functions with a structure similar to MTBDDs. In usual MTBDDs, each path from the root to a terminal node represents an input assignment, and the terminal node indicates a value of the function.

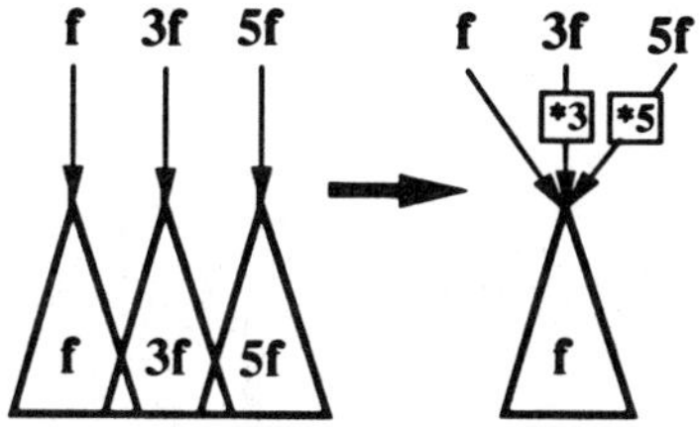

Figure 1.3.7 *BMDs.

Table 1.3.1 Complexity for bit-wise arithmetic functions.

Form	X	$X+Y$	$X \times Y$	X^2	c^X
MTBDD	$O(2^n)$	$O(2^n)$	$O(4^n)$	$O(2^n)$	$O(2^n)$
BDD-V.	$O(n)$	$O(n)$	$O(4^n)$	$O(2^n)$	$O(2^n)$
EVBDD	$O(n)$	$O(n)$	$O(4^n)$	$O(2^n)$	$O(2^n)$
BMD	$O(n)$	$O(n)$	$O(n^2)$	$O(n^2)$	$O(2^n)$
*BMD	$O(n)$	$O(n)$	$O(n)$	$O(n^2)$	$O(n)$

On the other hand, in a BMD, each path corresponds to a product term in the algebraic expression and each terminal node indicates a coefficient of the product term. An example is shown in Fig. 1.3.6. The multi-valued function $(3x_1 + x_2)$ contains two product terms x_1 and x_2, and their coefficients are 3 and 1, respectively.

BMDs are effective for representing bit-wise arithmetic functions. In particular, an n-bit multiplier can be represented by $O(n^2)$ BMD nodes, while conventional BDDs or MTBDDs require an exponential number. In addition, BMDs are improved by using attributed edges similarly to EVBDDs. They call the graph ***BMDs** (Fig. 1.3.7). In this technique, the multiplier functions are further reduced to $O(n)$ nodes. BMDs and *BMDs are useful for formal verification

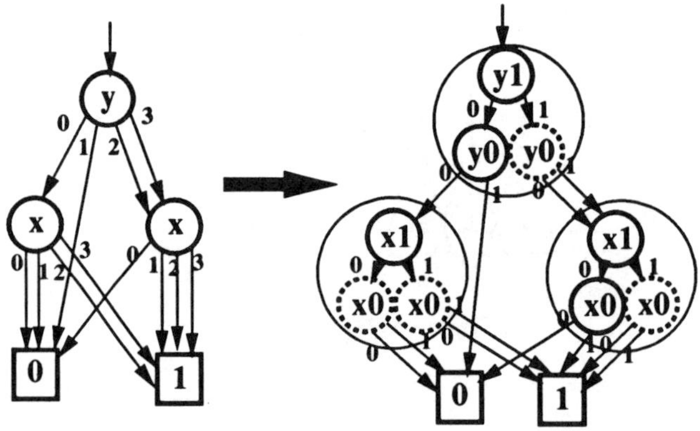

Figure 1.3.8 An MDD and its implementation.

of arithmetic circuits. The complexity of each representation is summarized in Table 1.3.1.

1.3.5 MDDs

When we deal with multi-valued input variables, each decision node should have multiple branches. For example, **TDDs (Ternary Decision Diagrams)** are such decision diagrams with three branches at each node. In general, they are called **MDDs (Multi-valued Decision Diagrams)** [19]. As shown in Fig. 1.3.8, MDDs can be represented using ordinary BDDs by encoding multi-valued inputs into binary-coded variables, and MDD operations can be translated to BDD operations. In this way, we can utilize BDD techniques for various applications in discrete variable problems.

1.4 REPRESENTATION OF CUBE SETS

We are sometimes faced with combinatorial problems which need to manipulate **sets of combinations**, or **cube sets**, rather than Boolean functions [23, 24]. Some variants of BDDs have been proposed for representing cube sets efficiently.

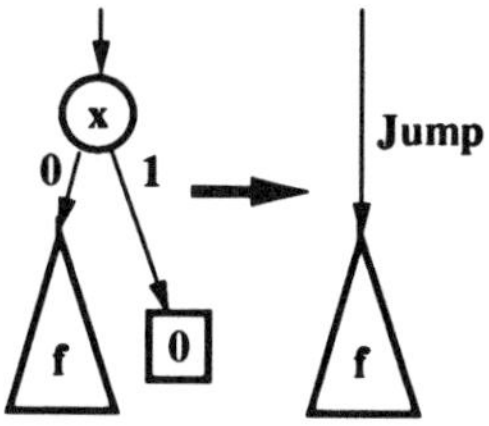

Figure 1.4.1 New reduction rule for ZBDDs.

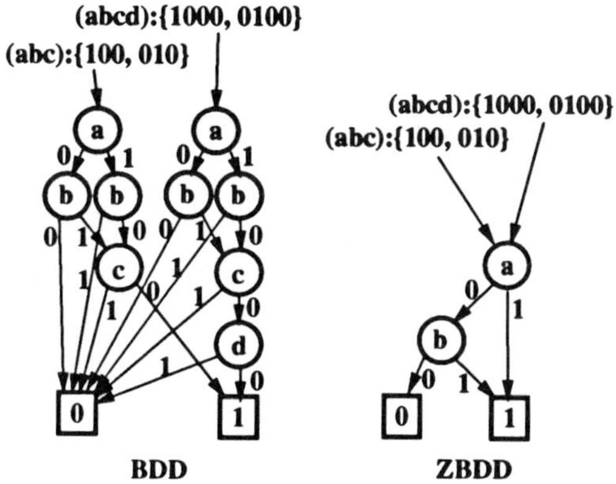

Figure 1.4.2 Effect of ZBDD reduction rule.

1.4.1 ZBDDs

ZBDDs (Zero-suppressed BDDs) [25] are a variant of BDDs [25] adapted for representing cube sets. They are based on the following reduction rules:

- Eliminate all nodes with the 1-edge pointing to the 0-terminal node. Then connect the edge to the other sub-graph directly (Fig. 1.4.1).
- Share all equivalent sub-graphs, as in BDDs.

Notice that, contrary to usual BDDs, we do not eliminate nodes whose two edges both point to the same node. When the number and order of the variables are fixed, ZBDDs provide canonical forms for Boolean functions as well as usual BDDs.

Fig. 1.4.2 illustrates a BDD and a ZBDD representing the same cube set. The ZBDD reduction rule automatically eliminates the nodes for irrelevant variables which do not appear in the cube set. To evaluate the effect, we generated a set of one hundred combinations each of which selects k out of 100 objects randomly. We then compared the size of both BDDs and ZBDDs representing these random combinations. The result in Fig. 1.4.3 shows that ZBDDs are much more compact than BDDs especially when k is small. Namely, ZBDDs

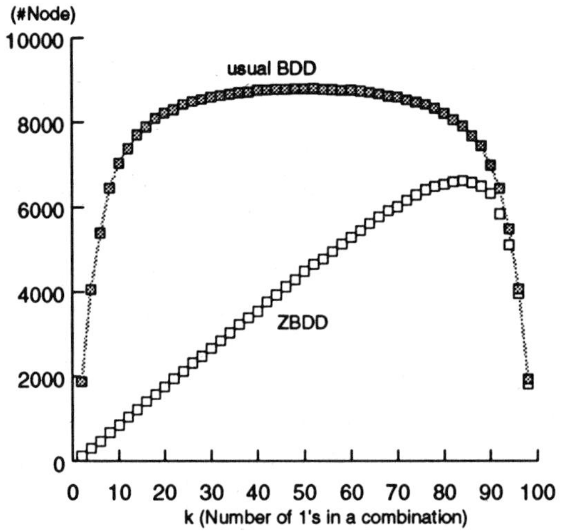

Figure 1.4.3 Comparison of sizes for BDDs and ZBDDs.

are remarkably effective for sparse combinations, such that each cube consists of a few variables chosen from many.

ZBDDs are manipulated by applying set operations, such as union, intersection, and difference. These operations can be performed in a time almost linear to the graph size, as for BDDs. Fig. 1.4.4 shows an example of generating a ZBDD.

In many applications of cube sets in digital system design, they commonly use **binate** cube sets containing both positive and negative literals: $x, \overline{x}$. To deal with binate literals, we use two different variable x_1 and x_0. For example, a binate cube set $(a\,\overline{b} + c)$ is expressed as a unate cube set $(a_1 b_0 + c_1)$. x_1 and x_0 never appear together in the same cube, so that the cubes become sparse and ZBDDs are effective.

ZBDDs provide an implicit cube set representation which is very efficient for manipulating large-scale cube sets. Minato presented a fast method of two-level logic optimization [26] and multi-level logic optimization [27], based on ZBDD representation.

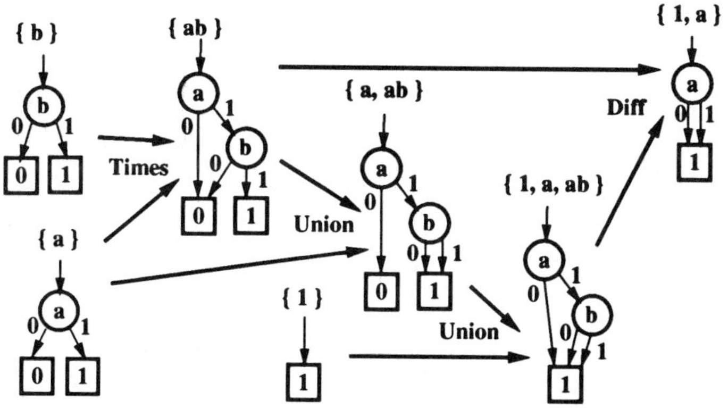

Figure 1.4.4 Generation of ZBDDs.

1.4.2 FDDs and KFDDs

Conventional BDDs are based on Shannon's expansion, which consists of AND-OR logic operation. However, there is another type of expansion using AND-EXOR operation:

$$f = f_{(x=0)} \oplus x \cdot (f_{(x=0)} \oplus f_{(x=1)}).$$

This expansion is called *(positive)* **Davio's expansion. FDDs (Functional Decision Diagrams)**, presented by Kebschull et al. [28], are a graph-based Boolean function representation based on Davio's expansion. A FDD is derived by reducing a binary tree graph which represents the **Reed-Muller expression** for a Boolean function, as shown in Fig. 1.4.5(a). In the binary tree, 0- and 1-terminal nodes represent logic values, and each node represents Davio's expansion, i.e. the 0-edge points to $f_{(x=0)}$, and the 1-edge points to $(f_{(x=0)} \oplus f_{(x=1)})$. The original paper uses the same reduction rules as BDDs; however, ZBDDs reduction rule is more appropriate for FDDs, since it can eliminate the nodes for variables irrelevant to the function. For example, the binary tree is reduced to Fig. 1.4.5(b) by the usual BDDs reduction rules, and more reduced to Fig. 1.4.5(c) by the ZBDDs rule.

Under a fixed variable order, FDDs provide canonical representation of Boolean functions as well as ordinary BDDs. FDDs can also be regarded as the implicit cube set representation of Reed-Muller expressions.

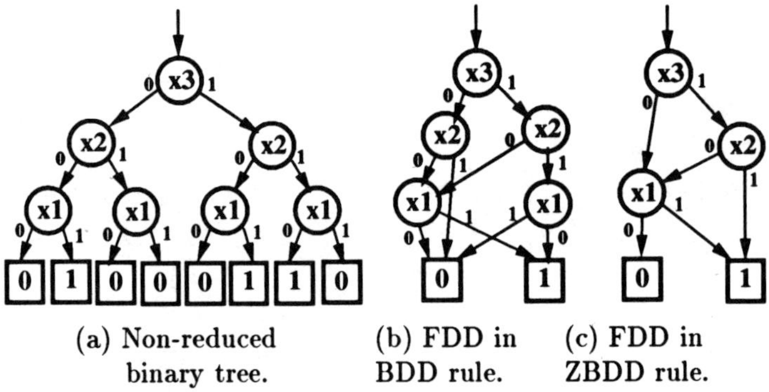

Figure 1.4.5 FDDs for $x_3x_2 \oplus x_3x_1 \oplus x_1$.

Drechsler et al. [29] presented **KFDDs (Kronecker FDDs)**, which are the generalization of FDDs. KFDDs are based on three types of expansions:

- $f = (\overline{x} \cdot f_{(x=0)}) \oplus (x \cdot f_{(x=1)})$ (Shannon's expansion)
- $f = f_{(x=0)} \oplus x \cdot (f_{(x=0)} \oplus f_{(x=1)})$ (positive Davio's expansion)
- $f = f_{(x=1)} \oplus \overline{x} \cdot (f_{(x=0)} \oplus f_{(x=1)})$ (negative Davio's expansion)

KFDDs mix the three type of expansions in one graph, provided that any two nodes with the same variable have the same type of expansion. In KFDDs, the Shannon type nodes are reduced by the usual BDD reduction rules, and the other node types follow ZBDDs rule. KFDDs give unique representation under a fixed order and types of variables. KFDDs can be regarded as the implicit cube set representation of **reduced Kronecker expressions**.

If we assign the positive Davio type only to all the variables, KFDDs become identical to FDDs. If the Shannon type is assigned to all, KFDDs will be BDDs. KFDDs may become more compact than FDDs and BDDs when a good type assignment is found. It is difficult to find the exactly optimal type assignment, and some heuristic methods are devised [29]. (See Chapter 2, [30] for more details).

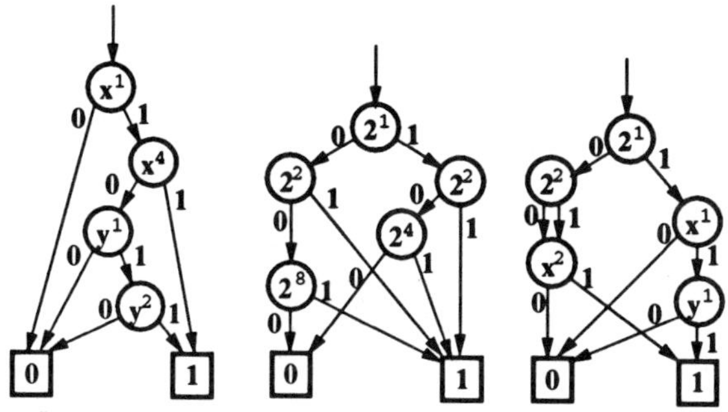

(a) $x^5 + x^1y^3$ (b) $2^8 + 2^5 + 2^3 + 2^2$ (c) $5x^2 + 2xy$

Figure 1.4.6 ZBDD-based polynomial representation.

1.4.3 Polynomial Representation Using ZBDDs

A polynomial formula is a basic model in mathematics. Minato [31] presented a method for representing polynomials using ZBDDs. It is similar to the cube set representation, except that they include higher degree variables and numerical coefficients.

The basic idea is based on binary coding of higher degrees and coefficients. A variable x^k can be broken down into:

$$x^k = x^{(k_1+k_2+\ldots+k_m)} = x^{k_1}x^{k_2}\ldots x^{k_m},$$

where $k_1, k_2, \ldots, k_m$ are different powers of 2. In this way, we can represent x^k as a combination of $x^1, x^2, x^4, x^8, \cdots, x^{2^{n-1}} (0 < k < 2^n)$. Such combinations can be dealt with efficiently using ZBDDs. For example, a polynomial $x^5 + xy^3$ can be written as $x^1x^4 + x^1y^1y^2$. It can be regarded as a cube set over x^1, x^4, y^1, and y^2. The formula, then, can be represented by using a ZBDD, as shown in Fig. 1.4.6(a).

Numerical coefficients can also be represented similarly. An integer c (> 1) can be written as a sum of powers of 2:

$$c = 2^{c_1} + 2^{c_2} + \ldots + 2^{c_m},$$

where $c_1, c_2, \ldots, c_m$ are different nonnegative integers. Then, regarding "2" as a variable, just like x, y, z, etc., it can be represented as a polynomial of variables

Table 1.4.1 Result for representing polynomials.

Expression	#Terms	ZBDD size	Time(s)
$(x+1)^{100}$	101	1,209	39.7
$(a+1)(b+1)...(j+1)$	1,024	10	0.1
$(a+1)(b+2)...(j+10)$	1,024	619	1.1
$(a+1)^4(b+1)^4...(j+1)^4$	9,765,625	2,053	2.8

using SPARC station 2 (128MB).

with degrees. Consequently, we can represent a constant number c as a cube set over $2^1, 2^2, 2^4, 2^8, \ldots, 2^{2^{n-1}} (0 < c < 2^{2^n})$ using ZBDDs. For example, the constant number $300 = 2^8 + 2^5 + 2^3 + 2^2$ can be written as $2^8 + 2^1 2^4 + 2^1 2^2 + 2^2$, and represented by a ZBDD as Fig. 1.4.6(b). When the number is used for a coefficient with other variables, we can regard the symbol "2" just as one sort of variable in the formula. Fig. 1.4.6(c) shows an example for $5x^2 + 2xy$, which is decomposed into $x^2 + 2^2 x^2 + 2^1 x^1 y^1$.

Experimental results are shown in Table 1.4.1. In this method, we can represent large-scale polynomials compactly and uniquely, and can manipulate them in practical time. Constructing canonical forms of polynomials immediately leads to equivalence checking of arithmetic expressions.

BMDs, discussed in the previous section, can also represent polynomials. One big difference is that BMDs assume binary-valued variables, so BMDs cannot deal with the higher-degree variables. Except for this difference, the two representations are similar to each other. Fig. 1.4.7(a) shows the ZBDD-based representation for $(4xy + 3x + 2y + 1)$. If we change the variable order such that the coefficient variables move from higher to lower position, the ZBDD becomes as shown in Fig. 1.4.7(b). In this graph, the sub-graphs with the coefficient variables correspond to the terminal nodes in the BMD. This observation indicates that the ZBDD-based representation and the BMD can be transformed into each other by changing the variable order.

1.5 SUMMARY

In this chapter, we surveyed the BDD techniques and their variants. In summary, we show a relationship chart of the different types of BDDs. In Fig. 1.5.1,

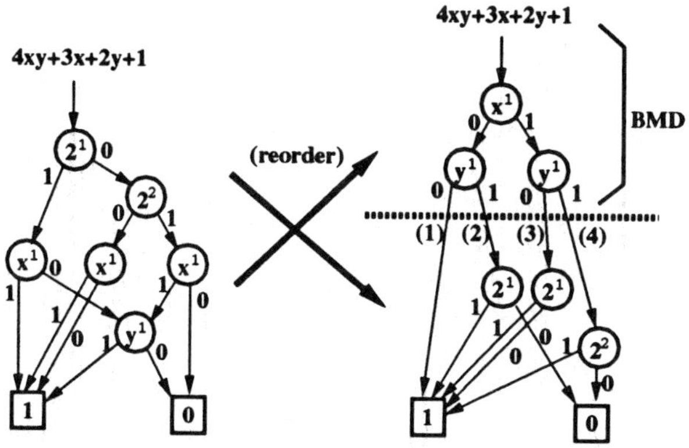

(a) Coefficients are higher. (b) Coefficients are lower.

Figure 1.4.7 Comparison of ZBDD-based method and BMD.

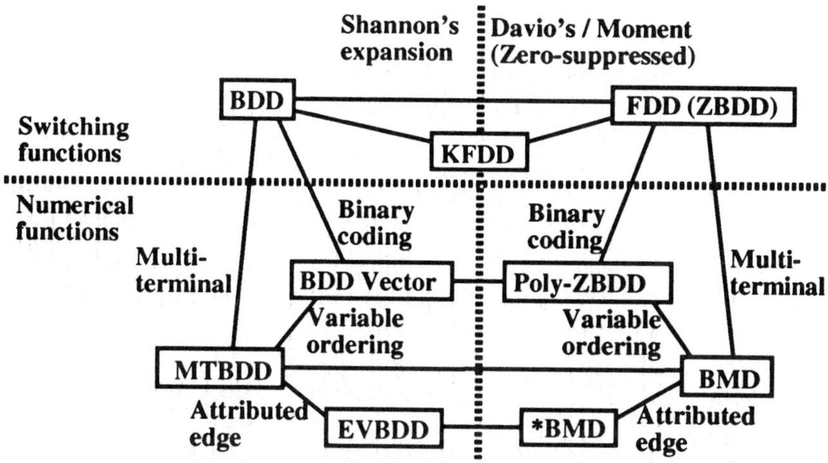

Figure 1.5.1 Relationship chart of BDD variants.

the switching function representations are shown in the upper portion, and the numerical data structures are in the lower. The left side ones are based on Shannon's expansion, and the right side ones are based on Davio's or moment decomposition, which requires the zero-suppressed reduction rule. This chart shows that BDD can be extended to MTBDD, BDD vector, and EVBDD by using techniques such as binary coding, multiple terminals, and attributed edges. Similarly, FDD (ZBDD) can be extended to Poly-ZBDD, BMD, and *BMD. KFDD is a hybrid method of BDD and FDD. This chart indicates that the relationship has a symmetric structure.

BDD techniques are based on the quick search of hash tables and linked list data structures. They greatly benefit from the property of the **random access machine model**, where any data in main memory can be accessed in constant time. The BDD representation is fairly sophisticated and adapted to the conventional computer model.

Presently, considerable research is in progress. BDDs and their improvements will become a key technique to problems not only in VLSI CAD, but also in various areas in Computer Science.

REFERENCES

[1] S. B. Akers, "Binary decision diagrams", *IEEE Trans. on Computers*, Vol. C-27, No. 6, pp. 509–516, June 1978.

[2] R. E. Bryant, "Graph-based algorithms for Boolean function manipulation", *IEEE Trans. on Computers*, Vol. C-35, No. 8, pp. 677–691, Aug. 1986.

[3] H.-T. Liaw and C.-S. Lin, "On the OBDD-representation of general Boolean functions", *IEEE Trans. on Computers*, Vol. C-41, No. 6, pp. 661–664, June 1992.

[4] N. Ishiura and S. Yajima, "A class of logic functions expressible by a polynomial-size binary decision diagrams", In *Proc. Synthesis and Simulation Meeting and International Interchange (SASIMI'90, Japan)*, pp. 48–54, Oct. 1990.

[5] S. Minato, N. Ishiura and S. Yajima, "Shared binary decision diagram with attributed edges for efficient Boolean function manipulation", *Proc. 27th IEEE/ACM DAC*, pp. 52-57, June 1990

[6] Bill Lin and Fabio Somenzi, "Minimization of symbolic relations", *Proc. IEEE/ACM ICCAD'90*, pp. 88-91, Nov. 1990.

[7] J. C. Madre and J. P. Billon, " Proving circuit correctness using formal comparison between expected and extracted behaviour", *Proc. 25th ACM/IEEE DAC*, pp. 205–210, June 1988.

[8] S. Tani, K. Hamaguchi, and S. Yajima, "The complexity of the optimal variable ordering of a shared binary decision diagram", *Technical Report 93-6, Department of Information Science, Faculty of Science, University of Tokyo*, Dec. 1993.

[9] N. Ishiura, H. Sawada, and S. Yajima, "Minimization of binary decision diagrams based on exchanges of variables", *Proc. IEEE/ACM ICCAD'91*, pp. 472–475, Nov. 1991.

[10] M. Fujita, H. Fujisawa and N. Kawato, "Evaluation and improvement of Boolean comparison method based on binary decision diagrams", *Proc. IEEE/ACM ICCAD'88*, pp. 2-5, Nov. 1988.

[11] S. Malik, A. R. Wang, R. K. Brayton, and A. L. Sangiovanni-Vincentelli, "Logic verification using binary decision diagrams in a logic synthesis environment", *Proc. IEEE/ACM ICCAD'88*, pp. 6–9, Nov. 1988.

[12] K. M. Butler, D. E. Ross, R. Kapur, and M. R. Mercer, "Heuristics to compute variable orderings for efficient manipulation of ordered binary decision diagrams", *Proc. of 28th ACM/IEEE DAC*, pp. 417–420, June 1991.

[13] M. Fujita, Y. Matsunaga, and T. Kakuda, "On variable ordering of binary decision diagrams for the application of multi-level logic synthesis", *Proc. IEEE EDAC'91*, pp. 50–54, February 1991.

[14] S. Minato, "Minimum-width method of variable ordering for binary decision diagrams", *IEICE Trans. Fundamentals*, Vol. E75-A, No. 3, pp. 392–399, Mar. 1992.

[15] R. Rudell, "Dynamic variable ordering for ordered binary decision diagrams", *Proc. IEEE/ACM ICCAD'93*, pp. 42–47, Nov. 1993.

[16] E. M. Clarke, K. L. McMillan, X. Zhao, M. Fujita, and J. Yang, "Spectral transforms for large Boolean functions with applications to technology mapping", *Proc. 30th ACM/IEEE DAC*, pp. 54–60, June 1993. (Also Chapter 4 of this book).

[17] R. I. Bahar, E. A. Frohm, C. M. Gaona, G. D. Hachtel, E. Macii, A. Pardo, and F. Somenzi, "Algebraic decision diagrams and their applications", *Proc. IEEE/ACM ICCAD'93*, pp. 188–191, Nov. 1993.

[18] J. Jain, J. Bitner, D. Fussell, and J. Abraham, "Probabilistic Design Verification", *Proc. IEEE/ACM ICCAD'91*, pp. 468–471, Nov. 1991. (Also Chapter 6 of this book).

[19] A. Srinivasan, T. Kam, S. Malik, and R. Brayton, "Algorithms for Discrete Function Manipulation", *Proc. IEEE/ACM ICCAD'90*, pp. 92–95, Nov. 1990.

[20] S. Minato, "BEM-II:an arithmetic Boolean expression manipulator using BDDs", *IEICE Trans. Fundamentals*, Vol. E76-A, No. 10, pp. 1721–1729, Oct. 1993.

[21] Y.-T. Lai, M. Pedram, and S. B. Vrudhula, "FGILP: An integer linear program solver based on function graphs", *Proc. IEEE/ACM ICCAD'93*, pp. 685–689, Nov. 1993. (Also Chapter 5 of this book).

[22] R. E. Bryant and Y.-A. Chen, "Verification of arithmetic functions with binary moment diagrams", *Proc. 32nd ACM/IEEE DAC*, June 1995.

[23] J. R. Burch, E. M. Clarke, K. L. McMillan and D. L. Dill, "Sequential circuit verification using symbolic model checking", *Proc. 27th ACM/IEEE DAC*, pp. 46-51, June 1990.

[24] O. Coudert and J. C. Madre, "Implicit and incremental computation of primes and essential implicant primes of Boolean functions", *Proc. 29th ACM/IEEE DAC*, pp. 36-39, June 1992.

[25] S. Minato, "Zero-suppressed BDDs for set manipulation in combinatorial problems", *Proc. 30th ACM/IEEE DAC*, pp. 272–277, June 1993.

[26] S. Minato, "Fast generation of irredundant sum-of-products forms from binary decision diagrams", *Proc. of Synthesis and Simulation Meeting and International Interchange (SASIMI'92, Japan)*, pp. 64–73, April 1992.

[27] S. Minato, "Fast weak-division method for implicit cube representation", *Proc. of Synthesis and Simulation Meeting and International Interchange (SASIMI'93, Japan)*, pp. 423–432, October 1993.

[28] U. Kebschull, E. Schubert and W. Rosenstiel, "Multilevel Logic Synthesis Based on Functional Decision Diagrams", *IEEE EDAC'92*, pp. 43-47, Mar. 1992.

[29] R. Drechsler, A. Sarabi, M. Theobald, B. Becker and M. Perkowski, "Efficient representation and manipulation of switching functions based on ordered Kronecker functional decision diagrams", *Proc. 31st ACM/IEEE DAC*, pp. 415-419, June 1994. (Also Chapter 7 of this book).

[30] T. Sasao, "Representation of Logic Functions using EXOR Operators", *IFIP WG. 10.5 Workshop on Applications of the Reed-Muller Expansions in Circuit Design* Aug., 1995. (Also Chapter 2 of this book).

[31] S. Minato, "Implicit manipulation of polynomials using zero-suppressed BDDs", *IEEE/ACM ED&TC'95*, pp. 449-454, Mar. 1995.

2

REPRESENTATIONS OF LOGIC FUNCTIONS USING EXOR OPERATORS

Tsutomu SASAO

Department of Computer Science and Electronics
Kyushu Institute of Technology
Iizuka, Japan

Abstract— Logic functions are usually represented by logical expressions or decision diagrams using AND and OR operators. However, some functions have more compact representations with EXOR operators. This chapter surveys representations of logic functions using EXOR operators.

2.1 INTRODUCTION

Various methods exist to represent logic functions. A truth table is a straightforward representation, but its size increases exponentially when n, the number of the input variables, increases.

Logical expressions and decision diagrams are useful to represent functions with many inputs, since they are often more compact than truth tables. The logical expressions and decision diagrams usually use AND and OR operators. However, some functions have more compact representation with EXOR operators. For example, to represent a parity function of n variables: $f = x_1 \oplus x_2 \oplus \cdots \oplus x_n$, a sum-of-products expression (SOP) requires 2^{n-1} products while an AND-EXOR expression requires only n products. Thus, for this function, the EXOR-based representation is much more compact than the SOP-based ones.

This chapter surveys various methods to represent logic functions by using EXOR operators.

The rest of this chapter is organized as follows:

Section 2.2 introduces various trees using EXOR operators.

Section 2.3 defines various AND-EXOR expressions, compares the relations among them, surveys optimization methods, and compares their complexities.

Section 2.4 shows various decision diagrams using EXORs.

Section 2.5 introduces EXOR ternary decision diagrams (ETDDs).

2.2 TREES USING EXOR OPERATORS

In this section, we survey tree representation of logic functions.

2.2.1 Three Types of Expansions

Consider the following three types of expansions using EXOR operators:

$$f = \bar{x} f_0 \oplus x f_1, \quad (2.2.1)$$

$$f = f_0 \oplus x f_2, \quad (2.2.2)$$

$$f = f_1 \oplus \bar{x} f_2, \quad (2.2.3)$$

where f_0 (f_1) is f with x replaced by 0 (1), and $f_2 = f_0 \oplus f_1$.

(2.2.1) is the **Shannon expansion**, where the EXOR operator is used instead of the inclusive OR operator. The Shannon expansion is denoted by S. (2.2.2) is the **positive Davio expansion**. This expansion uses only the positive literal, and is denoted by pD. (2.2.3) is the **negative Davio expansion**. This expansion uses only the negative literal, and is denoted by nD. Fig. 2.2.1 shows circuits corresponding to the three types of expansions.

These three expansions are the basis of the EXOR-based representation of logic functions.

2.2.2 Shannon Tree

By applying the Shannon expansions recursively to a logic function, we can represent a logic function by an expansion tree. Fig. 2.2.2 shows an example of

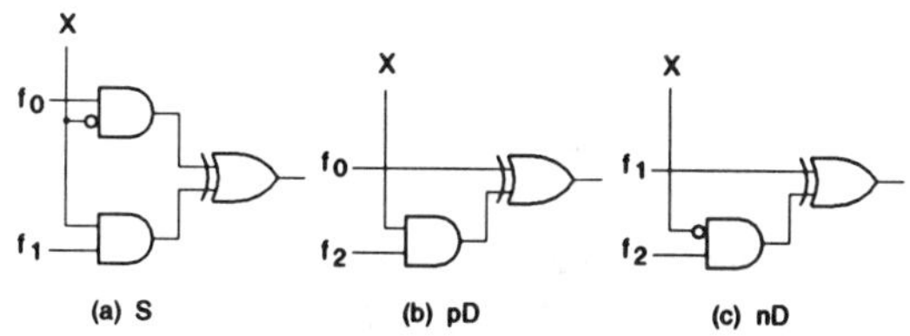

Figure 2.2.1 Circuits corresponding three types of expansions.

an expansion tree for a three-variable function f, where the symbol S denotes the Shannon expansion. This tree is called a **Shannon tree.** The terminal nodes represent binary constants. Each edge has a literal of a variable as a label. A product of the literals from the root node to a terminal node represents a product. For example, the leftmost path represents the product $\bar{x}_1\bar{x}_2\bar{x}_3$. This tree shows the expression for f:

$$\begin{aligned} f &= f_{000}\bar{x}_1\bar{x}_2\bar{x}_3 \oplus f_{001}\bar{x}_1\bar{x}_2x_3 \oplus f_{010}\bar{x}_1x_2\bar{x}_3 \oplus f_{011}\bar{x}_1x_2x_3 \\ &\quad \oplus f_{100}x_1\bar{x}_2\bar{x}_3 \oplus f_{101}x_1\bar{x}_2x_3 \oplus f_{110}x_1x_2\bar{x}_3 \oplus f_{111}x_1x_2x_3. \end{aligned}$$

A product term that includes all variable is called a **minterm**, and an expression consisting only of minterms is called a **minterm expansion.** Note that the function is unchanged if the exclusive OR is replaced with inclusive OR. The result is an another canonical expression for this set of variables. Note that the products having zero coefficients disappear. Thus, the number of non-zero coefficients equals to the number of products in the expression.

2.2.3 Positive Davio Tree

Alternatively, we can apply positive Davio expansions recursively to a logic function. Fig. 2.2.3 shows an example of an expansion tree for a three-variable function f, where the symbol pD denotes the positive Davio expansion. This produces the **positive Davio tree.** Note that each edge has either a constant one or a positive literal as a label. Also in this case, each path from the root node to a terminal node corresponds to a product.

The expression corresponding to this tree is

$$\begin{aligned} f &= f_{000}\cdot 1\cdot 1\cdot 1 \oplus f_{002}\cdot 1\cdot 1\cdot x_3 \oplus f_{020}\cdot 1\cdot x_2\cdot 1 \oplus f_{022}\cdot 1\cdot x_2x_3 \\ &\quad \oplus f_{200}x_1\cdot 1\cdot 1 \oplus f_{202}x_1\cdot 1\cdot x_3 \oplus f_{220}x_1x_2\cdot 1 \oplus f_{222}x_1x_2x_3. \end{aligned}$$

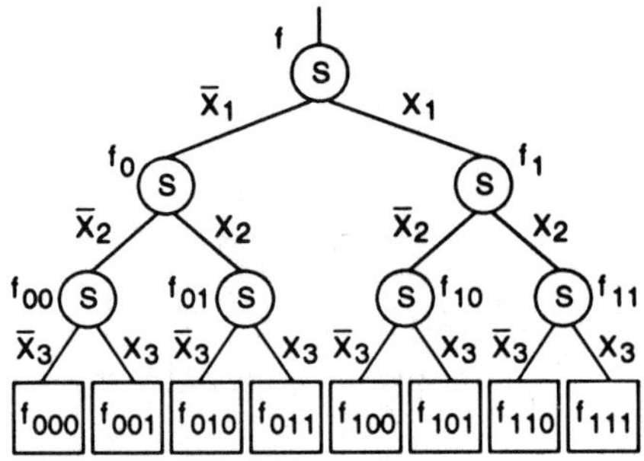

Figure 2.2.2 Shannon tree.

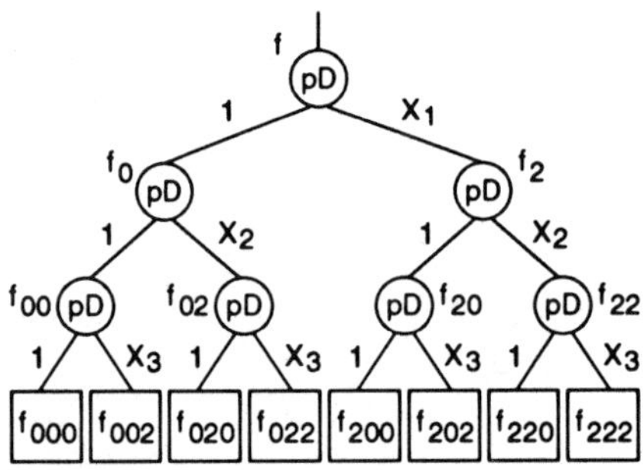

Figure 2.2.3 Positive Davio tree.

This expression uses only positive literals, and is called a **positive polarity Reed-Muller expression (PPRM)**. This is also a canonical expression.

2.2.4 Reed-Muller Tree

Suppose that we can use either the positive or the negative Davio expansion for each variable. Then, we have a Reed-Muller tree. Fig. 2.2.4 shows an example of a **Reed-Muller tree** for a three-variable function. In this tree, variable x_1 uses the positive Davio expansion, variable x_2 uses the negative Davio expansion, and variable x_3 uses the positive Davio expansion. The corresponding

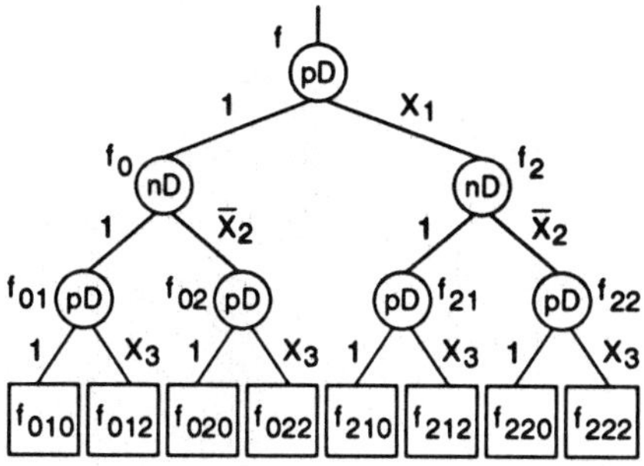

Figure 2.2.4 Reed-Muller tree.

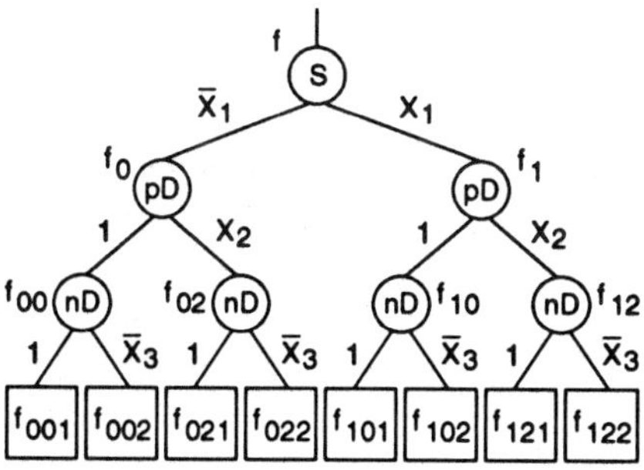

Figure 2.2.5 Kronecker tree.

expression is

$$\begin{aligned} f &= f_{010} \cdot 1 \cdot 1 \cdot 1 \oplus f_{012} \cdot 1 \cdot 1 \cdot x_3 \oplus f_{020} \cdot 1 \cdot \bar{x}_2 \cdot 1 \oplus f_{022} \cdot 1 \cdot \bar{x}_2 x_3 \\ &\quad \oplus f_{210} x_1 \cdot 1 \cdot 1 \oplus f_{212} x_1 \cdot 1 \cdot x_3 \oplus f_{220} x_1 \bar{x}_2 \cdot 1 \oplus f_{222} x_1 \bar{x}_2 x_3. \end{aligned}$$

We call this a **fixed polarity Reed-Muller expression (FPRM)**. Again, this expression is canonical for a given set of polarities. Note that there are 2^n different expansions for an n-variable function. Different expansions will produce expressions with different number of products.

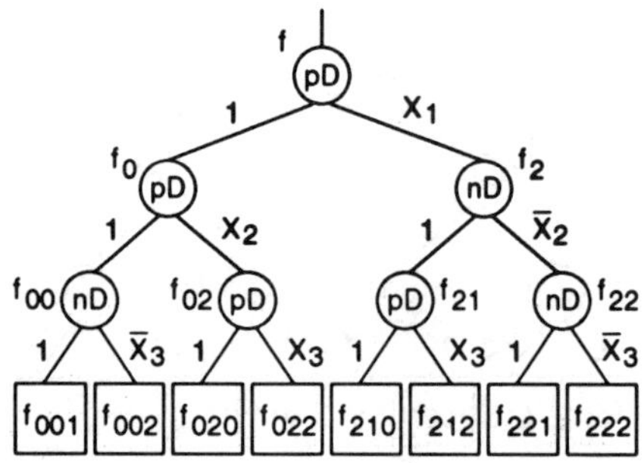

Figure 2.2.6 Pseudo-Reed-Muller tree.

2.2.5 Kronecker Tree

Suppose that we can use any of the three expansions for each variable, then we have a Kronecker tree. The three expansions are the Shannon expansion, the positive Davio expansion, and the negative Davio expansion. Fig. 2.2.5 shows an example of a **Kronecker tree** for a three-variable function. In this tree, variable x_1 uses the Shannon expansion, variable x_2 uses the positive Davio expansion, and variable x_3 uses the negative Davio expansion. The corresponding expression is

$$\begin{aligned} f &= f_{001}\bar{x}_1 \cdot 1 \cdot 1 \oplus f_{002}\bar{x}_1 \cdot 1 \cdot \bar{x}_3 \oplus f_{021}\bar{x}_1 x_2 \cdot 1 \oplus f_{022}\bar{x}_1 x_2 \bar{x}_3 \\ &\oplus f_{101}x_1 \cdot 1 \cdot 1 \oplus f_{102}x_1 \cdot 1 \cdot \bar{x}_3 \oplus f_{121}x_1 x_2 \cdot 1 \oplus f_{122}x_1 x_2 \bar{x}_3. \end{aligned}$$

This is called a **Kronecker expression (KRO)**. Again, this expression is canonical for a given way of expansion, and there are 3^n different expansions.

2.2.6 Pseudo-Reed-Muller Tree

Suppose that we can use either the positive or the negative Davio expansion for each node, then we have a **pseudo-Reed-Muller tree**. For example, Fig. 2.2.6 shows a pseudo Reed-Muller tree for a three-variable function, where variable x_1 uses the positive Davio expansion, variable x_2 and variable x_3 use both the positive and the negative Davio expansions. In this tree, there are 7 nodes. So 2^7 different expansions exist. In general, for n-variable functions, we have 2^{2^n-1} different expansions. The corresponding expression is

$$f = f_{001} \cdot 1 \cdot 1 \cdot 1 \oplus f_{002} \cdot 1 \cdot 1 \cdot \bar{x}_3 \oplus f_{020} \cdot 1 \cdot x_2 \cdot 1 \oplus f_{022} \cdot 1 \cdot x_2 x_3$$

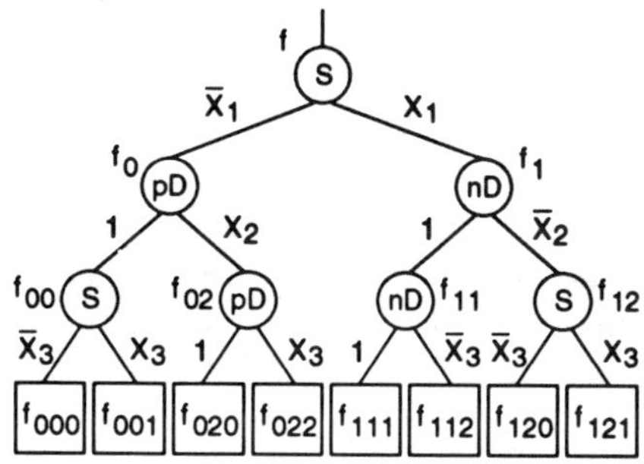

Figure 2.2.7 Pseudo-Kronecker tree.

$$\oplus f_{210} x_1 \cdot 1 \cdot 1 \oplus f_{212} x_1 \cdot 1 \cdot x_3 \oplus f_{221} x_1 \bar{x}_2 \cdot 1 \oplus f_{222} x_1 \bar{x}_2 \bar{x}_3.$$

We call this as **pseudo-Reed-Muller expression (PSDRM)**.

2.2.7 Pseudo-Kronecker Tree

Suppose that we can use any of the three expansions for each node, then we have a pseudo-Kronecker tree. Fig. 2.2.7 shows an example of a **pseudo-Kronecker tree** for three-variable function, where the first variable uses the Shannon expansion, the second variable uses both positive and negative Davio expansions, and the last variable uses all the three expansions. In this tree, there are 7 nodes. So 3^7 different expansions exist. For n-variable functions, we have 3^{2^n-1} different expansions. The corresponding expression is

$$\begin{aligned} f &= f_{000} \bar{x}_1 \cdot 1 \cdot \bar{x}_3 \oplus f_{010} \bar{x}_1 \cdot 1 \cdot x_3 \oplus f_{020} \bar{x}_1 x_2 \cdot 1 \oplus f_{022} \bar{x}_1 x_2 x_3 \\ &\oplus f_{111} x_1 \cdot 1 \cdot 1 \oplus f_{112} x_1 \cdot 1 \cdot \bar{x}_3 \oplus f_{120} x_1 \bar{x}_2 \bar{x}_3 \oplus f_{121} x_1 \bar{x}_2 x_3. \end{aligned}$$

We call this as **pseudo-Kronecker expression (PSDKRO)**.

2.3 VARIOUS AND-EXOR EXPRESSIONS

2.3.1 Expressions Generated by Expansion Trees

As shown in the previous section, different expansion trees generate different AND-EXOR expressions. A PPRM (Positive Polarity Reed-Muller expression) is generated from the positive Davio tree. When $n = 3$, it has a form

$$F = a_0 \oplus a_1x_1 \oplus a_2x_2 \oplus a_3x_3 \oplus a_4x_1x_2 \oplus a_5x_1x_3 \oplus a_6x_2x_3 \oplus a_7x_1x_2x_3.$$

Note that all the literals are positive. An FPRM (Fixed Polarity Reed-Muller expression) is generated from a Reed-Muller tree. It has a form

$$F = a_0 \oplus a_1x_1^* \oplus a_2x_2^* \oplus a_3x_3^* \oplus a_4x_1^*x_2^* \oplus a_5x_1^*x_3^* \oplus a_6x_2^*x_3^* \oplus a_7x_1^*x_2^*x_3^*,$$

where x_i^* occurs throughout as either x_i or $\bar{x}_i$. In similar ways, a KRO (Kronecker expression) is generated from a Kronecker tree; a PSDRM (Pseudo-Reed-Muller expression) is generated from a pseudo-Reed-Muller tree; and a PSDKRO (Pseudo-Kronecker expression) is generated from a pseudo-Kronecker tree. Note that the number of non-zero terminal nodes equals the number of products in the expression. Table 2.3.1 shows the relation of trees and expressions generated by the trees.

2.3.2 GRM and ESOP

So far, we have defined five classes of AND-EXOR expressions. In this part, we define two more classes. A **GRM (generalized Reed-Muller expression)** [7] is derived from a positive polarity Reed-Muller expression. For example, a GRM for a three-variable function can be represented as follows:

$$F = a_0 \oplus a_1x_1^* \oplus a_2x_2^* \oplus a_3x_3^* \oplus a_4x_1^*x_2^* \oplus a_5x_1^*x_3^* \oplus a_6x_2^*x_3^* \oplus a_7x_1^*x_2^*x_3^*,$$

where x_i^* denotes either the positive or the negative literal. In a GRM, both the positive and the negative literals may appear at the same time. For a 3-variable function, there are 12 literals, so $2^{12} = 4096$ different GRMs exist. In general, the total number of literals is $n2^{n-1}$, where n is the number of variables. Thus, $2^{n2^{n-1}}$ different expressions exist. Note that certain GRMs have no corresponding tree. For example, $\bar{x}_1\bar{x}_2 \oplus x_2\bar{x}_3 \oplus x_1x_3$ is a GRM, but cannot be generated from a pseudo Kronecker tree. In a GRM, no two products have the same set of variables. Thus, $x_1x_2x_3 \oplus \bar{x}_1\bar{x}_2\bar{x}_3$ is not a GRM. Note

Table 2.3.1 Relations between trees, AND-EXOR expressions, and decision diagrams.

Type of tree	Expression generated from the tree	DD generated from the tree
Shannon tree	Minterm expansion	BDD
Positive Davio tree	PPRM	FDD
Reed-Muller tree	FPRM	Reed-Muller DD
Kronecker tree	KRO	KDD
Pseudo-Reed-Muller tree	PSDRM	Pseudo Reed-Muller DD
Pseudo-Kronecker tree	PSDKRO	PKDD
	GRM	
EXOR Ternary decision tree	ESOP	ETDD

that many authors use term "GRM" to denote FPRM [17]. Some authors use other terminology for GRMs [6]. Thus, the terminology is not uniform.

The other class is **ESOPs** (**EXOR sum-of-products expressions**). This is the most general class of AND-EXOR expressions. For example, a three-variable function can be represented by

$$F = \sum \oplus x_1^* x_2^* x_3^*,$$

where x_i^* can be 1, x_i or $\bar{x}_i$, and $\sum \oplus$ represents the EXOR of the products. Note that in ESOPs, each occurrence of x_i^* can be chosen as 1, x_i or $\bar{x}_i$ independent of the other choice for x_i^*. Other classes of AND-EXOR expressions are defined in [18].

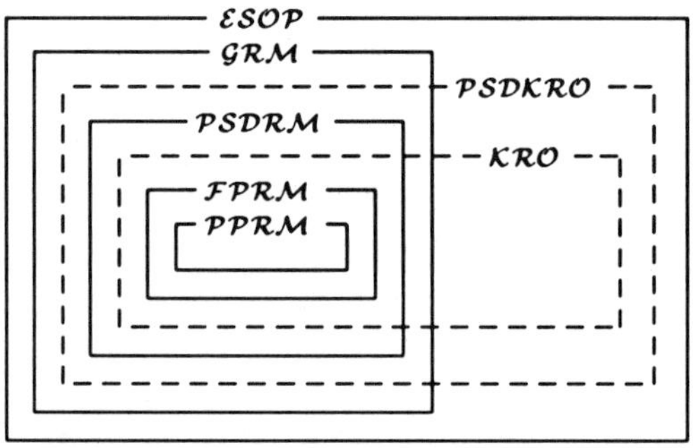

Figure 2.3.1 Relations among various classes of AND-EXOR expressions.

2.3.3 Comparison of Various Expressions

So far, we have defined 7 classes of AND-EXOR expression. The following examples show the difference of the classes.

1. $x_1x_2 \oplus x_2x_3 \oplus x_3x_1$ is a PPRM, since all the literals are positive.
2. $x_1\bar{x}_2 \oplus \bar{x}_2x_3 \oplus x_3x_1$ is an FPRM, but not a PPRM, since x_2 has the negative polarity.
3. $\bar{x}_1\bar{x}_2 \oplus x_2\bar{x}_3 \oplus x_3x_1$ is a GRM, since x_1 has the both polarities. It is not a PSDKRO, since there is no tree generating it.
4. $\bar{x}_1\bar{x}_2\bar{x}_3 \oplus x_1x_2x_3$ is an ESOP, but not a GRM. Since two products have the same set of variables.

Fig. 2.3.1 shows the relation among 7 classes of AND-EXOR expressions, where $\mathcal{PPRM}$, $\mathcal{FPRM}$, $\mathcal{PSDRM}$, $\mathcal{KRO}$, $\mathcal{PSDKRO}$, $\mathcal{GRM}$ and $\mathcal{ESOP}$ denote the set of expressions for PPRMs, FPRMs, PSDRMs, KROs, PSDKROs, GRMs and ESOPs, respectively. Note that $\mathcal{ESOP}$ is the most general class, and the $\mathcal{PPRM}$ is the most restricted class. This means that ESOPs require the fewest products to represent given functions among these seven classes of expressions.

Fig. 2.3.2 shows a **programmable logic array (PLA)** for PPRMs. Note that the number of literal lines is n. Fig. 2.3.3 shows a PLA for FPRMs. In this case, EXOR gates in the inputs are used to produce either the positive or the negative literals. Fig. 2.3.4 shows a PLA for PSDRMs, KROs, PSDKROs, and ESOPs. In this case the number of literal lines is $2n$. Fig. 2.3.5 shows an

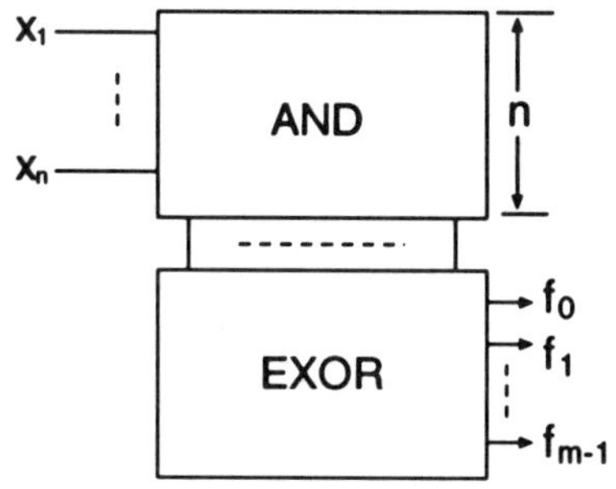

Figure 2.3.2 PLA for PPRMs.

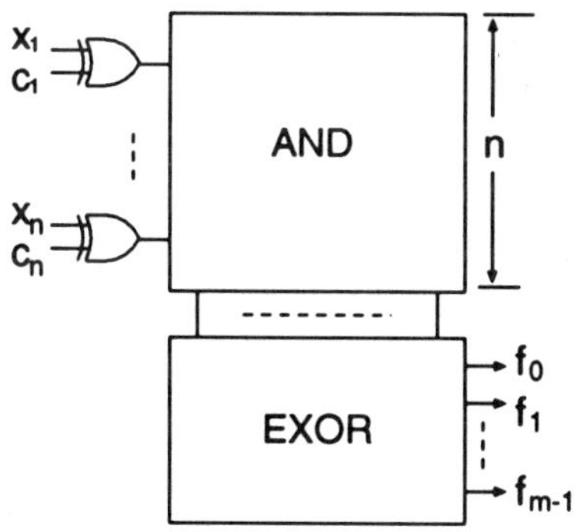

Figure 2.3.3 PLA for FPRMs.

easily testable PLA for GRMs [42]. In this PLA, in the normal mode, c is set to one, and the PLA realizes a GRM. In the test mode, c is set to zero, and the PLA realizes PPRMs, which are very easy to test [30, 15, 32]. Thus, this PLA has both features of GRMs and PPRMs, i.e., it requires fewer products and is easily testable.

In AND-EXOR two-level networks, tests that detect all detectable stuck-at faults can be generated in polynomial time of the number of the products. On the other hand, in AND-OR two-level networks, the test generation problem is not polynomial time solvable [48].

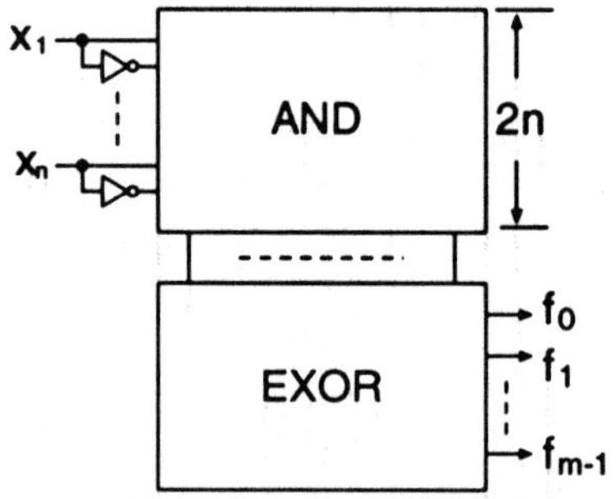

Figure 2.3.4 PLA for ESOPs.

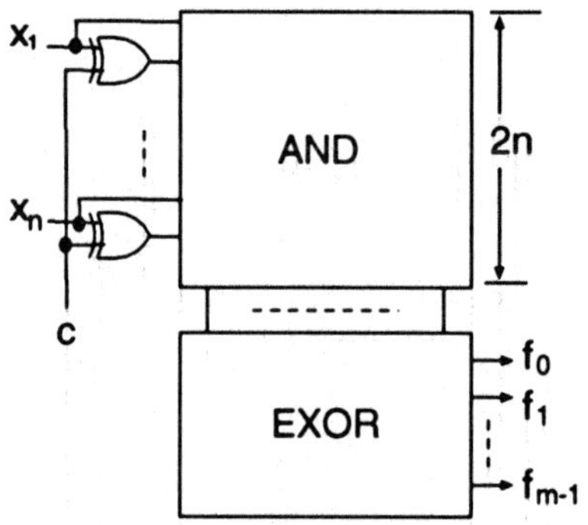

Figure 2.3.5 Easily testable PLA for GRMs.

ESOPs efficiently realize symmetric functions. An arbitrary symmetric function of $2r$ variables can be represented by using at most $2 \cdot 3^{r-1}$ products [14, 35]. ESOPs never require more products than SOPs to realize symmetric functions [31].

Table 2.3.2 Number of 4-variable functions requiring t products.

t	PPRM	FPRM	KRO	PSD RM	PSD KRO	GRM	ESOP	SOP
0	1	1	1	1	1	1	1	1
1	16	81	81	81	81	81	81	81
2	120	836	2268	2212	2268	2212	2268	1804
3	560	3496	8424	19160	21384	20856	21744	13472
4	1820	8878	15174	35150	36918	37818	37530	28904
5	4368	17884	19260	7404	3564	4512	3888	17032
6	8008	20152	19440	1480	1296	56	24	3704
7	11440	11600	864	24	0			512
8	12870	2336	0	24	24			26
9	11440	240	0					
10	8008	32	24					
11	4368							
12	1820							
13	560							
14	120							
15	16							
16	1							
av	8.00	5.50	4.73	3.80	3.70	3.68	3.66	4.13

av : average

2.3.4 Optimization of AND-EXOR Expressions

An expression for f is said to be **minimum** if it represents f and requires the minimum number of products among all expressions in its class.

For PPRM, no minimization is needed, since a PPRM is unique for a given function. The minimization of FPRMs corresponds to finding a Reed-Muller tree (Fig. 2.2.4) that minimizes the non-zero terminal nodes. There are 2^n different expressions. An efficient algorithm for FPRM optimization exists, which uses an extended truth vector and a weight vector [7, 23]. A straightforward implementation of this algorithm is practical for up to $n = 15$. The size of the memory and the computation time are both $O(3^n)$. Recently, an implementation using EXOR ternary decision diagram (ETDDs: will be described in section 2.5) has been developed. An FPRM with 94 inputs has been successfully minimized [44]. [50, 53] show optimization of FPRMs with don't cares.

The minimization of KROs is similar to that of FPRMs. There are 3^n different expressions. An algorithm that uses an extended truth vector and an extended weight vector requires the memory of $O(3^n)$ and $O(n3^n)$, respectively [7]. A straightforward implementation of this algorithm is practical up to $n = 15$. An ETDD based implementation has been also developed [44].

The minimization of PSDRMs is more complicated. For a given order of the input variables, 2^{2^n-1} different pseudo Reed-Muller trees (Fig. 2.2.6) exist. Because different orders of the input variables produce different expressions, we have to consider all possible orders of the input variables ($n!$). An efficient algorithm to optimize PSDRM for a given order of input variables is available [37]. To obtain an absolute minimum PSDRM, we have to consider all possible permutations of the input variables. Up to $n = 8$, the absolute minimum PSDRMs can be obtained in a reasonable computation time [37].

The minimization of PSDKRO is similar to that of PSDRMs. An ETDD based implementation has been developed. It obtains the minimum expression for a given order of the input variables [37].

The minimizations of GRMs and ESOPs are quite different from those of other expressions. BDD based algorithms have been developed [40, 43, 13]. However, they are practical for up to $n = 6$ or 7 only. Heuristic optimization programs have been developed for larger n: EXMIN2 [39] and EXORCISM-MV2 [47] are for ESOPs, and CANNES [6] and GRMIN [8] are for GRMs.

Table 2.3.2 compares the number of 4-variable functions requiring t products. There are 65,536 different functions. These functions are divided into 3984 P-equivalence classes, where each function in each class can be obtained from any other function by a permutation of variables. We minimized all the representative functions for each class. In the worst case, a PPRM requires 16 products, while an ESOP and a GRM require at most 6 products. SOPs (sum-of-products expressions) require at most 8 products. The last line in Table 2.3.2 shows the average numbers of products. On the average, GRMs and ESOPs require fewer products than SOPs. However, for some functions, ESOPs require more products than SOPs. For example, to represent $f = x_1y_1 \vee x_2y_2 \vee \cdots \vee x_ny_n$, an SOP requires only n products, while an ESOP requires $2^n - 1$ products. An average number of products in the PPRMs for n-variable functions is 2^{n-1} [42].

Table 2.3.3 Number of products to realize arithmetic functions.

Data	PPRM	FPRM	KRO	PSD RM	PSD KRO	GRM	ESOP	SOP
adr4	34	34	34	34	34	34	31	75
inc8	16	16	16	16	16	15	15	37
log8	253	193	171	163	128	105	96	123
mlp4	97	97	97	90	81	71	61	121
nrm4	216	185	157	150	105	96	69	120
rdm8	56	56	56	46	41	31	31	76
rot8	225	118	83	81	44	51	35	57
sqr8	168	168	168	164	146	121	112	178
sym9	210	173	173	127	90	126	51	84
wgt8	107	107	107	107	107	107	58	255

Table 2.3.3 shows the number of products for certain arithmetic functions [34]. For example, adr4 is the 4-bit adder. In this case, EXOR-based expressions require fewer products than the SOP. Also, for wgt8 (rd84 [52]), EXOR-based expressions require fewer products than the SOP. These functions all have 8 inputs. Multiple-output optimizations were done so that the total number of product terms in PLAs is minimized.

Table 2.3.4 compares the number of products for other benchmark functions [3, 52]. In this table, GRMs were simplified by GRMIN [8], ESOPs were simplified by EXMIN2 [39], and SOPs were simplified by MINI2 [33], so they are near minimum. For many functions, FPRMs require more products the SOPs, while ESOPs require fewer products than SOPs.

2.4 DECISION DIAGRAMS USING EXORS

In this section, we survey various decision diagrams using EXOR operators.

2.4.1 Decision Trees and Decision Diagrams

The decision diagrams (DDs) are obtained by simplifying decision trees using the following rules:

Table 2.3.4 Number of products for other benchmark functions.

Data	IN	OUT	Products					
			FPRM	KRO	PSD KRO	GRM	ESOP	SOP
amd	14	24	156	104	64	71	58	66
b9	16	5	105	105	81	81	81	119
bc0	26	11	1117	467	180	197	168	177
cordic	23	2	6438	1980	1036	1742	776	914
cps	24	109	291	249	154	175	137	172
duke2	22	29	255	209	102	87	81	86
ex7	16	5	105	105	81	81	81	119
gary	15	11	349	242	115	116	102	107
in2	19	10	355	262	117	125	108	134
in7	26	10	72	72	36	38	35	54
intb	15	7	1815	1605	492	375	327	629
m181	15	9	67	65	31	29	29	41
misg	56	23	104	104	67		66	69
mish	94	43	53	53	53		53	82
misj	35	14	17	17	16		16	35
misex3	14	14	3536	1421	689	764	553	690
rckl	32	7	32	32	32	32	32	32
ryy6	16	1	64	48	40	40	40	112
t1	21	23	232	209	101	94	90	103
t481	16	1	13	13	13	13	13	481
tial	14	8	3683	2438	790	1003	487	579
ts10	22	16	432	128	128	128	128	128
x6dn	39	5			104		95	81

1. If two sub-graphs are isomorphic, delete one, and connect the severed edge to the remaining sub-graph (Fig. 2.4.1).

2. Delete the Shannon node if its two descendent nodes are identical (Fig. 2.4.2). Delete the Davio node if the edge labeled with x^* is connected to the constant 0 (Fig. 2.4.3).

For example, consider the pseudo Kronecker tree shown in Fig. 2.4.5. In this tree, two sub-trees having Shannon nodes for x_3 are isomorphic. If we delete the right sub-tree, then we have a DD shown in Fig. 2.4.6. In this diagram, the positive Davio node for x_3 has an edge labeled with x_3 that is connected to the constant 0. Thus, we can delete the positive Davio node, yielding have the DD shown in Fig. 2.4.7. Because only two terminal nodes for constants 0 and

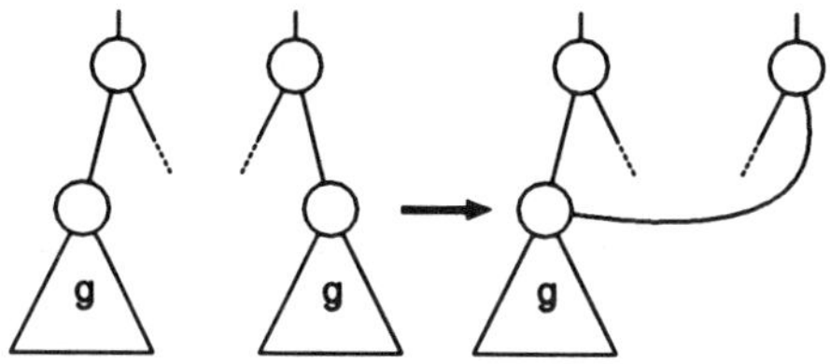

Figure 2.4.1 Merging isomorphic sub-graphs.

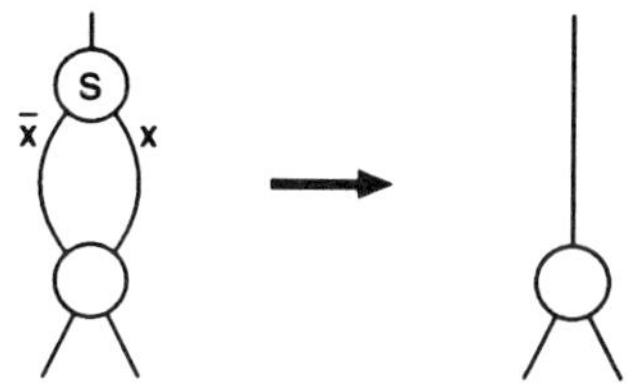

Figure 2.4.2 Elimination of the Shannon node.

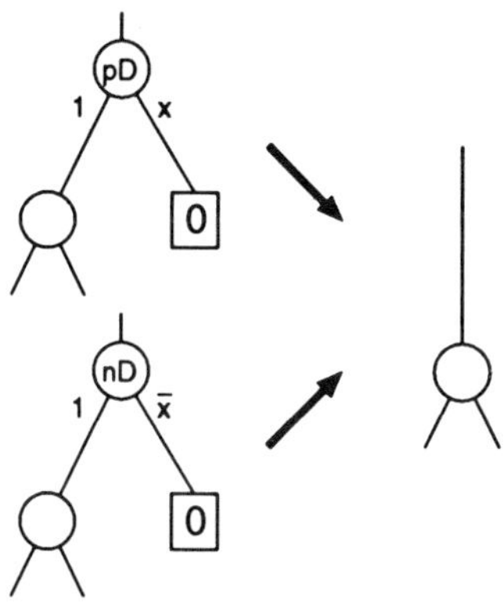

Figure 2.4.3 Elimination of the Davio nodes.

1 are necessary, we have the DD shown in Fig. 2.4.8. Here, we cannot simplify the graph anymore. Such a DD is called a **reduced DD**.

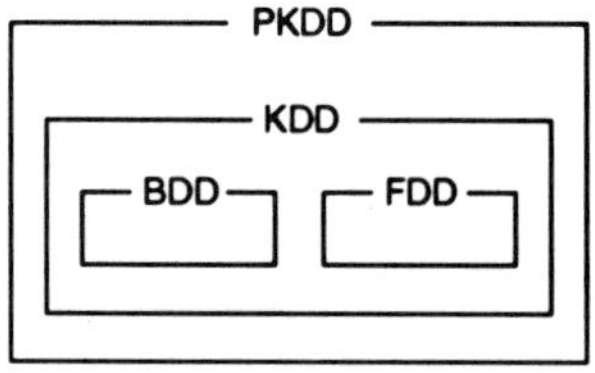

Figure 2.4.4 Relation among decision diagrams.

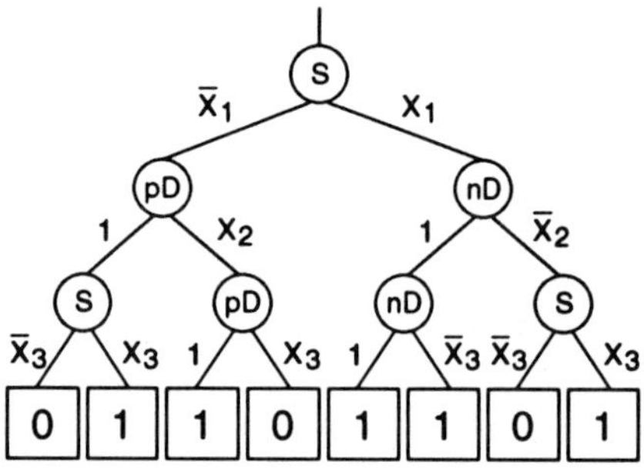

Figure 2.4.5 Pseudo Kronecker tree.

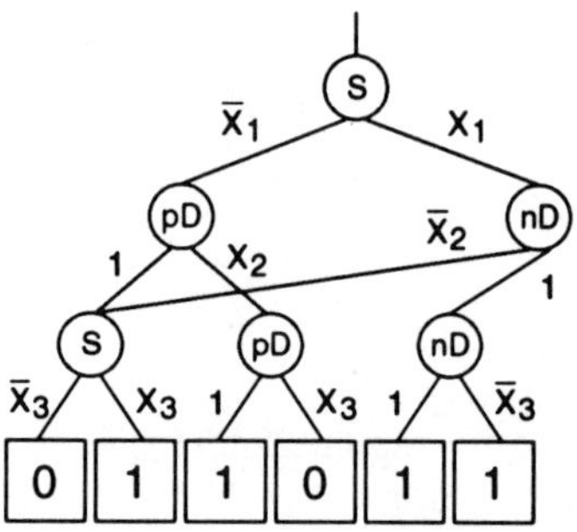

Figure 2.4.6 After deleting isomorphic sub-graph.

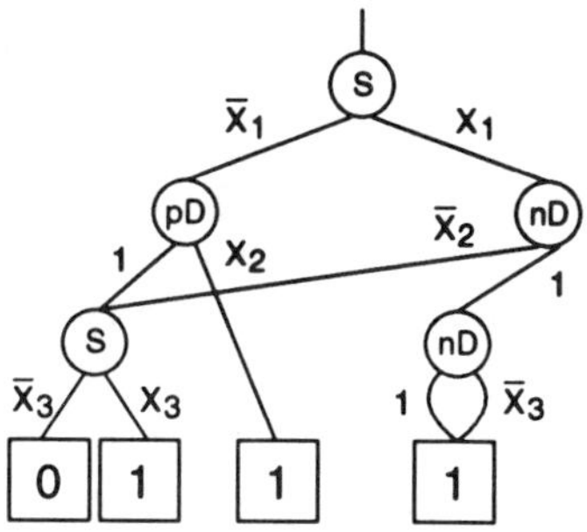

Figure 2.4.7 After deleting redundant positive Davio node.

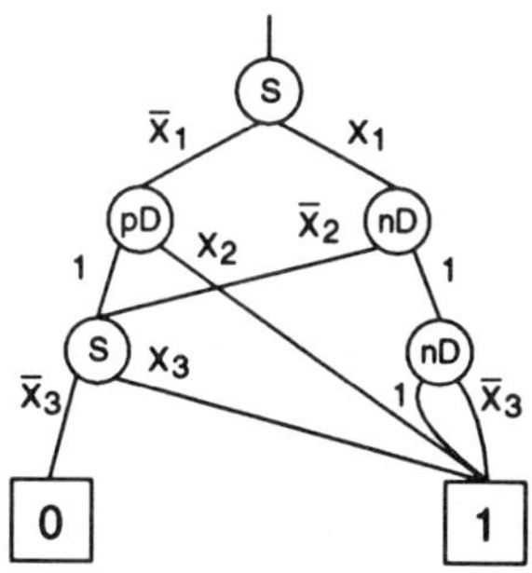

Figure 2.4.8 Reduced PKDD.

2.4.2 Relation Among Diagrams

Table 2.3.1 also shows the relations of trees and decision diagrams. Binary decision diagrams (BDDs) are derived from Shannon trees; functional decision diagrams (FDDs) are derived from positive Davio trees; Reed-Muller decision diagrams are derived from Reed-Muller trees; Kronecker decision diagrams (KDDs) are derived from Kronecker trees; and pseudo-Kronecker decision diagrams (PKDDs) are derived from pseudo-Kronecker trees. Note that some researchers define FDDs as those that are derived from Reed-Muller trees [21, 10]. KDDs are also called as OKFDDs (ordered Kronecker functional decision diagrams) [9].

PKDDs are the most general diagrams, while BDDs and FDDs are the most restricted ones. This means that PKDDs require the fewest nodes to represent given functions. Fig. 2.4.4 shows the relation among these decision diagrams.

There exists a class of functions whose BDDs have exponential size, while whose FDDs have polynomial size. On the other hand, there exist a class of functions whose FDDs have exponential size, while whose BDDs have polynomial size [2].

2.4.3 Optimization of Decision Diagrams

This part considers methods to represent given functions by decision diagrams with as few nodes as possible. In the case of PKDDs, two ways exist to make a choice from which one chooses the one with the fewest nodes:

1. Permutation of the input variables ($n!$ ways).
2. Selection of the expansion for each node (3^{2^n-1} ways).

Note that there are $n!3^{2^n-1}$ combinations to consider. Thus, it is virtually impossible to consider all these possibilities. Heuristic algorithms to optimize various decision diagrams exist [41]. Table 2.4.1 compares the number of nodes for various decision diagrams obtained by heuristic methods. It shows that, on the average, PKDDs require 30% fewer nodes than BDDs. Optimization of Reed-Muller decision diagrams are considered in [25, 1], and optimization of KDDs are considered in [9].

2.5 EXOR TERNARY DECISION DIAGRAMS

This part considers EXOR ternary decision diagrams (ETDDs), the decision diagram derived from ternary trees. ETDDs are classified as canonical and as non-canonical [29].

Table 2.4.1 Number of nodes for various decision diagrams.

function	in	out	# of nodes			
			BDD	FDD	KDD	PKDD
9sym	9	1	33	27	27	26
adr4	8	5	29	15	15	15
bc0	26	11	584	687	575	414
duke2	22	29	373	367	332	269
mlp4	8	8	141	108	108	99
tial	14	8	690	543	456	381
ratio			1.00	1.01	0.84	0.70

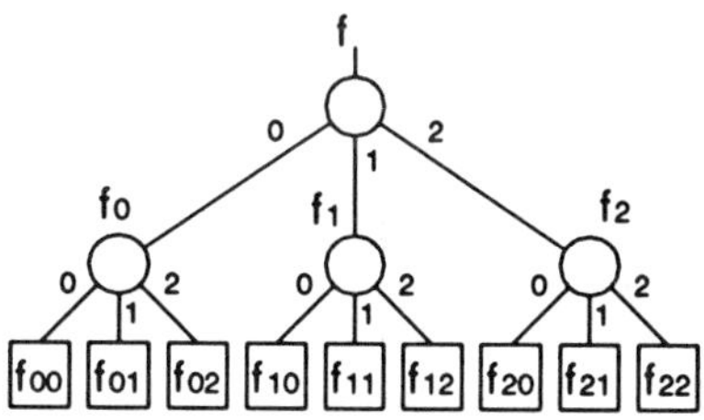

Figure 2.5.1 EXOR ternary decision tree.

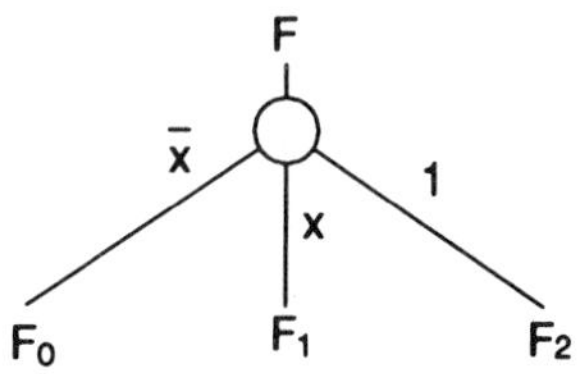

Figure 2.5.2 Ternary decision tree representing an ESOP.

2.5.1 Canonical ETDDs

Fig. 2.5.1 shows the EXOR ternary decision tree. Note that f_0 and f_1 appear in the Shannon expansion, and f_2 represents $f_0 \oplus f_1$. This data structure contains important information for optimizing AND-EXOR two-level expressions [37, 44, 26], as well as decision diagrams [45]. EXOR ternary decision diagrams (ETDDs) are obtained from EXOR ternary decision tree by using similar simplification rules to EXOR binary decision diagrams. The number of nodes for a reduced ETDD is $O(3^n/n)$ for arbitrary functions, and $O(n^3)$ for symmetric functions [37].

2.5.2 Non-Canonical ETDDs

Non-canonical ETDDs are used to represent logical expressions [19, 38, 51]. Fig. 2.5.2 shows an example of an EXOR ternary decision tree representing an ESOP

$$F = \bar{x}F_0 \oplus xF_1 \oplus 1 \cdot F_2.$$

Note that three edges have literals as labels. Again, ETDDs are obtained from EXOR ternary decision trees in a similar way. A path from the root node to a constant node represents a product term. For a given order of the input variables, a unique ETDD exists for an ESOP, while many ETDDs exist for a logic function.

This representation is similar to AND-OR ternary decision tree [38], where Fig. 2.5.2 represents an SOP:

$$F = \bar{x}F_0 \vee xF_1 \vee 1 \cdot F_2.$$

For a given order of the input variables, a unique AND-OR TDD exists for an SOP.

2.6 CONCLUSION AND COMMENTS

In this chapter, we surveyed various representations using EXOR operators. We defined seven classes of AND-EXOR expressions. Among them, ESOPs have compact representations for arithmetic and symmetric functions. Thus, ESOPs can be used as initial solutions of logic synthesis. Exact minimization of ESOPs is possible only for functions with a small number of inputs. Some heuristic simplification algorithms have been developed, but they are still immature compared with ones for SOPs.

We also defined various decision diagrams using EXOR operators. Among them, PKDDs require the fewest nodes. In a PKDD for a function f, if we replace each node with one of the circuits in Fig. 2.2.1, then we have a network for f. Thus, it can be used as an initial solution for a multi-level logic network for f [45]. Finally, we introduced ETDDs. Canonical ETDDs are useful for optimizing AND-EXOR expressions and EXOR based decision diagrams. Non-canonical ETDDs are used to represent and optimize ESOPs [19].

Acknowledgments

This work was supported in part by a Grant in Aid for Scientific Research of the Ministry of Education, Science and Culture of Japan. The comments of many reviewers improved the quality of presentation. Especially, Prof. Jon T. Butler's comments were useful.

REFERENCES

[1] B. Becker, R. Drechsler and M. Theobald, "On the implementation of a package for efficient representation and manipulation of functional decision diagrams," *IFIP 10.5 Workshop on Applications of the Reed-Muller Expansion in Circuit Design*, September 1993.

[2] B. Becker, R. Drechsler, and M. Theobald, "OKFDDs versus OBDDs and OFDDs," *International Workshop on Logic Synthesis*, Lake Tahoe, May 1995.

[3] R. K. Brayton, G. D. Hachtel, C. T. McMullen, and A. L. Sangiovanni-Vincentelli, *Logic Minimization Algorithms for VLSI Synthesis*, Boston, MA. Kluwer, 1984.

[4] R. E. Bryant, "Graph-based algorithms for Boolean function manipulation," *IEEE Trans. Comput.* Vol. C-35, No. 8, Aug. 1986, pp. 677-691.

[5] E. M. Clarke, M. Fujita, P.C. McGeer, K.L. McMillan and J. C. Yang, "Multi-terminal binary decision diagrams: An efficient data structure for matrix representation" *Int. Workshop on Logic Synthesis,* May 1993, pp. 6A 1-15. E. M. Clarke, K.L. McMillan, X. Zhao, M. Fujita, P.C. McGeer, and J. C. Yang, "Spectral transforms for large Boolean functions with application to technology mapping," *Int. Workshop on Logic Synthesis,* May 1993, pp. P6b 1-15. (Also Chapter 4 of this book).

[6] L. Csanky, M. A. Perkowski, and I. Schäfer, "Canonical restricted mixed polarity exclusive-or sums of products and the efficient algorithm for their minimization," *Proc. IEE,* Part. E, Vol. 140, No. 1, pp. 69-77, Oct. 1993.

[7] M. Davio, J-P. Deschamps, and A. Thayse, *Discrete and Switching Functions,* McGraw-Hill International, 1978.

[8] D. Debnath and T. Sasao, "GRMIN: a heuristic simplification algorithm for generalized Reed-Muller expressions," *IFIP WG 10.5 Workshop on Applications of the Reed-Muller Expansion in Circuit Design,* August 1995.

[9] R. Drechsler, A. Sarabi, M. Theobald, B. Becker and M.A. Perkowski, "Efficient representation and manipulation of switching functions based on ordered Kronecker functional decision diagrams," *Proc. 1994 Design Automation Conference,* June 1994, pp. 415-419.

[10] R. Drechsler, M. Theobald, and B. Becker, "Fast FDD based minimization of generalized Reed-Muller forms," *In European Design Automation Conf.,* 1994, pp.2-7.

[11] R. Drechsler, B. Becker and A. Jahnke, "On variable ordering and decomposition type choice in OKFDDs," *IFIP WG. 10.5 Workshop on Applications of the Reed-Muller Expansions in Circuit Design,* August 1995.

[12] R. Drechsler and B. Becker, "Dynamic minimization of OKFDDs," *IFIP WG. 10.5 Workshop on Applications of the Reed-Muller Expansions in Circuit Design,* August 1995. (Also Chapter 7 of this book).

[13] M. Escobar and F. Somenzi, "Synthesis of AND-EXOR expressions via satisfiability," *IFIP WG. 10.5 Workshop on Applications of the Reed-Muller Expansions in Circuit Design,* August 1995.

[14] S. Even, I. Kohavi and A. Paz, "On minimal modulo-2 sums of products for switching functions," *IEEE Trans. Electronic Computers,* Vol. EC-16, pp. 671-674, Oct. 1967.

[15] H. Fujiwara, *Logic Testing and Design for Testability,* The MIT Press, 1985.

[16] J. Gergov and C. Meinel, "Mod-2-OBDD's: a generalization of OBDD's and EXOR-sum-of-products," *IFIP 10.5 Workshop on Applications of the Reed-Muller Expansion in Circuit Design,* September 1993.

[17] D. Green, *Modern Logic Design,* Addison-Wesley Publishing Company, 1986.

[18] D. H. Green, "Families of Reed-Muller canonical forms," *International Journal of Electronics,* Vol. 63. No. 2, pp. 259-280, Jan. 1991.

[19] M.A. Heap, W.A. Rogers, and M.R. Mercer, "A synthesis algorithm for two-level XOR based circuits," *IEEE 1992 International Conference on Computer Design: VLSI in Computers and Processors,* ICCD'92, 459-62, 1992.

[20] U. Kebschull, E. Schubert and W. Rosenstiel, "Multilevel logic synthesis based on functional decision diagrams," *EDAC 92,* 1992, pp. 43-47.

[21] U. Kebschull and W. Rosenstiel, "Efficient graph based computation and manipulation of functional decision diagrams," *In Proc. EDAC'93,* pp. 43-47, 1993.

[22] N. Koda and T. Sasao, "LP equivalence class of logic functions," *IFIP 10.5 Workshop on Applications of the Reed-Muller Expansion in Circuit Design,* Hamburg, Sept. 16-17, 1993.

[23] P. K. Lui and J. Muzio, "Boolean matrix transforms for the parity spectrum and the minimization of modulo-2 canonical expansions," *Proc. IEE.* vol. 138, No. 6, pp. 411-417, Nov. 1991.

[24] A. Mukhopadhyay and G. Schmitz, "Minimization of exclusive OR and logical equivalence of switching circuits," *IEEE Trans. Comput.,* C-19, pp. 132-140, 1970.

[25] L. McKenzie, L. Xu and A. Almaini, "Graphical representation of generalized Reed-Muller expansions," *IFIP 10.5 Workshop on Applications of the Reed-Muller Expansion in Circuit Design,* Hamburg, Sept. 16-17, 1993.

[26] G. Papakonstantinou, "Minimization of modulo-2 sum of products," *IEEE Trans. on Comput.*, Vol. C-28, pp. 163-167, Feb. 1979.

[27] M. Perkowski and M. Chrzanowska-Jeske, "An exact algorithm to minimize mixed-radix exclusive sums of products for incompletely specified Boolean functions," *Proc. ISCAS,* pp. 1652-1655, June 1990.

[28] M. A. Perkowski, L. Csanky, A. Sarabi, and I. Schäfer, "Fast minimization of mixed-polarity AND-XOR canonical networks," *Proc. ICCD-92,* pp. 33-36, Oct. 1992.

[29] M. A. Perkowski, A. Sarabi, and I. Schäfer, "Multi-level logic synthesis based on Kronecker and Boolean ternary decision diagrams for incompletely specified functions," *(to appear VLSI journal).*

[30] S. M. Reddy, "Easily testable realization for logic functions," *IEEE Trans. Computers,* Vol. C-21, No. 11, pp. 1183-1188.

[31] U. Rollwage, "The complexity of mod-2 sum PLA's for symmetric functions," *IFIP 10.5 Workshop on Application of the Reed-Muller Expansion in Circuit Design,* Sept. 1993.

[32] K. K. Saluja and S. M. Reddy, "Fault detecting test set for Reed-Muller canonic networks," *IEEE Trans. on Comput.*, Vol. C-24, No. 10, pp. 995-998, 1975.

[33] T. Sasao, "Input variable assignment and output phase optimization of PLA's," *IEEE Trans. on Comput.*, Vol. C-33, No. 10, pp. 879-894, Oct. 1984.

[34] T. Sasao, "HART: A hardware for logic minimization and verification," *IEEE International Conference on Computer Design: VLSI in Computer, ICCD'85,* New York, pp. 713-718, Oct. 7-10, 1985.

[35] T. Sasao and P. Besslich, "On the complexity of MOD-2 Sum PLA's," *IEEE Trans. Comput.*, vol. 39, No. 2, pp. 262-266, Feb. 1990.

[36] T. Sasao, "Transformation of multiple-valued input two-valued output functions and its application to simplification of exclusive-or sum-of-products expressions," *Proc. International Symposium on Multiple-Valued Logic,* pp. 270-279, May 1991.

[37] T. Sasao (ed.), *Logic Synthesis and Optimization,* Kluwer Academic Publishers 1993-01.

[38] T. Sasao, "Ternary decision diagram and their applications," *International Workshop on Logic Synthesis,* Lake Tahoe, May 1993, pp. 6c, 1-11. (Also Chapter 12 of this book).

[39] T. Sasao, "EXMIN2:A simplification algorithm for exclusive-OR Sum-of-products expressions for multiple-valued input two-valued output functions, *IEEE Transactions on Computer-Aided Design of Integrated Circuits and Systems,* vol. 12, No. 5, May 1993, pp. 621-632.

[40] T. Sasao, "An exact minimization of AND-EXOR expressions using BDDs," *IFIP 10.5 Workshop on Application of the Reed-Muller Expansion in Circuit Design,* Sept. 1993.

[41] T. Sasao and J. T. Butler, "A design method for look-up table type FPGA by pseudo-Kronecker expansion," *Proc. International Symposium on Multiple-Valued Logic*, May 1994, pp. 97-106.

[42] T. Sasao, "Easily testable realization for generalized Reed-Muller expressions," *IEEE The 3rd Asian Test Symposium,* November 15-17, 1994, Nara Japan, pp. 157-162.

[43] T. Sasao and D. Debnath, "An exact minimization algorithm for generalized Reed-Muller expressions" *IEEE Asia-Pacific Conference on Circuits and Systems (APCCAS'94)*, December 5-8, 1994, Taipei, Taiwan, pp. 460-465.

[44] T. Sasao and F. Izuhara, "Exact minimization of AND-EXOR expressions using multi-terminal EXOR ternary decision diagram," *IFIP WG 10.5 Workshop on Applications of the Reed-Muller Expansion in Circuit Design,* August 1995. (Also Chapter 8 of this book).

[45] T. Sasao, S. Wada, H. Hamachi, and M. Matsuura, "Multi-level logic synthesis based on pseudo-Kronecker decision diagrams and logic transformation," *IFIP WG 10.5 Workshop on Applications of the Reed-Muller Expansion in Circuit Design,* August 1995.

[46] T. Sasao, "A design Method for AND-OR-EXOR three-level networks," *International Workshop on Logic Synthesis*, Lake Tahoe, May 1995.

[47] N. Song, and M. Perkowski, "EXORCISM-MV-2: Minimization of exclusive-sum-of-products expressions for multiple-valued input incompletely specified functions," *1993 International Symposium on Multiple-Valued Logic,* May 24, 1993. pp. 132-137.

[48] S. Toida and N. S. V. Rao, "On test generation for combinational circuits consisting of AND and EXOR gates," *Digest of Papers of 1992 IEEE VLSI Test Symposium. Design, Test and Application: ASICs and Systems-on-a-Chip,* pp. 113-118,1992.

[49] C.C. Tsai and M. Marek-Sadowska, "Efficient minimization algorithms for fixed polarity and/xor canonical networks," *In Great Lake Symposium VLSI,* pp. 76-79, 1993.

[50] D. Varma and E. A. Trachtenberg, "Computation of Reed-Muller expansions of incompletely specified Boolean functions from reduced representations," IEE Proc. E, 1991, vol. 138, pp. 85-92.

[51] K. Yasuoka,"Ternary decision diagrams to represent ringsum-of-products forms," *IFIP WG. 10.5 Workshop on Applications of the Reed-Muller Expansions in Circuit Design*, August 1995.

[52] S. Yang,"Logic synthesis and optimization benchmark user guide," version 3.0, *MCNC,* Jan. 1991.

[53] A. Zakrevskij, "Minimum polynomial implementation of systems of incompletely specified Boolean functions," *IFIP WG. 10.5 Workshop on Applications of the Reed-Muller Expansions in Circuit Design*, August 1995.

3

SPECTRAL TRANSFORM DECISION DIAGRAMS

Radomir S. STANKOVIĆ[1]
Tsutomu SASAO[2]
Claudio MORAGA[3]

[1]*Department of Computer Science*
Faculty of Electronics
18 000 Nis, Yugoslavia

[2]*Department of Computer Science and Electronics*
Kyushu Institute of Technology
Iizuka, Japan

[3]*Department of Computer Science*
Dortmund University
Germany

Abstract— This chapter proposes spectral decision diagrams (STDDs), that are graphical representations of spectral transforms of switching functions and integer-valued functions. Binary decision diagrams (BDDs) and functional decision diagrams (FDDs) are graphical representations for switching functions and their Reed-Muller transforms, respectively. Multi-terminal decision diagrams (MTBDDs), arithmetic transform decision diagrams (ACDDs), and Walsh transform decision diagrams (WDDs) are graphical representations for integer-valued functions, their arithmetic transforms, and their Walsh transforms, respectively. This chapter shows that an STDD represents a function and its spectral transform at the same time. As for n-bit adders, ACDDs and WDDs require $O(n)$ nodes while MTBDDs require $O(2^n)$ nodes. As for n-bit multipliers, ACDDs and WDDs require $O(n^2)$ nodes while MTBDDs require $O(4^n)$ nodes.

Table 3.1.1 Truth table for a half adder.

x_0	y_0	f_1	f_0
0	0	0	0
0	1	0	1
1	0	0	1
1	1	1	0

3.1 INTRODUCTION

An n-variable **switching function** is a mapping:

$$\{0,1\}^n \rightarrow \{0,1\}.$$

An **integer-valued function** in this chapter denotes a mapping:

$$\{0,1\}^n \rightarrow \{0,1,\ldots,p-1\}, \quad \text{where} \quad p \geq 2.$$

For example, consider the half adder shown in Table 3.1.1. This is a two-input two-output function. If we consider that the outputs represent decimal numbers $\{0,1,2\}$, then this multiple-output function is represented by the integer-valued function:

$$\{0,1\}^2 \rightarrow \{0,1,2\}.$$

In general, an n-input m-output function is represented by an integer-valued function:

$$\{0,1\}^n \rightarrow \{0,1,\ldots,2^m-1\}.$$

Various decision diagrams (DDs) to represent switching functions and integer-valued functions are proposed [6, 7, 25, 26, 27]. It is often more convenient to represent transformed functions instead of original functions [13, 19]. The coefficients of the transformed functions are called spectrum of the transform.

In this chapter, we will consider DDs that represent the spectra of the given functions. Such DDs are called spectral decision diagrams (STDDs). Binary decision diagrams (BDDs) [3] and functional decision diagrams (FDDs) [16] are STDDs for switching functions. Multi-terminal binary decision diagrams (MTBDDs) [6], algebraic decision diagrams (ADDs) [2], and binary moment diagrams (BMDs) [4] are STDDs for integer-valued functions.

This chapter is organized as follows: Section 3.2 reviews basic matrix theory. Section 3.3 first introduces Shannon trees, positive Davio trees, BDDs, and

FDDs. Then, it shows relationship between BDDs and FDDs. Section 3.4 presents STDDs as generalization of BDDs and FDDs. Sections 3.5 and 3.6 introduce arithmetic transform and Walsh transform, respectively. Section 3.7 introduces generalized BDD reduction rules for STDDs, and define arithmetic transform decision diagrams (ACDDs) and Walsh decision diagrams (WDDs). Section 3.8 extends STDDs to represent integer-valued functions. Section 3.9 shows that to represent n-bit adders and multipliers, ACDDs (WDDs) require $O(n)$ and $O(n^2)$ nodes, while MTBDDs require $O(2^n)$ and $O(4^n)$ nodes, respectively.

3.2 MATRIX THEORY

In this section, we review the matrix theory which is used in this chapter [13]. An n-tuple $\mathbf{a} = (a_1, a_2, \ldots, a_n)$, where a_i is a binary, integer, real, or complex number, is a **vector**. The set of all possible n-tuple vectors forms a **vector space**. The set of vectors V is **linearly dependent** iff some vector in V is represented as a linear combination of other vectors in V. V is **linearly independent** iff it is not linearly dependent.

Example 3.2.1 *Let* $\mathbf{a} = (1,0)$, $\mathbf{b} = (0,1)$, *and* $\mathbf{c} = (1,1)$. $V = \{\mathbf{a}, \mathbf{b}, \mathbf{c}\}$ *is linearly dependent since* $\mathbf{c}$ *is represented as* $\mathbf{a}+\mathbf{b} = (1,0)+(0,1) = (1,1)$. $\{\mathbf{a}, \mathbf{b}\}$ *is linearly independent. Also* $\{\mathbf{b}, \mathbf{c}\}$ *and* $\{\mathbf{a}, \mathbf{c}\}$ *are linearly independent.* ∎

Let $\mathbf{a} = (a_1, a_2, \ldots, a_n)$ and $\mathbf{b} = (b_1, b_2, \ldots, b_n)$ be vectors. The **inner product** of vectors $\mathbf{a}$ and $\mathbf{b}$ is

$$\mathbf{a} \cdot \mathbf{b} = a_1 b_1 + a_2 b_2 + \cdots + a_n b_n.$$

Two vectors $\mathbf{a}$ and $\mathbf{b}$ are **orthogonal** if $\mathbf{a} \cdot \mathbf{b} = 0$. If we consider the calculation in $GF(2)$, we use the symbol $\oplus$ instead of $+$. In this case $1 \oplus 1 = 0$.

Given a matrix $\mathbf{M}$, the **transposed matrix** $\mathbf{M}^t$ is obtained by interchanging the rows and columns. Let $\mathbf{M}$ be a real square matrix. $\mathbf{M}$ is orthogonal if $\mathbf{M}\mathbf{M}^t = \mathbf{M}^t\mathbf{M} = \mathbf{I}$, where $\mathbf{I}$ is the **identity matrix**. If $\mathbf{M}\mathbf{M}^t = \mathbf{M}^t\mathbf{M} = k\mathbf{I}$, then $\mathbf{M}$ is **orthogonal up to the constant** k. $\mathbf{M}^{-1}$ is the **inverse** of $\mathbf{M}$ if $\mathbf{M}\mathbf{M}^{-1} = \mathbf{M}^{-1}\mathbf{M} = \mathbf{I}$. $\mathbf{M}$ has its inverse iff the column vectors of $\mathbf{M}$ are linearly independent. If $\mathbf{M}$ is orthogonal up to the constant k, then $\mathbf{M}^{-1} = (\frac{1}{k})\mathbf{M}^t$. $\mathbf{M}$ is **symmetric** if $\mathbf{M} = \mathbf{M}^t$. $\mathbf{M}$ is **self-inverse** if $\mathbf{M} = \mathbf{M}^{-1}$. If $\mathbf{M} = k\mathbf{M}^{-1}$, then $\mathbf{M}$ is **self-inverse up to the constant** k. If $\mathbf{M}$ is orthogonal and symmetric, then $\mathbf{M}$ is self-inverse.

Example 3.2.2

1) Let $\mathbf{B} = \begin{bmatrix} 1 & 0 \\ 0 & 1 \end{bmatrix}$. *Since* $\mathbf{B} = \mathbf{B}^t$, $\mathbf{B}$ *is symmetric. Since*

$$\mathbf{B}\mathbf{B}^t = \begin{bmatrix} 1 & 0 \\ 0 & 1 \end{bmatrix} = \mathbf{I},$$

$\mathbf{B}$ *is orthogonal. Since* $\mathbf{B} = \mathbf{B}^{-1} = \mathbf{I}$, $\mathbf{B}$ *is self-inverse.*

2) Let $\mathbf{A} = \begin{bmatrix} 1 & 0 \\ -1 & 1 \end{bmatrix}$. $\mathbf{A}$ *is not symmetric. Since*

$$\begin{bmatrix} 1 & 0 \\ -1 & 1 \end{bmatrix}\begin{bmatrix} 1 & 0 \\ 1 & 1 \end{bmatrix} = \begin{bmatrix} 1 & 0 \\ 0 & 1 \end{bmatrix} = \mathbf{I}, \ \mathbf{A}^{-1} = \begin{bmatrix} 1 & 0 \\ 1 & 1 \end{bmatrix}.$$

Thus, $\mathbf{A} \neq \mathbf{A}^{-1}$, *i.e.,* $\mathbf{A}$ *is not self-inverse. Since*

$$\mathbf{A}\mathbf{A}^t = \begin{bmatrix} 1 & 0 \\ -1 & 1 \end{bmatrix}\begin{bmatrix} 1 & -1 \\ 0 & 1 \end{bmatrix} = \begin{bmatrix} 1 & -1 \\ -1 & 2 \end{bmatrix},$$

$\mathbf{A}$ *is not orthogonal.*

3) Let $\mathbf{W} = \begin{bmatrix} 1 & 1 \\ 1 & -1 \end{bmatrix}$. $\mathbf{W}$ *is symmetric. Since*

$$\mathbf{W}\mathbf{W}^t = \begin{bmatrix} 1 & 1 \\ 1 & -1 \end{bmatrix}\begin{bmatrix} 1 & 1 \\ 1 & -1 \end{bmatrix} = \begin{bmatrix} 2 & 0 \\ 0 & 2 \end{bmatrix} = 2\begin{bmatrix} 1 & 0 \\ 0 & 1 \end{bmatrix} = 2\mathbf{I},$$

$$\mathbf{W}^{-1} = \frac{1}{2}\begin{bmatrix} 1 & 1 \\ 1 & -1 \end{bmatrix} = \frac{1}{2}\mathbf{W}.$$

In this case, $\mathbf{W}$ *is self-inverse up to the constant 2.*

4) Let $\mathbf{R} = \begin{bmatrix} 1 & 0 \\ 1 & 1 \end{bmatrix}$. *Since* $\mathbf{R} \neq \mathbf{R}^t$, $\mathbf{R}$ *is not symmetric.* $\mathbf{R}$ *is not orthogonal in* $GF(2)$, *since*

$$\mathbf{R}\mathbf{R}^t = \begin{bmatrix} 1 & 0 \\ 1 & 1 \end{bmatrix}\begin{bmatrix} 1 & 1 \\ 0 & 1 \end{bmatrix} = \begin{bmatrix} 1 & 1 \\ 1 & 0 \end{bmatrix} \neq \mathbf{I}.$$

$\mathbf{R}$ *is self-inverse in* $GF(2)$ *since,*

$$\mathbf{R}\mathbf{R} = \begin{bmatrix} 1 & 0 \\ 1 & 1 \end{bmatrix}\begin{bmatrix} 1 & 0 \\ 1 & 1 \end{bmatrix} = \begin{bmatrix} 1 & 0 \\ 0 & 1 \end{bmatrix} = \mathbf{I}.$$

It follows that $\mathbf{R}^{-1} = \mathbf{R}$. ■

Let **A** and **B** be $(n \times n)$ square matrices, where

$$\mathbf{A} = \begin{bmatrix} a_{11} & a_{12} & \cdots & a_{1n} \\ a_{21} & a_{22} & \cdots & a_{2n} \\ \vdots & \vdots & & \vdots \\ a_{n1} & a_{n2} & \cdots & a_{nn} \end{bmatrix}.$$

The **Kronecker product** of A and B is

$$\mathbf{A} \otimes \mathbf{B} = \begin{bmatrix} a_{11}\mathbf{B} & a_{12}\mathbf{B} & \cdots & a_{1n}\mathbf{B} \\ a_{21}\mathbf{B} & a_{22}\mathbf{B} & \cdots & a_{2n}\mathbf{B} \\ \vdots & \vdots & & \vdots \\ a_{n1}\mathbf{B} & a_{n2}\mathbf{B} & \cdots & a_{nn}\mathbf{B} \end{bmatrix}.$$

Thus, if **A** and **B** are $(n \times n)$ matrices, then $\mathbf{A} \otimes \mathbf{B}$ is a $(n^2 \times n^2)$ matrix. The $\otimes$ operator is associative, but not commutative: $\mathbf{A} \otimes (\mathbf{B} \otimes \mathbf{C}) = (\mathbf{A} \otimes \mathbf{B}) \otimes \mathbf{C} = \mathbf{A} \otimes \mathbf{B} \otimes \mathbf{C}$. $\mathbf{A} \otimes \mathbf{B} \neq \mathbf{B} \otimes \mathbf{A}$.

Example 3.2.3

$$\textit{Let } \mathbf{A} = \begin{bmatrix} 1 & 0 \\ -1 & 1 \end{bmatrix} \textit{ and } \mathbf{B} = \begin{bmatrix} 1 & 0 \\ 0 & 1 \end{bmatrix},$$

$$\mathbf{A} \otimes \mathbf{B} = \begin{bmatrix} 1 & 0 & 0 & 0 \\ 0 & 1 & 0 & 0 \\ -1 & 0 & 1 & 0 \\ 0 & -1 & 0 & 1 \end{bmatrix}, \quad \mathbf{B} \otimes \mathbf{A} = \begin{bmatrix} 1 & 0 & 0 & 0 \\ -1 & 1 & 0 & 0 \\ 0 & 0 & 1 & 0 \\ 0 & 0 & -1 & 1 \end{bmatrix}.$$

■

3.3 BDDS AND FDDS

The discussion in this section is related to the theory of EXOR representations of logic functions (Chapter 2, [27]).

3.3.1 Shannon Trees

Shannon trees are defined by using the Shannon expansion $f = \bar{x}_i f_0 \oplus x_i f_1$. Recursive application of the Shannon expansion of f given by the truth vector

$\mathbf{F} = [f(0), \ldots, f(2^n - 1)]^t$ produce the matrix representation:

$$f = \mathbf{X}_b \mathbf{B}(n) \mathbf{F}, \tag{3.3.1}$$

where

$$\mathbf{X}_b = \bigotimes_{i=1}^{n} \left[\begin{array}{cc} \bar{x}_i & x_i \end{array} \right], \quad \mathbf{B}(n) = \bigotimes_{i=1}^{n} \mathbf{B}(1), \quad \mathbf{B}(1) = \left[\begin{array}{cc} 1 & 0 \\ 0 & 1 \end{array} \right],$$

and $\otimes$ denotes the Kronecker product. The matrix $\mathbf{B}(1)$ is called the basic Shannon matrix. The elements of $\mathbf{B}(1)$ and $\mathbf{B}(n)$ are logical 0 and 1, and the calculations are done in $GF(2)$.

The Shannon tree is a graphical representation of the **minterm expansion** (3.3.1) of f. The elements of that basis are the minterms generated by the Kronecker product in $\boldsymbol{X}_b$. For example, when $n = 2$, the minterm basis is

$$\begin{aligned} \mathbf{X}_b &= \left[\begin{array}{cc} \bar{x}_1 & x_1 \end{array} \right] \otimes \left[\begin{array}{cc} \bar{x}_2 & x_2 \end{array} \right] \\ &= \left[\bar{x}_1\bar{x}_2, \bar{x}_1 x_2, x_1 \bar{x}_2, x_1 x_2 \right]. \end{aligned}$$

In the matrix notation, these basic functions corresponds to the columns of the matrix $\mathbf{B}(n)$. Formally, the relations

$$\mathbf{B}_f = \mathbf{B}(n)\mathbf{F},$$

and

$$f = \mathbf{X}_b \mathbf{B}_f,$$

form the transform pair defined with respect to the minterm basis. In this case the spectrum $\mathbf{B}_f$ is identical to $\mathbf{F}$.

Example 3.3.1 *Consider the function $f(x_1, x_2) = \bar{x}_1 \vee x_2$. The truth vector is $\mathbf{F} = [1, 1, 0, 1]^t$. The spectrum $\mathbf{B}_f$ is given by*

$$\mathbf{B}_f = \mathbf{B}(2)\mathbf{F} = \left[\begin{array}{cccc} 1 & 0 & 0 & 0 \\ 0 & 1 & 0 & 0 \\ 0 & 0 & 1 & 0 \\ 0 & 0 & 0 & 1 \end{array} \right] \left[\begin{array}{c} 1 \\ 1 \\ 0 \\ 1 \end{array} \right] = \left[\begin{array}{c} 1 \\ 1 \\ 0 \\ 1 \end{array} \right].$$

f is represented by

$$\begin{aligned} f &= \mathbf{X}_b \mathbf{B}_f = \left[\bar{x}_1\bar{x}_2, \bar{x}_1 x_2, x_1 \bar{x}_2, x_1 x_2 \right] \left[\begin{array}{c} 1 \\ 1 \\ 0 \\ 1 \end{array} \right] \\ &= \bar{x}_1\bar{x}_2 \oplus \bar{x}_1 x_2 \oplus x_1 x_2. \end{aligned}$$

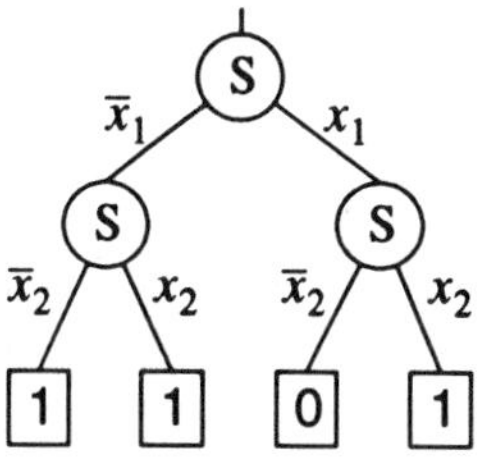

Figure 3.3.1 Shannon tree for $f = \bar{x}_1 \vee x_2$.

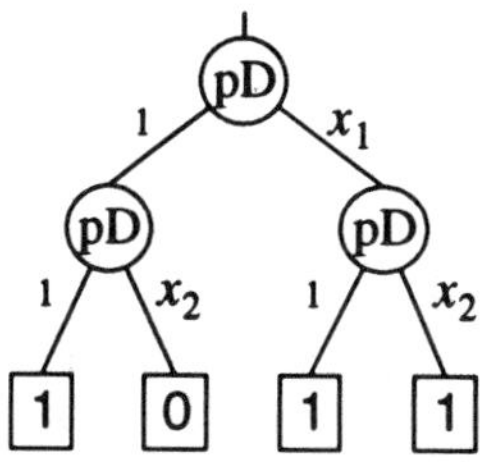

Figure 3.3.2 Positive Davio tree for $f = \bar{x}_1 \vee x_2$.

This is the ***minterm expansion*** *for f. Fig. 3.3.1 shows the Shannon tree for f, where S denotes the Shannon expansion.* ■

In a Shannon tree, each path from the root node up to a terminal node corresponds to a minterm. The values of the terminal nodes are the spectral coefficients, but since the minterm basis is used, they are equal to the function values. The I-mapping is defined by $\mathbf{B}(n)$.

3.3.2 Positive Davio Trees

Positive Davio trees are derived by using the positive Davio expansion $f = 1 \cdot f_0 \oplus x_i(f_0 \oplus f_1)$, where $f_0 = f(x_i = 0)$ and $f_1 = f(x_i = 1)$. Recursive application of the positive Davio expansion to f given by the truth-vector $\mathbf{F} = [f(0), \ldots, f(2^n - 1)]^t$ produce the matrix representation:

$$f = \mathbf{X}_r \mathbf{R}(n) \mathbf{F},$$

where

$$\mathbf{X}_r = \bigotimes_{i=1}^{n} \begin{bmatrix} 1 & x_i \end{bmatrix}, \quad \mathbf{R}(n) = \bigotimes_{i=1}^{n} \mathbf{R}(1), \quad \mathbf{R}(1) = \begin{bmatrix} 1 & 0 \\ 1 & 1 \end{bmatrix},$$

and $\otimes$ denotes the Kronecker product. The elements of $\mathbf{R}(1)$ and $\mathbf{R}(n)$ are logical 0 and 1, and the calculations are done in $GF(2)$.

The Kronecker product performed as a symbolic computation in $\mathbf{X}_r$ generates the product terms appearing in the positive polarity Reed-Muller expressions

(**PPRMs**). For example, when $n = 2$, we have

$$\begin{aligned} \mathbf{X}_r &= \begin{bmatrix} 1 & x_1 \end{bmatrix} \otimes \begin{bmatrix} 1 & x_2 \end{bmatrix} \\ &= \begin{bmatrix} 1 & x_2 & x_1 & x_1x_2 \end{bmatrix}. \end{aligned}$$

These product terms represent particular switching functions whose truth-vectors correspond to the columns of the **Reed-Muller transform matrix** $\mathbf{R}(n)$. The function generated in this way are called the Reed-Muller (RM) functions.

Definition 3.3.1 *The* ***Reed-Muller spectrum*** $\mathbf{R}_f = [R_f(0), \ldots, R_f(2^n-1)]^t$ *of a switching function* f *with the truth-vector* $\mathbf{F}$ *is*

$$\mathbf{R}_f = \mathbf{R}(n)\mathbf{F}. \tag{3.3.2}$$

For each transform, we need the inverse transform to get back from the spectral domain. We consider the transform pair consisting of the forward and the inverse transform. Since, the Reed-Muller transform matrix $\mathbf{R}(n)$ is a self-inverse matrix over $GF(2)$, the forward and inverse transforms are given by the same matrix.

The elements of $\mathbf{X}_r$ are symbolic representation for columns of $\mathbf{R}(n)$, and therefore, the inverse Reed-Muller transform can be written as

$$f = \mathbf{X}_r\mathbf{R}_f. \tag{3.3.3}$$

Relations (3.3.2) and (3.3.3) form the **Reed-Muller transform pair.**

Example 3.3.2 *Consider the function* $f(x_1, x_2) = \bar{x}_1 \vee x_2$. *The truth vector is* $\mathbf{F} = [1, 1, 0, 1]^t$. *The Reed-Muller spectrum is given by*

$$\mathbf{R}_f = \mathbf{R}(2)\mathbf{F} = \begin{bmatrix} 1 & 0 & 0 & 0 \\ 1 & 1 & 0 & 0 \\ 1 & 0 & 1 & 0 \\ 1 & 1 & 1 & 1 \end{bmatrix} \begin{bmatrix} 1 \\ 1 \\ 0 \\ 1 \end{bmatrix} = \begin{bmatrix} 1 \\ 0 \\ 1 \\ 1 \end{bmatrix}.$$

f is represented by

$$\begin{aligned} f &= \mathbf{X}_r\mathbf{R}_f = [1, x_2, x_1, x_1x_2] \begin{bmatrix} 1 \\ 0 \\ 1 \\ 1 \end{bmatrix} \\ &= 1 \oplus x_1 \oplus x_1x_2. \end{aligned}$$

This is the PPRM for f. Fig. 3.3.2 shows the positive Davio tree for f, where pD denotes the positive Davio expansion. ■

The positive Davio tree is the graphical representation of the PPRM. Each path from the root node up to a terminal node corresponds to a product term from $\mathbf{X}_r$, thus, to a particular Reed-Muller function determined by the product of labels at edges. The values of terminal nodes in the positive Davio tree are the Reed-Muller coefficients defined by (3.3.2). The RM-mapping is defined by $\mathbf{R}(n)$.

3.3.3 Decision Diagrams

DDs are derived from decision trees by deleting redundant nodes, and by sharing equivalent sub-graphs.

Definition 3.3.2 (BDD reduction rules)

1. *If two descendent nodes of a node are identical (Fig. 3.3.3(a)), then delete the node and connect the incoming edges of the deleted node to the corresponding successor.*
2. *Share equivalent sub-graphs (Fig. 3.3.3(b)).*

Zero-suppressed BDDs (ZBDDs) [21] are useful to represent sets of combinations.

Definition 3.3.3 (ZBDD reduction rules)

1. *If the outgoing edge of a node labeled with x_i or $\bar{x}_i$, points to the constant zero, then delete the node and connect the edge to the other sub-graph directly (Fig. 3.3.3(c)).*
2. *Share equivalent sub-graphs (Fig. 3.3.3(b)).*

The BDD reduction rules can be applied to both Shannon and Davio trees. They produce the reduced BDDs (RBDDs) and the quasi-reduced FDDs (QRFDDs), respectively. The ZBDD reduction rules produce reduced FDDs

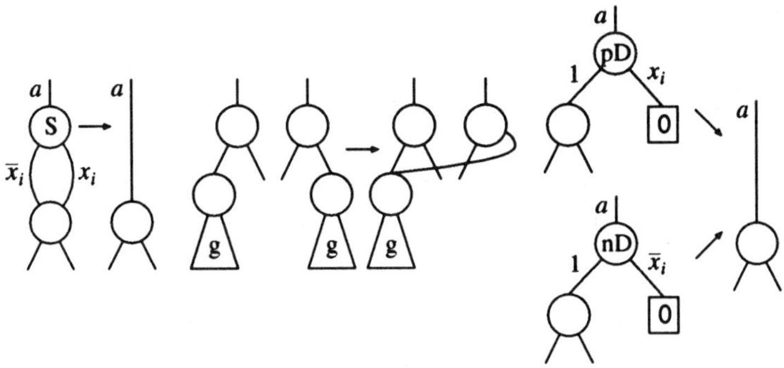

(a) (b) (c)

Figure 3.3.3 Reduction rules, (*a*) Elimination of the Shannon node, (*b*) Sharing of the isomorphic sub-graphs, (*c*) Elimination of the Davio nodes.

(RFDDs). They are not so effective in the cases of DDs for integer-valued functions, since in these cases zero is one of many possible values of the terminal nodes. The ZBDD reduction rules make the interpretation of various DDs more complicated, and therefore, will not be used in the reduction of STDDs. However, the presented theory is true also for DDs reduced by the ZBDD reduction rules. For the reduction of STDDs, the generalized BDD reduction rules will be introduced in Definition 3.7.1.

The **size** of a DD is defined as the number of nodes. However, it does not express all properties of the DD. Therefore, we define the shape of the DD as follows. The **shape** of DD is defined by the number of nodes, their place and distribution over the levels of DDs and the connections between the nodes. In a decision tree, the **length** of an edge connecting the nodes at successive levels is defined to be one. In a reduced decision diagram (RDD), edges longer than one, i.e., connecting nodes at non-successive levels can appear. For example, the length of an edge connecting a node at $(i-1)$-th level with a node at $(i+1)$-th level is two.

Definition 3.3.4 *In an RDD, a* **cross point** *is a point where an edge longer than one crosses a level in the DD.*

Note that the introduction of the cross points does not extend the DDs into the complete trees. Two cross points are shown in Fig. 3.3.4(*b*).

3.3.4 Relation Between BDDs and FDDs

$\mathbf{B}(n)$ and $\mathbf{R}(n)$ matrices show the relationship between BDDs and FDDs.

Statement 3.3.1 *Since $\mathbf{R}(n)\mathbf{B}(n) = \mathbf{R}(n)$, the FDD for f is the BDD of its Reed-Muller spectrum if the meaning of nodes and the labels of edges are correspondingly changed: The positive Davio nodes, pD, are replaced by the Shannon nodes, S, and $1 \rightarrow \bar{x}_i$, $x_i \rightarrow x_i$ in the labels of edges. Conversely, since $\mathbf{B}(n)\mathbf{R}(n) = \mathbf{R}(n)$, the BDD of the Reed-Muller spectrum $\mathbf{R}_f$ of f is the FDD of f, if the reverse procedure to nodes and labels of edges is performed.*

Statement 3.3.2 *A BDD and an FDD represent f and $\mathbf{R}_f$, the Reed-Muller spectrum of f, at the same time.*

Statement 3.3.1 shows that the change of interpretation of nodes and labels will change the meaning of a DD. Since the shape of the DD remains invariant, Statement 3.3.2 applies directly to RBDDs and QRFDDs.

In reading f from an FDD, by following the labels at edges, we perform the Reed-Muller transform starting from the terminal nodes, since the labels at the edges represent the columns of $\mathbf{R}(n)$. The values of terminal nodes represent the Reed-Muller coefficients and, therefore, we read f, since the Reed-Muller matrix is self-inverse.

Example 3.3.3 *Fig. 3.3.4(a) shows the FDD for $f(x_1, x_2, x_3) = x_3 \oplus x_1x_2 \oplus x_1x_3$ reduced by the BDD reduction rules. This DD represents*

$$\begin{aligned} f &= 1 \cdot (1 \cdot (1 \cdot 0 \oplus x_3 \cdot 1) \oplus x_2 \cdot 0) \\ &\quad \oplus x_1(1 \cdot (1 \cdot 0 \oplus x_3 \cdot 1) \oplus x_2(1 \cdot 1 \oplus x_3 \cdot 0)) \\ &= x_3 \oplus x_1x_3 \oplus x_1x_2. \end{aligned}$$

However, if we relabel edges as $1 \rightarrow \bar{x}_i$, then we will perform the $\mathbf{B}(n)$ mapping at each node and we will read the Reed-Muller spectrum $\mathbf{R}_f$ as follows:

$$R_f = \bar{x}_1(\bar{x}_2(\bar{x}_3 \cdot 0 \oplus x_3 \cdot 1) \oplus x_2 \cdot 0)$$

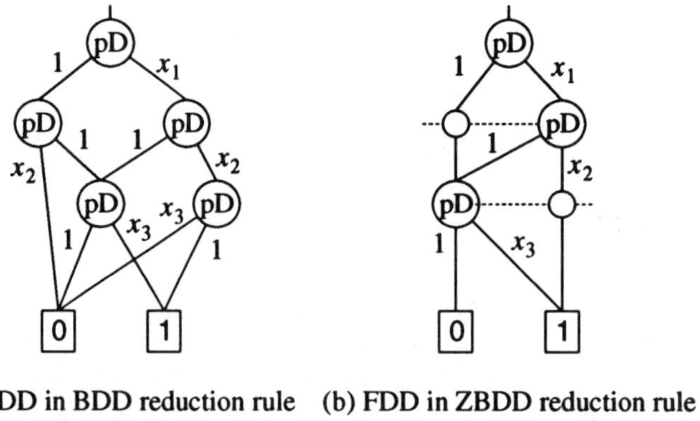

Figure 3.3.4 FDDs for f in Example 3.3.3.

$$
\begin{aligned}
& \oplus\, x_1(\bar{x}_2(\bar{x}_3 \cdot 0 \oplus x_3 \cdot 1) \oplus x_2(\bar{x}_3 \cdot 1 \oplus x_3 \cdot 0)) \\
= \;& \bar{x}_1\bar{x}_2x_3 \oplus x_1\bar{x}_2x_3 \oplus x_1x_2\bar{x}_3.
\end{aligned}
$$

Thus, $\mathbf{R}_f = [0, 1, 0, 0, 0, 1, 1, 0]^t$. ■

Statement 3.3.2 extends also to the FDDs reduced by the ZBDD reduction rules if the effect of the deleted nodes is properly taken into account.

In the ZBDD reduction rules, we delete a node whose edge labeled by x_i points to the constant zero. Thus, we delete the node for which $f_0 \oplus f_1 = 0$, but the edge labeled by 1 remains in the DD. Therefore, the effect of the term $1 \cdot f_0$ must be taken into account in reading the Reed-Muller spectrum from the FDD reduced by the ZBDD reduction rules. That can be done through the cross points to which the Shannon expansion should be also applied. However, since in the deleted node the value associated to x_i is zero, the Shannon expansion in the cross points reduces to $\bar{x}_i f_0$.

Example 3.3.4 *Fig. 3.3.4(b) shows the FDD for f in Example 3.3.3 reduced by the ZBDD reduction rules. We can read f and the Reed-Muller spectrum $\mathbf{R}_f$ from this DD by using the positive Davio and the Shannon expansion, respectively.*

$$
\begin{aligned}
f &= 1 \cdot (1 \cdot 0 \oplus x_3 \cdot 1) \oplus x_1(1 \cdot (1 \cdot 0 \oplus x_3 \cdot 1) \oplus x_2 \cdot 1) \\
&= x_3 \oplus x_1x_3 \oplus x_1x_2,
\end{aligned}
$$

$$
\begin{aligned}
R_f &= \bar{x}_1(\bar{x}_2(\bar{x}_3 \cdot 0 \oplus x_3 \cdot 1)) \oplus x_1(\bar{x}_2(\bar{x}_3 \cdot 0 \oplus x_3 \cdot 1) \oplus x_2(\bar{x}_3 \cdot 1)) \\
&= \bar{x}_1\bar{x}_2x_3 \oplus x_1\bar{x}_2x_3 \oplus x_1x_2\bar{x}_3.
\end{aligned}
$$

Thus, $\mathbf{R}_f = [0,1,0,0,0,1,1,0]^t$. ∎

3.4 GENERALIZATION

Interpretation of the positive Davio expansions as the partial Reed-Muller transform relates FDDs to the spectral representations of switching functions, and suggests for further generalizations of DDs to integer-valued functions. We can generalize concept of FDDs and BDDs to spectral transforms representations.

Definition 3.4.1 *Let Q be a mapping defined by a $(2^n \times 2^n)$ matrix $\mathbf{Q}(n)$ whose columns are linearly independent. For a function f given by the truth-vector $\mathbf{F}$, the Q-spectrum $\mathbf{Q}_f = [Q_f(0), \ldots, Q_f(2^n-1)]^t$ is*

$$
\mathbf{Q}_f = \mathbf{Q}(n)\mathbf{F}. \tag{3.4.1}
$$

Definition 3.4.2 *The Q-decision tree for a function f is the decision tree associated with f. In the Q-decision tree, each path from the root node to a terminal node corresponds to a basic function, i.e., to a column of $\mathbf{Q}^{-1}(n)$. The values of terminal nodes of the Q-decision tree represent the Q-spectrum of f.*

Definition 3.4.3 *A* ***Q-decision diagram (QDD)*** *for a function f is the DD derived from the corresponding Q-decision tree.*

Suppose that $\mathbf{Q}(n)$ is represented as the Kronecker product of n identical factors $\mathbf{Q}(1)$, i.e.,

$$
\mathbf{Q}(n) = \bigotimes_{i=1}^{n} \mathbf{Q}(1).
$$

For such a mapping Q, the basic matrix $\mathbf{Q}(1)$ defines an expansion of f with respect to the i-th variable:

$$
f = \mathbf{Q}^{-1}(1)\mathbf{Q}(1)\begin{bmatrix} f_0 \\ f_1 \end{bmatrix}. \tag{3.4.2}
$$

This expansion is a generalization of the Shannon, the positive Davio and the negative Davio expansions. It is denoted as the Q-expansion. A mapping inverse to the Q-expansion is defined by the columns of $\mathbf{Q}^{-1}(1)$.

Thus, the function expansion described in (3.4.2) is performed in the nodes of a QDD at the i-th level, and the terminal nodes of a QDD represent the Q-spectrum $\mathbf{Q}_f$ of f. This statement determines the way of drawing a QDD for a given f. Each path from the root node to a terminal node represents a column of $\mathbf{Q}^{-1}(n)$. This statement shows the way to determine f from a given QDD.

If Q is an RM-mapping, then $\mathbf{Q}(1) = \mathbf{Q}^{-1}(1) = \mathbf{R}(1)$ and the columns of $\mathbf{Q}^{-1}(1)$ are expressed by 1 and x_i. In this case, QDD is an FDD. If Q is an I-mapping, then $\mathbf{Q}(1) = \mathbf{Q}^{-1}(1) = \mathbf{B}(1)$ and the columns of $\mathbf{Q}^{-1}(1)$ are expressed by $\bar{x}_i$ and x_i. In this case, QDD is a BDD.

Therefore, a QDD includes a BDD and an FDD as special cases and permits to define DDs for an arbitrary mapping Q, including spectral transform decision diagrams (STDDs) associated with spectral transforms [13]. In the following sections, we will give two examples of STDDs.

3.5 ARITHMETIC TRANSFORM

3.5.1 Arithmetic Transform and Arithmetic Spectrum

Let $\mathbf{A}(n)$ be the arithmetic transform matrix defined by

$$\mathbf{A}(n) = \bigotimes_{i=1}^{n} \mathbf{A}(1), \quad \mathbf{A}(1) = \begin{bmatrix} 1 & 0 \\ -1 & 1 \end{bmatrix},$$

where the addition and the multiplication are done in integer. In symbolic representation

$$\mathbf{A}(1) = \begin{bmatrix} 1 - 2x_i & x_i \end{bmatrix}.$$

Definition 3.5.1 *For a function f given by the truth-vector $\mathbf{F}$, the* ***arithmetic spectrum*** $\mathbf{A}_f = [a_0, \ldots, a_{2^n-1}]^t$ *is*

$$\mathbf{A}_f = \mathbf{A}(n)\mathbf{F}. \tag{3.5.1}$$

$\mathbf{A}(1)$ is the basic arithmetic transform matrix. This matrix at the i-th position in the Kronecker product generating $\mathbf{A}(n)$ performs the partial arithmetic transform of f with respect to the i-th variable. The inverse mapping $\mathbf{A}^{-1}(n)$ is given by

$$\mathbf{A}^{-1}(n) = \bigotimes_{i=1}^{n} \mathbf{A}^{-1}(1), \quad \mathbf{A}^{-1}(1) = \begin{bmatrix} 1 & 0 \\ 1 & 1 \end{bmatrix},$$

or in symbolic representation

$$\mathbf{A}^{-1}(1) = \begin{bmatrix} 1 & x_i \end{bmatrix}.$$

Therefore, the inverse arithmetic transform is defined as

$$f = \mathbf{X}_a \mathbf{A}_f, \quad \mathbf{X}_a = \bigotimes_{i=1}^{n} \begin{bmatrix} 1 & x_i \end{bmatrix}. \tag{3.5.2}$$

Relations (3.5.1) and (3.5.2) form the arithmetic transform pair. The matrix $\mathbf{A}^{-1}(n)$ has the same form as the matrix $\mathbf{R}(n)$. However, in $\mathbf{R}(n)$ the entries are logical values 0 and 1, while in $\mathbf{A}^{-1}(n)$ these are the integers 0 and 1. The calculations with $\mathbf{R}(n)$ are performed in $GF(2)$, while calculations with $\mathbf{A}^{-1}(n)$ are performed in integers. Therefore, $\mathbf{A}^{-1}(n)$ can be considered as the integer Reed-Muller matrix and their columns represent the integer Reed-Muller functions. The matrix $\mathbf{A}(n)$ is inverse to $\mathbf{A}^{-1}(n)$ over integers and, therefore, the arithmetic transform is the Reed-Muller transform over integers. Thus, the arithmetic transform is also called the **integer Reed-Muller transform.**

Example 3.5.1 *Consider the function* $f(x_1, x_2) = \bar{x}_1 \vee x_2$. *The truth vector is* $\mathbf{F} = [1, 1, 0, 1]^t$. *The arithmetic spectrum is*

$$\mathbf{A}_f = \mathbf{A}(2)\mathbf{F}.$$

$$\mathbf{A}_f = \begin{bmatrix} 1 & 0 & 0 & 0 \\ -1 & 1 & 0 & 0 \\ -1 & 0 & 1 & 0 \\ 1 & -1 & -1 & 1 \end{bmatrix} \begin{bmatrix} 1 \\ 1 \\ 0 \\ 1 \end{bmatrix} = \begin{bmatrix} 1 \\ 0 \\ -1 \\ 1 \end{bmatrix}.$$

f is represented by

$$\begin{aligned} f &= \mathbf{X}_a \mathbf{A}_f = [1, x_2, x_1, x_1 x_2] \begin{bmatrix} 1 \\ 0 \\ -1 \\ 1 \end{bmatrix} \\ &= 1 - x_1 + x_1 x_2. \end{aligned}$$

■

Any switching function is represented as an ordinary arithmetic sum of partial product of uncomplemented variables by using the following identities:

$$\begin{aligned} \bar{x} &= 1 - x, \\ x \cdot y &= xy, \\ x \vee y &= x + y - xy, \\ x \oplus y &= x + y - 2xy. \end{aligned} \tag{3.5.3}$$

Such an expression is called an **arithmetic expression**, or a **real expansion** [19].

Example 3.5.2 *A switching function $f(x_1, x_2) = \bar{x}_1 \vee x_2$ is represented by the arithmetic expression.*

$$\begin{aligned} f &= (1 - x_1) + x_2 - (1 - x_1)x_2 \\ &= 1 - x_1 + x_2 - x_2 + x_1x_2 \\ &= 1 - x_1 + x_1x_2. \end{aligned}$$

■

The arithmetic transform was introduced for the replacement of the logical operations NOT, AND, OR and EXOR with the corresponding arithmetic operations in (3.5.3). From that approach, the term "arithmetic" was introduced [20, 12]. Formerly, we called "algebraic transform," etc., but we follow Heidtmann's terminology. The relationship to the integer Reed-Muller transform was shown later. This is the same transform, but derived from different motivations and, therefore, there is a sense to use different name.

3.5.2 Arithmetic Transform Decision Tree

From (3.4.2), the basic arithmetic transform matrices $\mathbf{A}^{-1}(1)$ and $\mathbf{A}(1)$ define the arithmetic transform expansion:

$$\begin{aligned} f &= \begin{bmatrix} 1 & x_i \end{bmatrix} \begin{bmatrix} 1 & 0 \\ -1 & 1 \end{bmatrix} \begin{bmatrix} f_0 \\ f_1 \end{bmatrix} = \begin{bmatrix} 1 & x_i \end{bmatrix} \begin{bmatrix} f_0 \\ f_1 - f_0 \end{bmatrix} \\ &= 1 \cdot f_0 + x_i(f_1 - f_0), \end{aligned}$$

which is also called *A*-**expansion**. Note that $+$ is the addition in normal arithmetic. Arithmetic transform decision trees are defined by using the *A*-expansion.

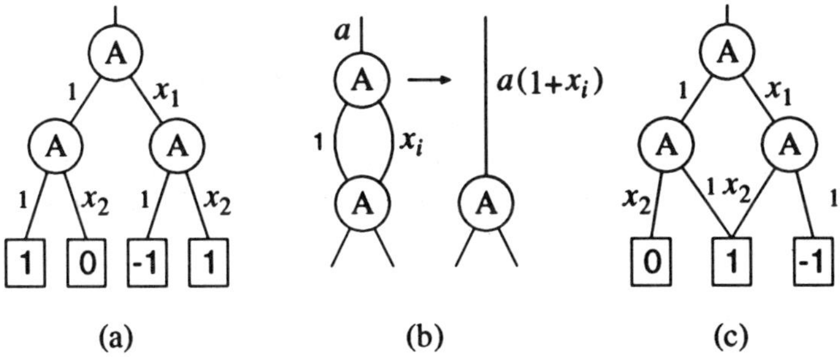

Figure 3.5.1 Arithmetic transform decision tree of $f(x_1, x_2) = \bar{x}_1 \vee x_2$, reduction rule, and RACDD.

Definition 3.5.2 *An* ***arithmetic transform decision tree*** *is a decision tree associated with a given function f by using the expansion $f = 1 \cdot f_0 + x_i(f_1 - f_0)$.*

Each path from the root node to a terminal node of the arithmetic transform decision tree corresponds to a product in the **arithmetic expression**. The values of terminal nodes in an arithmetic transform decision tree are the coefficients of the arithmetic spectrum. Therefore, to determine the function f represented by a given arithmetic transform decision tree, we should use a mapping inverse to the A-expansion. In the arithmetic transform decision tree, this mapping is expressed by the labels at the edges in the tree. Therefore, to read f we should simply follow the labels at the edges.

Example 3.5.3 *An arithmetic transform decision tree for $f(x_1, x_2) = \bar{x}_1 \vee x_2$ is shown in Fig. 3.5.1(a). This tree represents the function:*

$$\begin{aligned} f &= 1 \cdot (1 \cdot 1) + 0 \cdot (1 \cdot x_2) + (-1) \cdot (x_1 \cdot 1) + 1 \cdot (x_1 x_2) \\ &= 1 - x_1 + x_1 x_2. \end{aligned}$$

In Fig. 3.5.1, the symbol A denotes the A-expansion. ∎

3.6 WALSH TRANSFORM

3.6.1 Walsh Transform and Walsh Spectrum

An arithmetic transform is an example of a non-self-inverse transform. If we look for a self-inverse transform in rational numbers, the **Walsh transform** is a good candidate [13].

Let $\mathbf{W}(n)$ be the Walsh transform matrix defined by

$$\mathbf{W}(n) = \bigotimes_{i=1}^{n} \mathbf{W}(1), \quad \mathbf{W}(1) = \begin{bmatrix} 1 & 1 \\ 1 & -1 \end{bmatrix},$$

or in symbolic representation

$$\mathbf{W}(1) = \begin{bmatrix} 1 & 1 - 2x_i \end{bmatrix}.$$

Note that the calculation will be done in normal arithmetic.

Definition 3.6.1 *For a function f given by the truth-vector* $\mathbf{F}$, *the* ***Walsh spectrum*** $\mathbf{W}_f = [w_0, \ldots, w_{2^n-1}]^t$ *is*

$$\mathbf{W}_f = 2^{-n}\mathbf{W}(n)\mathbf{F}. \tag{3.6.1}$$

The Walsh transform matrix is self-inverse matrix up to the constant 2^n. Therefore, the inverse Walsh transform is defined by

$$f = \mathbf{X}_w \mathbf{W}_f, \quad \mathbf{X}_w = \bigotimes_{i=1}^{n} \begin{bmatrix} 1 & 1 - 2x_i \end{bmatrix}. \tag{3.6.2}$$

Relations (3.6.1) and (3.6.2) form the Walsh transform pair.

$\mathbf{W}(1)$ is the basic Walsh transform matrix. This matrix at i-th position in the Kronecker product generating $\mathbf{W}(n)$ performs the partial Walsh transform with respect to the i-th variable.

Example 3.6.1 *Consider the function* $f(x_1, x_2) = \bar{x}_1 \vee x_2$. *The truth vector is* $\mathbf{F} = [1, 1, 0, 1]^t$. *The Walsh spectrum is*

$$\mathbf{W}_f = 2^{-n}\mathbf{W}(n)\mathbf{F}$$

$$= 2^{-2}\begin{bmatrix} 1 & 1 & 1 & 1 \\ 1 & -1 & 1 & -1 \\ 1 & 1 & -1 & -1 \\ 1 & -1 & -1 & 1 \end{bmatrix}\begin{bmatrix} 1 \\ 1 \\ 0 \\ 1 \end{bmatrix} = 2^{-2}\begin{bmatrix} 3 \\ -1 \\ 1 \\ 1 \end{bmatrix}.$$

f is represented by

$$\begin{aligned} f &= \mathbf{X}_w \mathbf{W}_f = \frac{1}{4}[1, 1-2x_2, 1-2x_1, (1-2x_1)(1-2x_2)]\begin{bmatrix} 3 \\ -1 \\ 1 \\ 1 \end{bmatrix} \\ &= \frac{1}{4}[3-(1-2x_2)+(1-2x_1)+(1-2x_1)(1-2x_2)]. \end{aligned}$$

■

The last expression in the above example is called **Walsh expression.** In Walsh expressions, functions are represented by 1 and $(1-2x_i)$. If we expand and simplify the Walsh expression, then we have an arithmetic expression.

3.6.2 Walsh Transform Decision Trees

From (3.4.2), the Walsh transform defines the Walsh expansion:

$$\begin{aligned} f &= \frac{1}{2}\begin{bmatrix} 1 & 1-2x_i \end{bmatrix}\begin{bmatrix} 1 & 1 \\ 1 & -1 \end{bmatrix}\begin{bmatrix} f_0 \\ f_1 \end{bmatrix} \\ &= \frac{1}{2}\begin{bmatrix} 1 & 1-2x_i \end{bmatrix}\begin{bmatrix} f_0+f_1 \\ f_0-f_1 \end{bmatrix} \\ &= \frac{1}{2}[(f_0+f_1)+(1-2x_i)(f_0-f_1)], \end{aligned}$$

which is also called **W-expansion.** Walsh transform decision trees are defined by using the W-expansion.

Definition 3.6.2 *The* ***Walsh transform decision tree*** *is a decision tree associated with a given function f by using the expansion*

$$f = \frac{1}{2}[(f_0+f_1)+(1-2x_i)(f_0-f_1)].$$

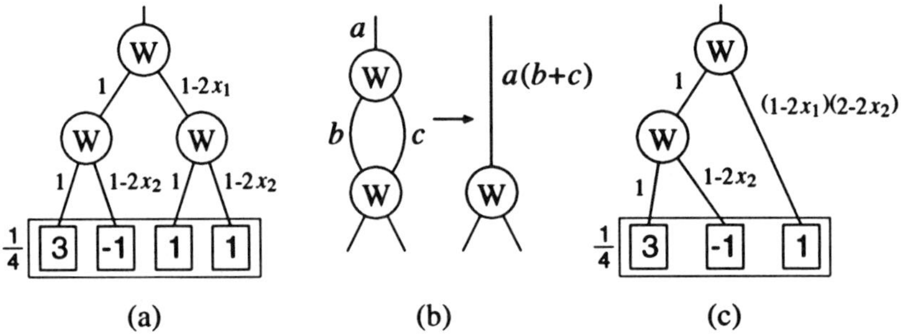

Figure 3.6.1 Walsh transform decision tree for $f(x_1, x_2) = \bar{x}_1 \vee x_2$, reduction rule and RWDD.

Example 3.6.2 *The Walsh transform decision tree in Fig. 3.6.1(a) represents* $f(x_1, x_2) = \bar{x}_1 \vee x_2$, *since*

$$\begin{aligned} f &= \frac{3}{4}(1 \cdot 1) - \frac{1}{4}[1 \cdot (1 - 2x_2)] + \frac{1}{4}[(1 - 2x_1) \cdot 1] + \frac{1}{4}(1 - 2x_1)(1 - 2x_2) \\ &= 1 - x_1 + x_1 x_2. \end{aligned}$$

In Fig. 3.6.1(a), the values of the terminal nodes are multiplied by $\frac{1}{4}$, *and the symbol W denotes the W-expansion.* ■

Each path from the root node to a terminal node in the Walsh transform decision tree corresponds to a particular Walsh function in the Hadamard ordering [13]. The values of terminal nodes of the Walsh transform decision tree are the Walsh spectral coefficients. Given a Walsh transform decision tree, to read f we should perform the W^{-1}-mapping derived from (3.6.2). This mapping is expressed by the labels at the edges.

3.7 REDUCED STDDS

3.7.1 Reduction Rules

The BDD reduction rule 1 (Fig. 3.3.3(a)) says that a node with the outgoing edges denoted by $\widetilde{x}_i \in \{\bar{x}_i, x_i\}$ can be deleted if they points to the same node.

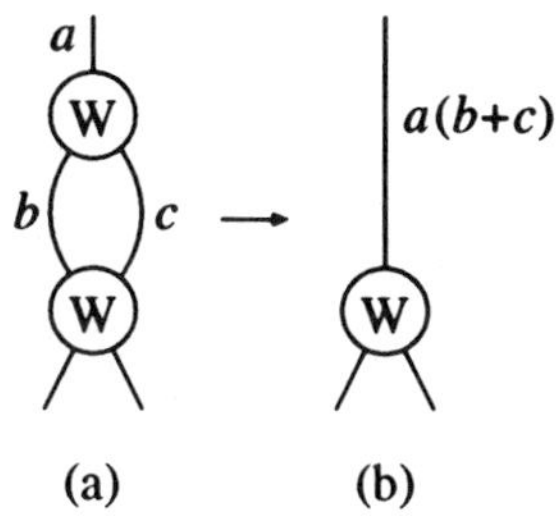

Figure 3.7.1 Generalized BDD reduction rules.

The manipulation with labels at edges is quite simple, and it is described by $a(\bar{x}_i \oplus x_i) = a$.

In the ZBDD reduction rule 1 (Fig. 3.3.3(c)), we have the similar rule, since $a \cdot (1 \oplus x_i \cdot 0) = a$ and $a \cdot (1 \oplus \bar{x}_i \cdot 0) = a$. With this reduction rule, we used the properties $\tilde{x}_i \cdot 0 = 0$ and $1 \cdot a = a$ for any a. Both properties are true in $GF(2)$ and integers.

In integer-valued functions, zero is one of many possible values, and the ZBDD reduction rules are not so effective as in the case of switching functions where only zeros and ones appear at the terminal nodes. Therefore, we use the BDD reduction rules in the reduction of STDDs, where terminal nodes take the values in integers. Note that, in STDDs the labels at edges may differ from 1, x_i, or $\bar{x}_i$.

Definition 3.7.1 (Generalized BDD reduction rules)

1. *If two outgoing edges of a node point to the same node, then delete the node and connect the incoming edge of the deleted node to the corresponding successor. Relabel this edge as is shown in Fig. 3.7.1.*
2. *Share equivalent sub-graphs (Fig. 3.3.3(b)).*

For a given f, reduced STDD (RSTDD) with a given order of variables is a canonical representation.

3.7.2 Arithmetic Transform Decision Diagrams

The **arithmetic transform decision diagrams (ACDDs)** are introduced by an analogy to the BDDs and FDDs. ACDDs should be distinguished from the algebraic decision diagrams (ADDs) [2], and the AND ternary decision diagrams (ATDDs) [24].

Definition 3.7.2 ***Arithmetic transform decision diagrams (ACDDs)*** *are DDs derived from arithmetic transform decision trees by using the generalized BDD reduction rules. An ACDD is reduced (RACDD) if further reduction with the same rules is impossible.*

In Section 3.8.2, we will show that ACDDs are the same as binary moment diagrams (BMDs) [4] except for the reduction rules.

Example 3.7.1 *Fig. 3.5.1(c) is the RACDD for* $f(x_1, x_2) = \bar{x}_1 \vee x_2$. *It represents an arithmetic expression:*

$$\begin{aligned} f &= 1 \cdot (1 \cdot 1) + 0 \cdot (1 \cdot x_2) + (-1) \cdot (x_1 \cdot 1) + 1 \cdot (x_1 x_2) \\ &= 1 - x_1 + x_1 x_2. \end{aligned}$$

■

3.7.3 Walsh Transform Decision Diagrams

In a similar way, we can define Walsh transform decision diagrams.

Definition 3.7.3 ***Walsh transform decision diagrams (WDDs)*** *are the DDs derived from Walsh transform decision trees by using the generalized BDD reduction rules. A WDD is reduced (RWDD) if further reduction with the same rules is impossible.*

Example 3.7.2 *Fig. 3.6.1(c) is the RWDD for* $f = \bar{x}_1 \vee x_2$. *The labels at edges are determined by the generalized BDD reduction rules. This RWDD represents the same function as the complete Walsh transform decision tree in*

Fig. 3.6.1(a):

$$\begin{aligned} f &= \frac{3}{4}\cdot 1 + (-\frac{1}{4})\cdot(1-2x_2) + \frac{1}{4}\cdot(1-2x_1)(2-2x_2) \\ &= 1 - x_1 + x_1x_2. \end{aligned}$$

■

3.8 RELATION BETWEEN STDDS AND OTHER DDS

3.8.1 STDDs and BDDs

Statement 3.3.2 establishes the relationship between BDDs and FDDs. Since $\mathbf{B}(n)\mathbf{Q}(n) = \mathbf{Q}(n)\mathbf{B}(n) = \mathbf{Q}(n)$, the same relationship exists between STDDs and BDDs. Note that Statement 3.3.2 is derived from Statement 3.3.1, where the change of labels at the edges and meaning of nodes is required. The algebraic structures for STDDs and BDDs are different. The I-mapping for switching functions is expressed by the Shannon expansion $f = \bar{x}_i f_0 \oplus x_i f_1$. On the other hand, the I-mapping for integer-valued functions is defined in terms of the addition and multiplication in integers, thus, by $f = \bar{x}_i f_0 + x_i f_1$. Therefore, we can extend Statement 3.3.2 to:

Statement 3.8.1 *Given an RBDD, we can read the Q-spectrum of f by the Q-mapping (3.4.1) to the all nodes and cross points of RBDD.*

The following examples illustrate how to read arithmetic and Walsh spectra.

Example 3.8.1 *To read the arithmetic spectrum from an RBDD, we should use the A-mapping $A = (1-2x_i)f_0 + x_i f_1$ derived from (3.5.1) with $n = 1$. Thus, the edges must be relabeled as follows: $\bar{x}_i \rightarrow 1-2x_i$. From the RBDD in Fig. 3.8.1, we have the expression:*

$$\begin{aligned} A_f &= (1-2x_1)x_2 + x_1(1-2x_2+x_2) \\ &= x_2 - 2x_1x_2 + x_1 - x_1x_2 \\ &= x_1 + x_2 - 3x_1x_2 \\ &= 0\cdot(\bar{x}_1\bar{x}_2) + 1\cdot(\bar{x}_1x_2) + 1\cdot(x_1\bar{x}_2) + (-1)\cdot(x_1x_2). \end{aligned}$$

Thus, the arithmetic spectrum of f is $\mathbf{A}_f = [0,1,1,-1]^t$. ■

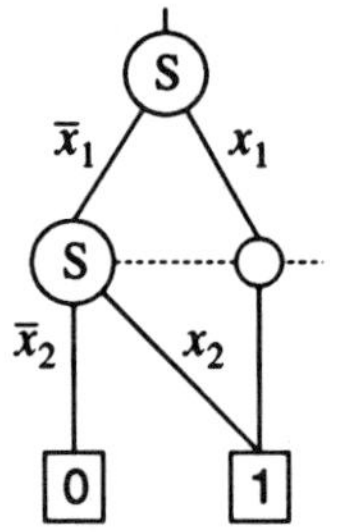

Figure 3.8.1 RBDD of $f(x_1, x_2) = x_1 \vee x_2$.

Example 3.8.2 *To read the Walsh spectrum from an RBDD, we should use the W-mapping* $W = \frac{1}{2}[f_0 + (1-2x_i)f_1]$ *derived from (3.6.1) with* $n = 1$. *Thus, the edges must be relabeled as follows:*

$$\begin{aligned} \bar{x}_i &\rightarrow \frac{1}{2} \\ x_i &\rightarrow \frac{1}{2}(1-2x_i). \end{aligned}$$

From the RBDD in Fig. 3.8.1, we have the expression:

$$\begin{aligned} W_f &= \frac{1}{2}[\frac{1}{2}\cdot 0 + \frac{1}{2}(1-2x_2)\cdot 1] + \frac{1}{2}(1-2x_1)\frac{1}{2}[(1+1-2x_2)\cdot 1] \\ &= \frac{1}{4}(3 - 4x_2 - 4x_1 + 4x_1x_2) \\ &= \frac{1}{4}[3\cdot \bar{x}_1\bar{x}_2 - 1\cdot \bar{x}_1x_2 - 1\cdot x_1\bar{x}_2 - 1\cdot x_1x_2]. \end{aligned}$$

Thus, the Walsh spectrum of f *is* $\mathbf{W}_f = \frac{1}{4}[3, -1, -1, -1]^t$. ∎

Note that W and W^{-1} mappings differ only in the normalization constant 2^{-1}.

Statement 3.8.2 *Given a reduced STDD representing* f *defined in terms of a basis* Q. *We can read the* Q*-spectrum of* f *from this RSTDD by using the* I*-mapping.*

Example 3.8.3 *To read the arithmetic spectrum from an RACDD, we should use the* I*-mapping* $f = \bar{x}_i f_0 + x_i f_1$. *Thus, the edges must be relabeled as follows:*

Table 3.8.1 Relation of transforms and decision diagrams.

Elements of spectrum	Transform	Basic matrix	Transform matrix	Decision diagram
$\{0,1\}$	Identical	$\begin{matrix}1 & 0\\ 0 & 1\end{matrix}$	$\mathbf{B}(n)$	BDD
	Reed-Muller	$\begin{matrix}1 & 0\\ 1 & 1\end{matrix}$	$\mathbf{R}(n)$	FDD
Integer	Identical	$\begin{matrix}1 & 0\\ 0 & 1\end{matrix}$	$\mathbf{B}(n)$	MTBDD
	Arithmetic	$\begin{matrix}1 & 0\\ -1 & 1\end{matrix}$	$\mathbf{A}(n)$	ACDD, BMD
Rational number	Walsh	$\begin{matrix}1 & 1\\ 1 & -1\end{matrix}$	$2^{-n}\mathbf{W}(n)$	WDD

$1 \rightarrow \bar{x}_i$. *From Fig. 3.5.1(c) we have the arithmetic spectrum:*

$$A_f = 1 \cdot \bar{x}_1\bar{x}_2 + 0 \cdot \bar{x}_1 x_2 - 1 \cdot x_1\bar{x}_2 + 1 \cdot x_1 x_2.$$

Thus, $\mathbf{A}_f = [1, 0, -1, 1]^t$. ∎

Example 3.8.4 *To read the Walsh spectrum from an RWDD, we should use the I-mapping* $f = \bar{x}_i f_0 + x_i f_1$. *Thus, the edges must be relabeled as follows:*

$$\begin{aligned} 1 &\rightarrow \bar{x}_i \\ 1 - 2x_i &\rightarrow x_i. \end{aligned}$$

From Fig. 3.6.1(c), we have the Walsh spectrum:

$$W_f = \frac{3}{4} \cdot (\bar{x}_1\bar{x}_2) - \frac{1}{4} \cdot (\bar{x}_1 x_2) + \frac{1}{4} \cdot (x_1\bar{x}_2) + \frac{1}{4} \cdot (x_1 x_2).$$

Thus, $\mathbf{W}_f = \frac{1}{4}[3, -1, 1, 1]^t$. ∎

3.8.2 STDDs for Integer-Valued Functions

STDDs are defined through the spectral interpretation of BDDs and FDDs. The same interpretation includes other EXOR-based DDs for switching functions. Generalization of the approach to integer-valued functions and related DDs is possible. In this setting, if the logical 0 and 1 are considered as integers 0 and 1, then MTBDDs are defined with respect to the integer counterpart of the basis used in the definition of the BDDs.

ACDDs are defined in terms of the integer Reed-Muller functions, thus, with respect to the integer counterpart of the basis used in the definition of the FDDs.

In the same way, the moment decomposition of f used in the definition of BMDs [4], is the same as the expansion of f with respect to the integer Reed-Muller basis. Thus, BMD and ACDD are the same concept derived from different interpretations. Note that ACDDs use generalized BDD reduction rules, while BMDs use ZBDD reduction rules. Thus, ACDDs and BMDs are different.

WDDs are defined in terms of the discrete Walsh functions. Discrete Walsh functions are the integer counterparts of the Reed-Muller functions in the coding $(0,1) \rightarrow (1,-1)$. Therefore, WDDs are integer counterparts of FDDs in the $(1,-1)$ coding. Thus, they also relate to FDDs and other EXOR-based DDs.

Terminal nodes in the multi-terminal binary decision tree represent elements of the truth-vector of f. Terminal nodes in the arithmetic transform decision tree show the arithmetic spectrum. Similarly, in the Walsh transform decision tree, terminal nodes represent the Walsh coefficients in the Hadamard ordering.

Table 3.8.1 shows relation of various transforms and decision diagrams. Fig. 3.8.2 illustrates the relationship among MTBDDs, ACDDs and WDDs for an integer-valued function [4]. From Fig. 3.8.2, we can see the followings:

1) The MTBDD for the arithmetic spectrum of f is identical to the ACDD of f.
2) The MTBDD for the Walsh spectrum of f is identical to the WDD of f.
3) The MTBDD for f is identical to the WDD for the Walsh spectrum of f if we ignore the constant factor. This is because $\mathbf{W}(2)\mathbf{W}(2) = 2\mathbf{I}(2)$.
4) The WDD for the arithmetic spectrum of f is different from the ACDD for the Walsh spectrum of f. This is because $\mathbf{W}(2)\mathbf{A}(2) \neq \mathbf{A}(2)\mathbf{W}(2)$.

Thus, we have the following:

Statement 3.8.3 *An STDD is an MTBDD for corresponding spectrum.*

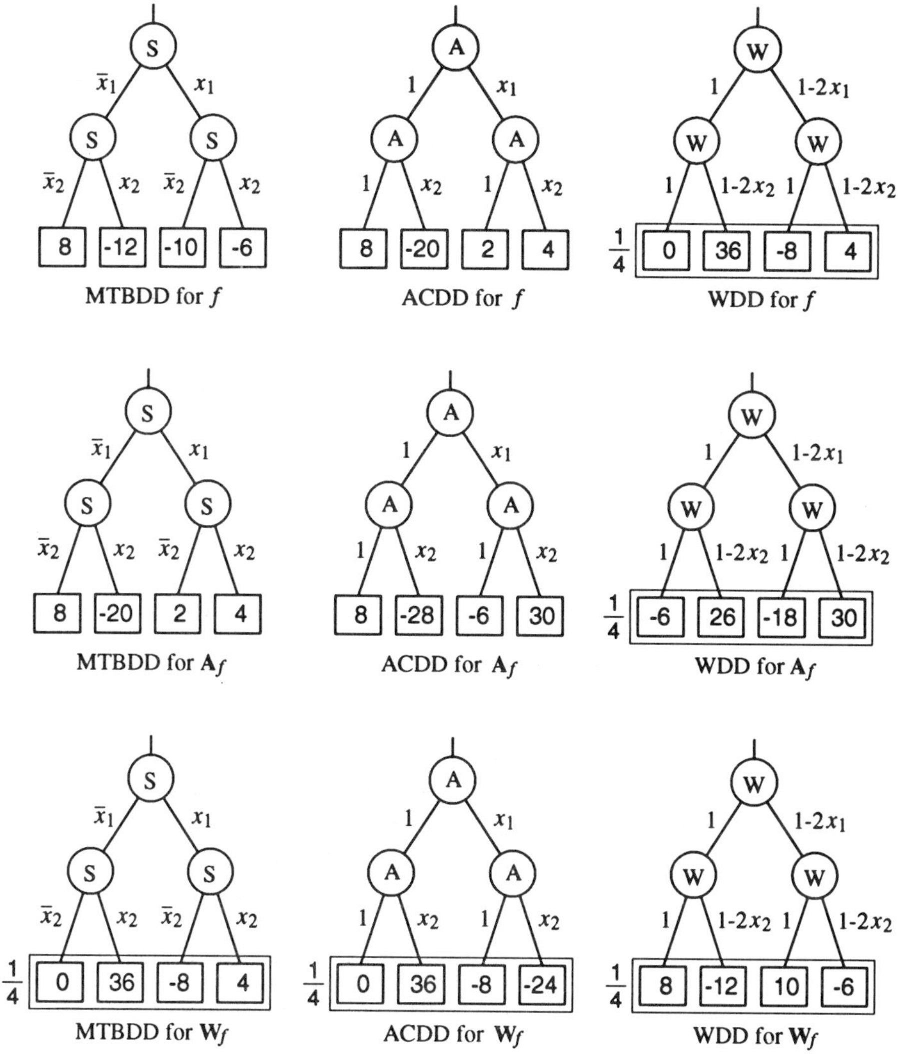

Figure 3.8.2 Relation of MTBDDs, ACDDs and WDDs.

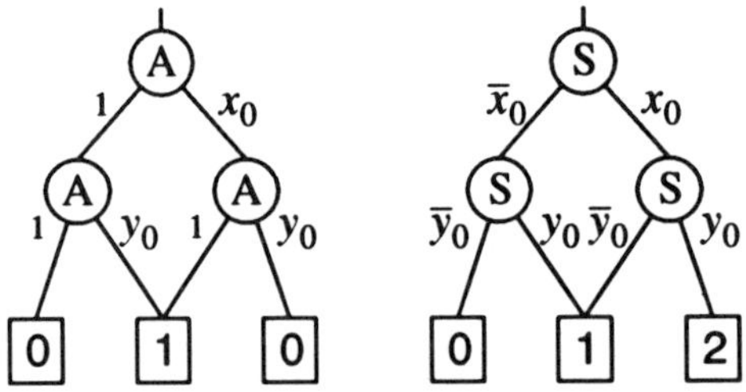

(a) ACDD (b) MTBDD

Figure 3.9.1 ACDD and MTBDD for a half adder.

3.9 STDDS FOR ARITHMETIC FUNCTIONS

3.9.1 ACDDs

ACDDs can represent **multiple-output switching function,** $f_{m-1}, \ldots, f_0$, through the integer-valued function:

$$f(z) = 2^{m-1} f_{m-1} + \cdots + 2^0 f_0. \tag{3.9.1}$$

Example 3.9.1 *Consider the half adder shown in Table 3.1.1. The multiple-output function is represented by expressions:*

$$\begin{aligned} f_1(x_0, y_0) &= x_0 y_0, \\ f_0(x_0, y_0) &= x_0 \oplus y_0. \end{aligned}$$

This multiple-output function is represented by the integer-valued function $f(z) = 2f_1 + f_0$. *The truth-vector of* $f(z)$ *is* $\mathbf{Z} = [0, 1, 1, 2]^t$. *The arithmetic spectrum is* $\mathbf{A}_z = [0, 1, 1, 0]^t$. *The ACDD and the MTBDD for* $f(z)$ *are shown in Fig. 3.9.1(a) and (b), respectively.* ■

The arithmetic transform decision tree for a multiple-output function represents the arithmetic spectra of the associated integer-valued functions. If $A_j(x_1, \ldots, x_n)$ is the arithmetic expression representing $f_j(x_1, \ldots, x_n)$, then the multiple-output function can be represented by the arithmetic expression [20]:

$$D = \sum_{j=0}^{m-1} 2^j A_j.$$

The representation of a multiple-output function is reduced to that of an arithmetic expression. The i-th bit in the binary representation of the resulting number shows the value of the i-th function [20].

Example 3.9.2 *The multiple-output function f_1, f_0 in Example 3.9.1 is represented by:*

$$\begin{aligned} A_1 &= x_0 y_0, \\ A_0 &= x_0 + y_0 - 2x_0 y_0. \end{aligned}$$

It is represented by the arithmetic expression:

$$D = 2A_1 + A_0 = x_0 + y_0.$$

The coefficients of this arithmetic expression are the values of the terminal nodes in Fig. 3.9.1(a). ■

In a similar way, the arithmetic expression for the 2-bit adder is

$$D = 4A_2 + 2A_1 + A_0 = 2(x_1 + y_1) + (x_0 + y_0).$$

The arithmetic expression for the n-bit adder is

$$D = \sum_{i=0}^{n-1} 2^i (x_i + y_i).$$

The arithmetic transform decision tree for the n-bit adder has $2n$ non-zero terminal nodes, and the values of them are $1, 2, 4, 8, \ldots, 2^{n-1}$. Thus, we can prove the followings:

Theorem 3.9.1 *The number of nodes to represent an n-bit adder by an ACDD is $O(n)$.*

(The proof is left to the readers as an exercise).

Note that the Shannon tree for the n-bit adder has $2^{2n} - 1$ non-zero terminal nodes, and the values of them are $1, 2, \ldots, 2^{n+1} - 2$. The number of nodes to represent the n-bit adder by an MTBDD is $O(2^n)$.

Example 3.9.3 *Consider the 2-bit multiplier:*

$$\begin{array}{cccc} & & x_1 & x_0 \\ & \times) & y_1 & y_0 \\ \hline f_3 & f_2 & f_1 & f_0 \end{array}$$

In this case, sum-of-products expressions for f_3, f_2, f_1*, and* f_0 *are*

$$\begin{aligned} f_0 &= x_0 y_0, \\ f_1 &= \bar{x}_1 x_0 y_1 \vee x_1 \bar{x}_0 y_0 \vee x_0 y_1 \bar{y}_0 \vee x_1 \bar{y}_1 y_0, \\ f_2 &= x_1 y_1 \bar{y}_0 \vee x_1 \bar{x}_0 y_1, \\ f_3 &= x_0 y_0 x_1 y_1. \end{aligned}$$

The arithmetic expression for the 2-bit multiplier is

$$\begin{aligned} D &= (2x_1 + x_0)(2y_1 + y_0) \\ &= 4x_1 y_1 + 2(x_0 y_1 + x_1 y_0) + x_0 y_0. \end{aligned}$$

The arithmetic transform decision tree has only four non-zero terminals, and the number of different values is only three. On the other hand, the Shannon tree has 9 non-zero terminals, and the number of different values is 6. We can also see that the arithmetic expression is much simpler than the sum-of-products expressions. ∎

For the n-bit multiplier, the arithmetic expression is

$$D = (\sum_{i=0}^{n-1} 2^i x_i)(\sum_{j=0}^{n-1} 2^j y_j).$$

Thus, the total number of non-zero terminals in the arithmetic transform decision tree is $n \times n = n^2$. Suppose that in the ACDD, the order of the variables is $x_0, x_1, \ldots, x_{n-1}, y_0, y_1, \ldots, y_{n-1}$, then we have the following:

Theorem 3.9.2 *The number of nodes to represent an* n*-bit multiplier by an ACDD is* $O(n^2)$.

(The proof is left to the readers as an exercise).

In the similar way, we can show that the number of nodes to represent X^k is $O(n^k)$. Note that the number of nodes to represent the n-bit multiplier by an MTBDD is $O(4^n)$.

MTBDDs for multiple-output functions are based on the point-wise composition of $f_{m-1}, \ldots, f_0$. However, ACDDs are based on a linear composition of the corresponding arithmetic expression $A_{m-1}, \ldots, A_0$. This is the major difference between ACDDs and MTBDDs. And ACDDs (BMDs) represent arithmetic functions more efficiently than MTBDDs.

3.9.2 WDDs

WDDs also represent multiple-output functions through $f(z)$. However, some advantages can be achieved by using an alternative coding. In the Walsh processing of switching functions, the coding $(0,1) \to (1,-1)$ of the function values is done to make the function more compatible to the transform [13]. In that coding, the Walsh spectral coefficients scaled by 2^n for an n-variable switching function are even numbers in the range -2^n to 2^n and their sum is 2^n. Thus, this coding excludes the odd integers in the terminal nodes in WDDs.

Theorem 3.9.3 *Let $\mathbf{Q}_f$ and $\mathbf{Q}_g$ be spectra for f and g, respectively. Then $\mathbf{Q}_f + \mathbf{Q}_g$ is the spectrum for $f + g$.*

(Proof) Let $\mathbf{F}$ and $\mathbf{G}$ be the truth vectors for f and g, respectively. Let $\mathbf{Q}$ be the transform matrix. Then, we have $\mathbf{Q}_f = \mathbf{QF}$ and $\mathbf{Q}_g = \mathbf{QG}$. Thus, $\mathbf{Q}_f + \mathbf{Q}_g = \mathbf{Q}(\mathbf{F} + \mathbf{G})$. □

Example 3.9.4 *Consider the half adder in Example 3.9.1. The multiple-output function is represented by an integer-valued function $f(z)$ by using the mapping (3.9.1). In WDDs, the component functions f_0 and f_1 can be in the $(0,1)$ or the $(1,-1)$ coding.*

In the case of the $(0,1)$ coding, we have $f(z) = 2f_1 + f_0$ and the truth-vector is given by $\mathbf{Z} = [0,1,1,2]^t$. The Walsh spectrum is $\mathbf{W}_f(z) = \frac{1}{4}\mathbf{W}(2)\mathbf{Z} = \frac{1}{4}[4,-2,-2,0]^t$. The Walsh spectrum is also obtained as the composition of the spectra for f_0 and f_1: $\mathbf{W}_{f_0} = \frac{1}{4}[2,0,0,-2]^t$, $\mathbf{W}_{f_1} = \frac{1}{4}[1,-1,-1,1]^t$, $\mathbf{W}_f(z) = 2\mathbf{W}_{f_1} + \mathbf{W}_{f_0} = \frac{1}{4}[4,-2,-2,0]^t$. The Walsh decision tree for $f(z)$ in the $(0,1)$ coding is shown in Fig. 3.9.2.

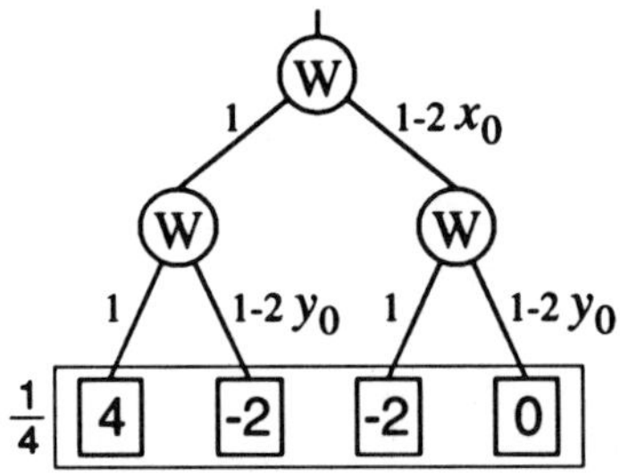

Figure 3.9.2 Walsh transform decision tree for a half adder with $(0, 1)$ coding.

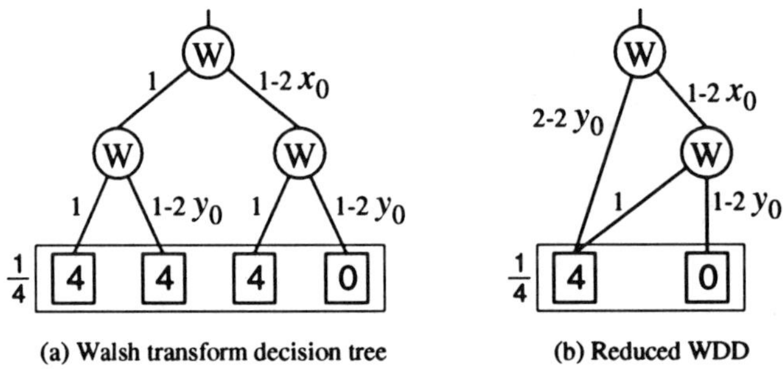

Figure 3.9.3 Walsh transform decision tree and RWDD for a half adder with $(1, -1)$ coding.

In the case of $(1, -1)$ *coding,* $\mathbf{F}_0 = [1, -1, -1, 1]^t$ *and* $\mathbf{F}_1 = [1, 1, 1, -1]^t$. *Thus,* $f_{(1,-1)}(z) = 2f_{1(1,-1)} + f_{0(1,-1)}$ *and the truth-vector is* $\mathbf{Z}_{(1,-1)} = [3, 1, 1, -1]^t$. *The Walsh spectrum is* $\mathbf{W}_{f_{(1,-1)}} = \frac{1}{4}\mathbf{W}(2)\mathbf{Z}_{(1,-1)} = \frac{1}{4}[4, 4, 4, 0]^t$. *The Walsh spectrum is obtained also as the composition of the spectra for* f_1 *and* f_0*:* $\mathbf{W}_{f_{(1,-1)}}(z) = 2\mathbf{W}_{f_{1(1,-1)}} + \mathbf{W}_{f_{0(1,-1)}}$, *where* $\mathbf{W}_{f_{0(1,-1)}} = \frac{1}{4}[0, 0, 0, 4]^t$ *and* $\mathbf{W}_{f_{1(1,-1)}} = \frac{1}{4}[2, 2, 2, -2]^t$. *The Walsh decision tree of* f *in the* $(1, -1)$ *coding and the corresponding RWDD are shown in Fig. 3.9.3.* ∎

Example 3.9.5 *Let us consider the half adder in Example 3.9.1. In the case of the* $(0,1)$ *coding, the arithmetic expressions are*

$$\begin{aligned} A_1 &= x_0 y_0, \\ A_0 &= x_0 + y_0 - 2x_0 y_0. \end{aligned}$$

Let $\alpha_0 = 1 - 2x_0$ *and* $\beta_0 = 1 - 2y_0$. *Then, we have*

$$\begin{aligned} W_1 &= \frac{1}{4}(1-\alpha_0)(1-\beta_0), \\ W_0 &= \frac{1}{4}[2(1-\alpha_0) + 2(1-\beta_0) - 2(1-\alpha_0)(1-\beta_0)]. \end{aligned}$$

The multiple-output function is represented by the expression:

$$\begin{aligned} D = 2W_1 + W_0 &= \frac{1}{4}[2(1-\alpha_0) + 2(1-\beta_0)] \\ &= \frac{1}{4}[4 - 2(\alpha_0 + \beta_0)]. \end{aligned}$$

In the case of the $(1,-1)$ *coding, we have to do the domain transform* $1 - 2f_j$:

$$\begin{aligned} W_1 &= 1 - 2A_1 = 1 - 2x_0 y_0, \\ W_0 &= 1 - 2A_0 = 1 - 2(x_0 + y_0 - 2x_0 y_0). \end{aligned}$$

Thus,

$$\begin{aligned} D &= 2W_1 + W_0 = 2(1 - 2x_0 y_0) + [1 - 2(x_0 + y_0 - 2x_0 y_0)] \\ &= 3 - 2(x_0 + y_0) = 3 - [(1-\alpha_0) + (1-\beta_0)] = \frac{1}{4}[4 + 4(\alpha_0 + \beta_0)]. \end{aligned}$$

■

By using the relation $x_i + y_i = 1 - \frac{1}{2}(\alpha_i + \beta_i)$, we can prove the following:

Theorem 3.9.4 *The number of nodes to represent an n-bit adder by a WDD is* $O(n)$.

Example 3.9.6 *The Walsh expression for the 2-bit multiplier is derived as follows: First, the integer-valued function is represented by* $D = (2x_1 + x_0)(2y_1 + y_0)$. *Let* $\alpha_i = 1 - 2x_i$, $\beta_i = 1 - 2y_i$. *Then, we have*

$$\begin{aligned} D &= [(1-\alpha_1) + \frac{1}{2}(1-\alpha_0)][(1-\beta_1) + \frac{1}{2}(1-\beta_0)] \\ &= \frac{1}{4}(3 - \alpha_0 - 2\alpha_1)(3 - \beta_0 - 2\beta_1) \\ &= \frac{1}{4}[9 - 3(\alpha_0 + \beta_0) - 6(\alpha_1 + \beta_1) + \alpha_0\beta_0 + 2(\alpha_0\beta_1 + \alpha_1\beta_0) + 4\alpha_1\beta_1]. \end{aligned}$$

Thus, the number of non-zero terminals is 9, and the number of different coefficients is only 6. ■

For the n-bit multiplier, the Walsh expression is

$$D = \frac{1}{4}[\sum_{i=0}^{n-1} 2^i(1-\alpha_i)][\sum_{j=0}^{n-1} 2^j(1-\beta_j)].$$

Thus, the total number of non-zero terminals in the Walsh decision tree is $(n+1)\times(n+1) = (n+1)^2$. Suppose that in the WDD, the order of the variables is $x_0, x_1, \ldots, x_{n-1}, y_0, y_1, \ldots, y_{n-1}$, then we have the following:

Theorem 3.9.5 *The number of nodes to represent an n-bit multiplier by a WDD is $O(n^2)$.*

In the similar way, we can show that the number of nodes to represent X^k is $O(n^k)$.

3.9.3 WDDs with $n = 3$

The use of negative edges is often useful for compact representations. WDDs with the $(1,-1)$ coding are often more compact than ones with the $(0,1)$ coding.

When we use the $(1,-1)$ coding, switching functions with $n = 3$ have three distinct sets of coefficients, which are scaled by 2^3 in order to have integer values: $\{0,8\}$, $\{0,4,-4\}$, $\{2,-2,6\}$ [13]. Thus, the maximal number of terminal nodes is three. Note that some combinations within these distinct sets are not allowed in the Walsh spectra of switching functions. For example, the vector $\frac{1}{8}[-2,-2,2,2,-6,-2,2,-2]^t$ does not correspond to any switching function. Among the Walsh spectra for the switching functions, the most complex case appears when the values and the sign of the spectral coefficients alternate. The simplest cases are 2^n switching functions corresponding to the discrete Walsh functions after the recoding $(1,-1) \rightarrow (0,1)$. In the Boolean domain, they are linear functions and their spectra contain a single non-zero coefficient 2^n.

RWDDs of three different sizes exist for the switching functions with $n = 3$ except for linear functions. Besides the root node and three terminal nodes, an RWDD for $n = 3$ can contain 5, 3 or 2 non-terminal nodes. RWDDs for linear functions have only two terminal nodes.

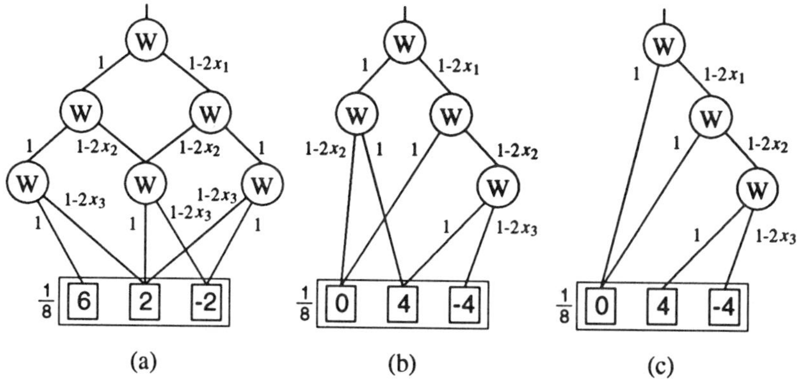

Figure 3.9.4 Possible shapes of RWDDs for three-variable functions under the $(1,-1)$ coding.

Example 3.9.7 *Fig. 3.9.4 shows RWDDs for* $f = x_2x_3 \oplus x_1x_2x_3$, $f = x_2x_3 \oplus x_1x_3$ *and* $f = x_1 \oplus x_2x_3$. *All WDDs are in the* $(1,-1)$ *coding.* ■

A relevant property of RWDDs is that a considerable subset of functions can be realized by an RWDD of a given shape. By Walsh spectral coefficients, 64 out of 256 functions for three variables can be represented by RWDDs of the shape shown in Fig. 3.9.4(*c*) with different assignments of labels at the edges and values to terminal nodes.

Example 3.9.8 *Fig. 3.9.5 shows RWDDs for* $f = x_2 \oplus x_1x_3$ *with different orders of the input variables. The RWDDs are in the* $(1,-1)$ *coding. This illustrates that the ordering of the variables influences the size of WDD.* ■

3.10 CONCLUSIONS AND COMMENTS

In this chapter, we generalized concepts of BDDs and FDDs. STDDs are graphical representation of spectral transform for switching functions and integer-valued functions. In BDDs and FDDs, the additions are done in $GF(2)$, while in MTBDDs, ACDDs, and WDDs, the additions are done in ordinary arithmetic. We introduced a generalized BDD reduction rules for STDDs. BMDs [4]

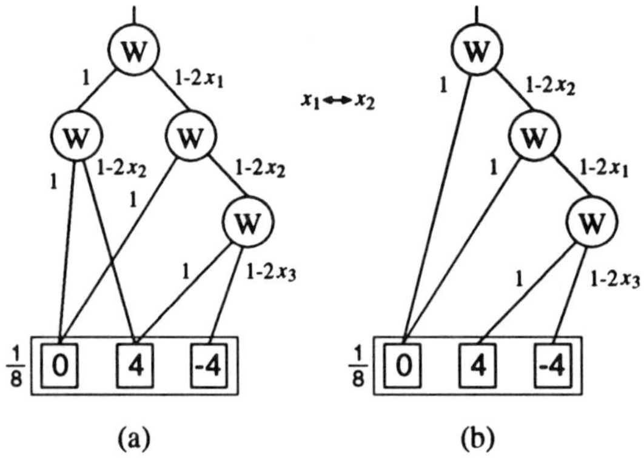

Figure 3.9.5 RWDD for $f = x_2 \oplus x_1 x_3$ with different orderings of variables.

are graphical representation of arithmetic spectra, and are equivalent to ACDDs except for the reduction rules. We showed methods to read a function and its spectrum from a given STDD. We also showed that ACDDs and WDDs for adders and multipliers are much smaller than MTBDDs.

The applications of STDDs are:

- Classification of switching functions (WDDs) [13, 19].
- Technology mapping (WDDs) [6].
- Verification (ACDDs) [15].
- Test generation (ACDDs) [12].

Extensions of STDDs are:

- Transforms using various bases (HDDs) [6].
- Multiple-place DDs (MDDs) [28].
- FDDs for multi-valued functions [30].
- Edge valued binary decision diagrams (EVBDDs) [18, 31].

REFERENCES

[1] Ph. W. Besslich, "Efficient computer method for XOR logic design", *IEE Proc., Part E*, Vol. 129, 1982, pp. 15-20.

[2] R. I. Bahar, E. A. Frohm, C. M. Gaona, G. D. Hachtel, E. Macii, A. Pardo, F. Somenzi, "Algebraic decision diagrams and their applications", *Int. Conf. on CAD*, Nov. 1993, pp. 188-191.

[3] R. E. Bryant, "Graph-based algorithms for Boolean functions manipulation", *IEEE Trans. Comput.*, Vol. C-35, No.8, 1986, pp. 667-691.

[4] R. E. Bryant, Y-A. Chen, "Verification of arithmetic functions with binary moment diagrams", May 1994, CMU-CS-94-160. Also *Proc. 32nd ACM/IEEE DAC*, June 1995.

[5] P. Calingaert, "Switching function canonical forms based on commutative and associative binary operations", *Trans. Amer. Inst. Elect. Eng.*, Part I, 52, 1961, pp. 808-814.

[6] E. M. Clarke, K. L. McMillan, X. Zhao, M. Fujita, "Spectral transforms for extremely large Boolean functions", *Proc. IFIP WG 10.5 Workshop on Applications of the Reed-Muller Expansion in Circuit Design*, Sept. 1993, Hamburg, pp. 86-90. (Also Chapter 4 of this book.)

[7] R. Drechsler, A. Sarabi, M. Theobald, B. Becker, M.A. Perkowski, "Efficient representation and manipulation of switching functions based on ordered Kronecker functional decision diagrams", *Proc. 31st ACM/IEEE DAC*, June 1994, pp. 415-419.

[8] B. J. Falkowski, M. A. Perkowski, "One more way to compute the Hadamard-Walsh spectrum of completely and incompletely specified Boolean functions", *Int. J. Electronics*, Vol. 69, No.5, 1990, pp. 595-602.

[9] J. E. Gibbs, "Instant Fourier transform," *Electron. Lett.*, Vol. 13, No.5, 1977, pp. 122-123.

[10] I. J. Good, "The interaction algorithm and practical Fourier analysis", *J. Roy. Statist. Soc.*, ser. B, Vol. 20, 1958, 361-372, Addendum, Vol. 22, 1960, pp. 372-375.

[11] B. Harking, C. Moraga, "Efficient derivation of Reed-Muller expansions in multiple-valued logic systems", *Proc. 22nd Int. Symp. on Multiple-valued Logic*, May 1992, Sendai, Japan, pp. 436-441.

[12] K. D. Heidtmann, "Arithmetic spectra applied to stuck-at-fault detection for combinatorial networks", *Proc. 2nd Technical Workshop New Directions in IC Testing*, 4.1-4.13, Winnipeg, April 1987. Also, *IEEE Trans. on Computers*, vol. 40, No. 3, March 1991, pp. 320-324.

[13] S. L. Hurst, D. M. Miller, J. C. Muzio, *Spectral Techniques in Digital Logic*, Academic Press, Bristol, 1985.

[14] M. G. Karpovsky, *Finite Orthogonal Series in the Design of Digital Devices*, Wiley and JUP, New York, 1976.

[15] J. Jain, "Arithmetic transform of Boolean functions," Chapter 6 of this book.

[16] U. Kebschull, E. Schubert, W. Rosenstiel, "Multilevel logic synthesis based on functional decision diagrams", *EDAC 92*, 1992, pp. 43-47.

[17] G.A. Kukharev, V.P. Shmerko, S.N. Yanushkievich, *Technique of Binary Data Parallel Processing for VLSI*, Vysheyshaja shcola, Minsk, 1991, Belaruss.

[18] Y. F. Lai, M. Pedram, S. B. K. Vrudhula, "EVBDD-based algorithms for integer linear programming, spectral transformation, and functional decomposition", *IEEE Trans. Computer-Aided Design of Integrated Circuits and Systems*, Vol. 13, No.8, 1994, pp. 959-975. (Also Chapter 5 of this book.)

[19] R. J. Lechner, "Harmonic analysis of switching functions", in A. Mukhopadhay, Ed., *Recent Development in Switching Theory*, Academic Press, 1971.

[20] V. D. Malyugin, "On a polynomial realization of a cortege of Boolean functions", *Repts. of the USSR Academy of Sciences*, Vol. 265, No. 6, 1982.

[21] S. Minato, "Graph-based representations of discrete functions", *Proc. IFIP WG 10.5 Workshop on Applications of the Reed-Muller Expansion in Circuit Design*, August 1995, Chiba, Japan, pp. 1-10. (Also Chapter 1 of this book.)

[22] J. C. Muzio, T. C. Wesselkamper, *Multiple-valued Switching Theory*, Adam Hilger, Bristol, 1986.

[23] M. A. Perkowski, "The generalized orthonormal expansion of functions with multiple-valued inputs and some of it applications", *Proc. 22nd Int. Symp. on Multiple-Valued Logic*, May 1992, Sendai, Japan.

[24] T. Sasao, "Ternary decision diagrams and their applications", *Int. Workshop on Logic Synthesis,* Lake Tahoe, May 1993. (Also Chapter 12 of this book.)

[25] T. Sasao, J. T. Butler, "A design method for look-up table type FPGA by pseudo-Kronecker expansions", *Proc. 24th Int. Symp. on Multiple-valued Logic*, Boston, May 1994, pp. 97-104.

[26] T. Sasao, D. Debnath, "An exact minimization algorithm for generalized Reed-Muller expansions", *Proc. IEEE Asia-Pacific Conference on Circuits and Systems*, Dec. 1994, Taipei.

[27] T. Sasao, "Representation of logic functions using EXOR operators", *Proc. IFIP WG 10.5 Workshop on Applications of the Reed-Muller Expansion in Circuit Design*, August 1995, Chiba, Japan, pp. 11-20. (Also Chapter 2 of this book.)

[28] A. Srinivasan, T. Kam, S. Malik, R. K. Brayton, "Algorithms for discrete function manipulation", in: *Proc. Int. Conf. on CAD*, 1990, pp. 92-95.

[29] R. S. Stanković, C. Moraga, "Reed-Muller-Fourier representations of multiple-valued functions over Galois fields of prime cardinality", *Proc. IFIP WG 10.5 Workshop on Applications of the Reed-Muller Expansion in Circuit Design*, Sept. 1993, Hamburg, pp. 115-124.

[30] R. S. Stanković, "Functional decision diagrams for multiple-valued functions", *Proc. 25th Int. Symp. on Multiple-Valued Logic*, Indiana, May 1995, pp. 284-289.

[31] R. S. Stanković, "Some remarks about spectral transform interpretation of MTBDDs and EVBDDs", *Proc. ASP-DAC'95*, August 1995, Chiba, Japan, pp. 385-390.

4

MULTI-TERMINAL BINARY DECISION DIAGRAMS AND HYBRID DECISION DIAGRAMS

Edmund M. CLARKE[1]
Masahiro FUJITA[2]
Xudong ZHAO[1]

[1] *School of Computer Science*
Carnegie Mellon University
Pittsburgh, PA 15213, USA

[2] *Fujitsu Labs of America Inc.*
3350 Scott Blvd. Bldg. #34
Santa Clara, CA 95054-3104, USA

Abstract— Functions that map vectors with binary values into the integers are important for the design and verification of arithmetic circuits. We demonstrate how multi-terminal binary decision diagrams (MTBDDs) can be used to represent such functions concisely. The Walsh transform and Reed-Muller transform have numerous applications in computer-aided design, but the usefulness of these techniques in practice has been limited by the size of the binary valued functions that can be transformed. We show how to compute the MTBDD representations of the Walsh transform and Reed-Muller transform for functions with several hundred variables. Bryant and Chen have proposed binary moment diagrams (BMDs) for representing the class of functions that we have considered. We discuss the relationship between these methods and describe a generalization called hybrid decision diagrams which is often much more concise.

4.1 INTRODUCTION

Large integer matrices arise naturally in the design and verification of arithmetic circuits. In this chapter, we describe how to represent and manipulate such matrices efficiently using **multi-terminal binary decision diagrams** (MTBDDs) [7]. An MTBDD is like an ordinary binary decision diagram except that the terminal nodes can be arbitrary integer values instead of just 0 and 1. Previously, we have demonstrated how MTBDDs can be used to represent functions that map vectors with binary values into the integers. Our representation for integer matrices is based on this technique. An integer matrix with dimensions $2^m \times 2^n$ can be treated as a function that maps vectors with binary values of length $m + n$ into the integers. Various matrix operations can be performed by operations on the corresponding integer functions.

The Walsh transform and the Reed-Muller transform [10] have numerous applications in computer aided design, particularly in synthesis and testing of circuits. Unfortunately, the usefulness of these techniques in practice has been limited by the size of the binary valued functions that can be handled by the transform. Since these transforms are given as vectors with length of 2^n where n is the number of variables in the function, currently available techniques limit the functions to less than 20 variables. Since the Walsh matrix and the Reed-Muller matrix have simple recursive definitions, they can be encoded efficiently by MTBDDs. In this manner, we can compute concise representations for the transforms of functions with several hundred variables.

Recently, Bryant and Chen [5] have proposed binary moment diagrams (BMDs) for representing functions that map vectors with binary values into the integers. We show that the BMD of a function is the MTBDD that results from applying the inverse integer Reed-Muller transformation [11] to the function. The transformation can be computed using the techniques that we have developed for manipulating large matrices. The transformation matrix in this case is the Kronecker product [2] of a number of identical 2×2 matrices. We show that the Kronecker products of other 2×2 matrices behave in a similar way. In fact, the transformations obtained from Kronecker products of other matrices will in many cases be more concise than the BMD. We have further generalized this idea so that the transformation matrix can be the Kronecker product of different matrices. In this way, we obtain a representation, called hybrid decision diagram (HDD), that is more concise than either the MTBDD or the BMD.

Our chapter is organized as follows: Section 4.2 gives the basic properties of MTBDDs that are used in the remainder of the chapter. Section 4.3 shows how the results of the previous section can be used to implement standard operations like addition and multiplication of very large integer matrices. Section 4.4 describes how BDDs can be obtained for recursively defined integer matrices and shows how to compute the spectral transforms for binary valued functions. In Section 4.4 we also illustrate the power of this representation by computing the transforms of several very large binary valued functions. Section 4.5

describes the relationship between BMDs and the inverse integer Reed-Muller transformation. This section also introduces Kronecker product and shows how it can be used to generalize BMDs. The next section introduces hybrid decision diagrams and provides experimental evidence to show the usefulness of this representation. The chapter concludes in Section 4.7 with a brief summary and a discussion of directions for future research.

4.2 MULTI-TERMINAL BINARY DECISION DIAGRAMS

Ordered binary decision diagrams (BDDs) are a canonical representation for binary valued functions proposed by Bryant [4]. They are often substantially more compact than traditional normal forms such as conjunctive normal form and disjunctive normal form. They can also be manipulated very efficiently. Hence, BDDs have become widely used for a variety of CAD applications, including symbolic simulation, verification of combinational logic and, more recently, verification of sequential circuits.

A BDD is similar to a binary decision tree, except that its structure is a directed acyclic graph (DAG) rather than a tree, and there is a strict total order placed on the occurrence of variables as one traverses the graph from root to leaf. Algorithms of linear complexity exist for computing BDD representations of $\neg f$ and $f \vee g$ from the BDDs for the functions f and g.

Let $f : B^m \to Z$ be a function that maps vectors with binary values of length m into integers. Suppose $n_1, \ldots, n_N$ are the possible values of f. The function f partitions the space B^m of vectors with binary values into N sets $\{S_1, \cdots, S_N\}$, such that $S_i = \{\bar{x} \mid f(\bar{x}) = n_i\}$. Let f_i be the characteristic function of S_i. We say that f is in *normal form* if $f(\bar{x})$ is represented as $\sum_{i=1}^{N} f_i(\bar{x}) \cdot n_i$. This sum can be represented as a BDD with integers as its terminal nodes. We call such DAGs **Multi-terminal BDDs** (MTBDDs) [1, 7].

Any arithmetic operation $\odot$ on MTBDDs can be performed in the following way.

$$\begin{aligned} h(\bar{x}) &= f(\bar{x}) \odot g(\bar{x}) \\ &= \sum_{i=1}^{N} f_i(\bar{x}) \cdot n_i \odot \sum_{j=1}^{N'} g_j(\bar{x}) \cdot n'_j \end{aligned}$$

$$= \sum_{i=1}^{N} \sum_{j=1}^{N'} f_i(\bar{x}) g_j(\bar{x}) (n_i \odot n'_j)$$

$$= \sum_{k=1}^{N''} \left(\sum_{n_i \odot n'_j = n''_k} f_i(\bar{x}) g_j(\bar{x}) \right) n''_k.$$

Since the f_i's are mutually disjoint and the g_j's are mutually disjoint, the $f_i g_j$'s are also mutually disjoint. Therefore, the summations $\sum_{n_i \odot n'_j = n''_k} f_i(\bar{x}) g_j(\bar{x})$ are mutually disjoint binary valued functions.

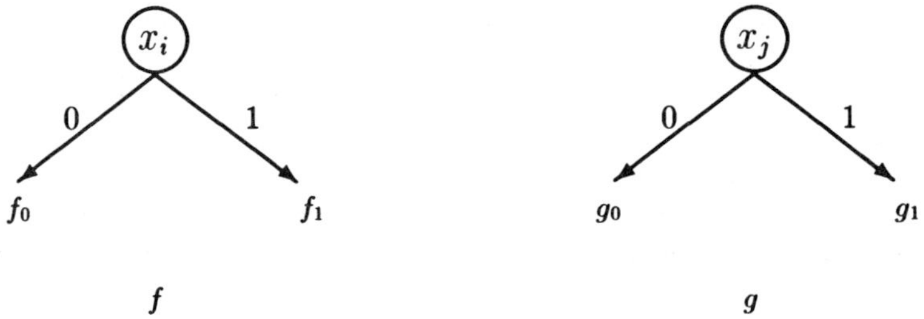

Figure 4.2.1 BDDs for f and g.

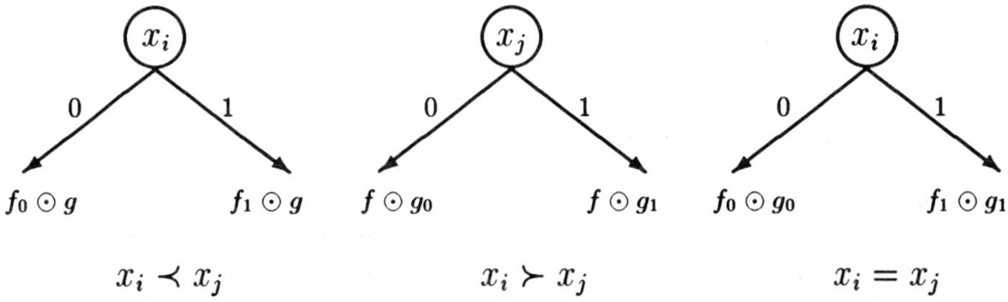

Figure 4.2.2 BDD of $f \odot g$.

We now give an efficient algorithm for computing $f(\bar{x}) \odot g(\bar{x})$.

- If f is a leaf, then for each leaf of g, apply $\odot$ with f as the first argument.
- If g is a leaf, then for each leaf of f, apply $\odot$ with g as the second argument.

- Otherwise, f and g have the form in Figure 4.2.1, and the BDD for $f \odot g$, depending on the relative order of x_i and x_j, is given in Figure 4.2.2. We use the notation $x_i \prec x_j$ to mean that variable x_i precedes variable x_j in the total ordering for the BDD variables.

The resulting diagram may not be in normal form. In order to convert it into normal form, a *reduction* phase is needed. The algorithm for this phase is essentially identical to the reduction phase in Bryant's algorithm for constructing BDDs [4] .

Functions that map vectors with binary values into the integers can also be represented as arrays of BDDs. Each BDD corresponds to one bit of the binary representation of the function value. In general, it is quite expensive to perform operations using this representation.

4.3 MATRIX OPERATIONS

Let M be a $2^k \times 2^l$ matrix over Z. It is easy to see that M can be represented as a function $M : B^{k+l} \rightarrow Z$, such that $M_{ij} = M(\bar{x}, \bar{y})$, where $\bar{x}$ is the binary vector encoding i and $\bar{y}$ is the binary vector encoding j. Therefore, matrices with integer values can be represented as integer valued functions using the MTBDD representation in Section 4.2. We need the following operations for integer matrices for computing the spectral transforms: *absolute value*, *scalar multiplication*, *addition*, *sorting a vector of integers*, *summation over one dimension*, and *matrix multiplication*. The first three operations are trivial and will not be discussed in this chapter.

- Summing matrices over one dimension

  It is sometimes desirable to obtain a 2^n vector from a $2^n \times 2^m$ matrix that each element in the vector is the summation of the corresponding column, i.e. $M'_i = \sum_{j=0}^{2^m-1} M_{ij}$. When the matrices are expressed in terms of integer valued functions, the equation becomes $M'(\bar{x}) = \sum_{\bar{y}} M(\bar{x}, \bar{y})$, where $\sum_{\bar{y}}$ means "sum over all possible assignments to $\bar{y}$". In practice, $\sum_{\bar{y}} M(\bar{x}, \bar{y})$ can be computed as:

$$\sum_{y_1 y_2 \cdots y_m} M(\bar{x}, y_1, y_2, \ldots, y_m)$$

$$
\begin{aligned}
&= \sum_{y_1 y_2 \cdots y_{m-1}} \sum_{y_m} M(\bar{x}, y_1, y_2, \ldots, y_m) \\
&= \sum_{y_1 y_2 \cdots y_{m-1}} (M(\bar{x}, y_1, y_2, \ldots, y_{m-1}, 0) \\
&\qquad + M(\bar{x}, y_1, y_2, \ldots, y_{m-1}, 1)).
\end{aligned}
$$

In this way, each variable in $\bar{y}$ is eliminated by performing an addition.

This operation can also be used to sum the elements of a vector and to obtain a two dimensional matrix from a three dimensional matrix by summing over one dimension. Although this operation works well in many cases, the worst case complexity can be exponential in the number of variables.

- Sorting vectors

  Frequently, it is useful to rearrange the elements in a vector so that they are in non-decreasing order. When the number of different values in the vector is not very large, the sorted vector can be represented concisely without using MTBDDs. In order to uniquely determine a sorted vector, we only need to know the set of different values and the number of occurrences of each value. Thus, the sorted vector can be represented as a list with length m, where m is the number of different values. Each element in the list contains the value and number of its occurrences.

  It is easy to find the set of different values, since it is only necessary to collect all of the terminal nodes in the MTBDD. The number of occurrences N_k of a possible value c_k can be calculated as $N_k = \sum_{i=0}^{2^n-1}(\text{if } V_i = c_k \text{ then } 1 \text{ else } 0)$, where V_i is the ith element of the vector. The operation of summation over a vector discussed previously can be applied to compute this sum. Although, in general, the complexity of the summation operation does not have a satisfactory upper bound, summation over a vector takes time linear in the size of the MTBDD representing the vector. Thus the complexity of the sorting operation is linear in both the number of distinct values in the vector and the size of the MTBDD representation of the vector.

- Matrix multiplication

  Suppose that two matrices A and B have dimensions $2^k \times 2^l$ and $2^l \times 2^m$, respectively. Let $C = A \times B$ be the product of A and B, C will have dimension $2^k \times 2^m$. If we treat A and B as integer valued functions, we can compute the product matrix C as

$$C(\bar{x}, \bar{z}) = \sum_{\bar{y}} A(\bar{x}, \bar{y}) B(\bar{y}, \bar{z}),$$

using the summation operation discussed above. In general, the complexity of this operation can also be exponential in the number of variables.

4.4 SPECTRAL TRANSFORMATIONS OF BINARY VALUED FUNCTIONS

Two of the most commonly used transformations in digital circuit design are the Walsh transform and the Reed-Muller transform [10]. In this section, we will show how the MTBDD based techniques described previously can be used to compute concise representations of the spectra for these transformations.

The Walsh matrix W_n has the recursive definition:

$$W_0 = 1 \qquad W_n = \begin{bmatrix} W_{n-1} & W_{n-1} \\ W_{n-1} & -W_{n-1} \end{bmatrix}.$$

Each element of the matrix is determined by its row and column coordinates. We will encode the 2^n columns by variables $y_n, \ldots, y_1$ and the 2^n rows by the variables $x_n, \ldots, x_1$. W_n can be represented as an integer valued function:

$$\begin{aligned} & W_n(y_n, \ldots, y_1, x_n, \ldots, x_1) \\ = & \begin{cases} W_{n-1}(y_{n-1}, \ldots, y_1, x_{n-1}, \ldots, x_1) & \text{if } (x_n y_n \neq 1) \\ -W_{n-1}(y_{n-1}, \ldots, y_1, x_{n-1}, \ldots, x_1) & \text{if } (x_n y_n = 1) \end{cases} \\ = & W_{n-1}(y_{n-1}, \ldots, y_1, x_{n-1}, \ldots, x_1) \\ & \cdot(\text{if } x_n y_n = 1 \text{ then } -1 \text{ else } 1). \end{aligned}$$

The above recursive definition can be expressed by an MTBDD as shown in Figure 4.4.1.

The Walsh transform maps a binary valued vector f with length 2^n to an integer vector of length 2^n, denoted by W_f, in which each component is between -2^n to 2^n. The transform can be easily expressed using the Walsh matrix, $W(f) = W_n \times (1 - 2f)$ [10]. For example, the vector $[0,1,1,1,1,0,0,0]^T$ is mapped into $[0,0,0,0,-4,4,4,4]^T$.

Likewise, the Reed-Muller matrix has the recursive definition:

$$R_0 = 1 \qquad R_n = \begin{bmatrix} R_{n-1} & 0 \\ R_{n-1} & R_{n-1} \end{bmatrix}$$

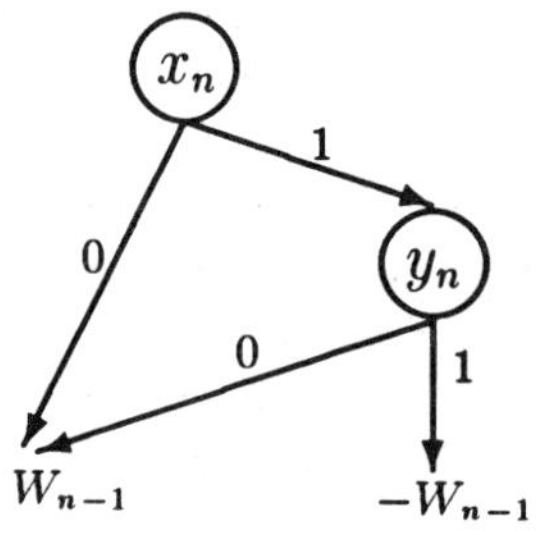

Figure 4.4.1 MTBDD for W_n.

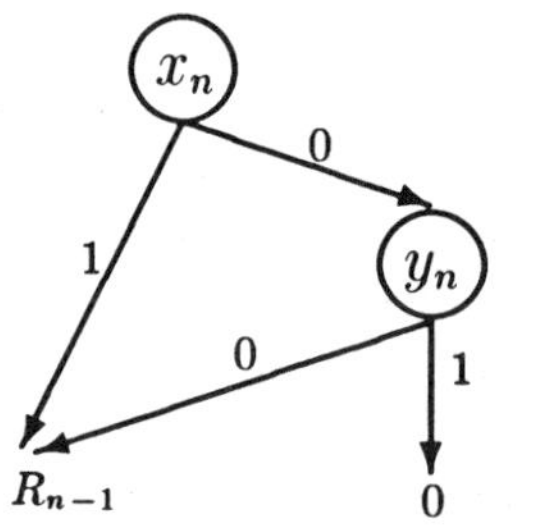

Figure 4.4.2 MTBDD for R_n.

which can be expressed by

$$\begin{aligned} & R_n(y_n, \ldots, y_1, x_n, \ldots, x_1) \\ = \ & \text{if } ((\neg x_n) \cdot y_n) \text{ then } 0 \\ & \text{else } R_{n-1}(y_{n-1}, \ldots, y_1, x_{n-1}, \ldots, x_1) \end{aligned}$$

and has the MTBDD representation in Figure 4.4.2.

The Reed-Muller transformation [11] maps a binary valued vector of length 2^n into another binary valued vector of the same length. This transformation can be expressed by the matrix multiplication $R(f) = R_n \times f$. However, during the matrix multiplication, integer addition is replaced by XOR in order to perform the modulo 2 arithmetic. For example, the vector $[0, 1, 1, 1, 1, 0, 0, 0]^T$ is mapped into $[0, 1, 1, 1, 1, 0, 0, 0]^T$.

Table 4.4.1 Experimental results for spectral transformations.

example circuit					Walsh coef.		R-M coef.	
circuit	# of inputs	output name	# of gates	|BDD|	|MTBDD|	time	|MTBDD|	time
c1908	33	9	880	3607	1850	44	27748	184
c3540	50	361	1669	520	15985	171	4679	8.2
c5315	178	813	2307	1397	7069	328	2647	25
$adder_{50}$	100	C_{50}	250	151	7456	23	249	2.3
$adder_{100}$	200	C_{100}	500	301	29906	128	499	11

When the number of variables is large, the transformations can be computed by representing the matrices and the vectors as MTBDDs and matrix operations can be performed as described in Section 4.3 and Section 4.4.

To illustrate the power of these techniques, we have computed the Walsh transformation and Reed-Muller transformation for some large combinatorial circuits, including two adders and some of the ISCAS benchmarks (Table 4.4.1). The examples were run on a DEC-5000 and run time is shown in seconds. We use the notation |BDD| to indicate the size of a BDD. A similar convention is used for MTBDDs.

4.5 KRONECKER TRANSFORMATIONS

Recently, Bryant and Chen[5] have developed a new representation for functions that map vectors with binary values to integer values. This representation is called the binary moment diagram (BMD) of the function. Instead of the Shannon expansion $f = xf_1 + (1-x)f_0$, they use the expansion $f = f_0 + xf'$, where f' is equal to $f_1 - f_0$. After merging the common subexpressions, a DAG representation for the function is obtained. They prove in their paper that this gives a compact representation for certain functions which have exponential size if represented by MTBDDs directly.

There is a close relationship between this representation and the inverse integer Reed-Muller transformation. The matrix for the inverse integer Reed-Muller transformation is defined recursively by

$$S_0 = 1 \qquad S_n = \begin{bmatrix} S_{n-1} & 0 \\ -S_{n-1} & S_{n-1} \end{bmatrix}$$

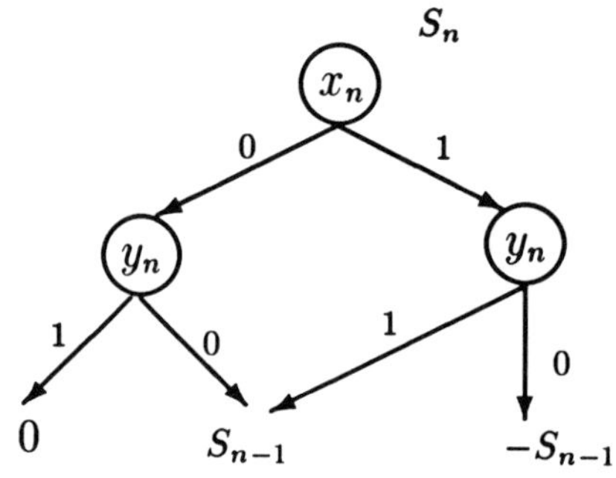

Figure 4.5.1 MTBDD for inverse integer Reed-Muller transformation matrix.

which has a linear MTBDD representation as shown in Figure 4.5.1. Let $\bar{i} \in B^n$ be the binary representation of the integer $0 \leq i < 2^n$. A function $f : B^n \rightarrow Z$ can be represented as a column vector where the value of the i-th entry is $f(\bar{i})$. We will not distinguish between a function and its corresponding column vector. The inverse integer Reed-Muller transformation can be obtained by multiplying the transformation matrix and the column vector $S(f) = S \times f$ using the technique described in the previous section.

Theorem 4.5.1 *The MTBDD of the inverse integer Reed-Muller transform of f is isomorphic to the BMD of f.*

The Kronecker product of two matrices is defined as follows:

$$A \otimes B = \begin{pmatrix} a_{11} & \dots & a_{1m} \\ \vdots & & \vdots \\ a_{n1} & \dots & a_{nm} \end{pmatrix} \otimes B = \begin{pmatrix} a_{11}B & \dots & a_{1m}B \\ \vdots & & \vdots \\ a_{n1}B & \dots & a_{nm}B \end{pmatrix}.$$

The inverse integer Reed-Muller matrix can be represented as the Kronecker product of n identical 2×2 matrices:

$$S_n = \begin{pmatrix} S_{n-1} & 0 \\ -S_{n-1} & S_{n-1} \end{pmatrix} = \begin{pmatrix} 1 & 0 \\ -1 & 1 \end{pmatrix} \otimes S_{n-1}$$

$$= \underbrace{\begin{pmatrix} 1 & 0 \\ -1 & 1 \end{pmatrix} \otimes \ldots \otimes \begin{pmatrix} 1 & 0 \\ -1 & 1 \end{pmatrix}}_{n}.$$

The inverse integer Reed-Muller transformation is not the only method that can be used to reduce the size of the BDD representation. Other transformations that are defined as Kronecker products of 2×2 matrices may also provide concise representations for functions mapping vectors with binary values into integers. In particular, Reed-Muller matrix R_n and Walsh matrix W_n can be represented as Kronecker products shown below:

$$\begin{aligned} R_n &= \begin{pmatrix} R_{n-1} & 0 \\ R_{n-1} & R_{n-1} \end{pmatrix} = \begin{pmatrix} 1 & 0 \\ 1 & 1 \end{pmatrix} \otimes R_{n-1} \\ &= \underbrace{\begin{pmatrix} 1 & 0 \\ 1 & 1 \end{pmatrix} \otimes \ldots \otimes \begin{pmatrix} 1 & 0 \\ 1 & 1 \end{pmatrix}}_{n} \quad \text{and} \\ W_n &= \begin{pmatrix} W_{n-1} & W_{n-1} \\ W_{n-1} & -W_{n-1} \end{pmatrix} = \begin{pmatrix} 1 & 1 \\ 1 & -1 \end{pmatrix} \otimes W_{n-1} \\ &= \underbrace{\begin{pmatrix} 1 & 1 \\ 1 & -1 \end{pmatrix} \otimes \ldots \otimes \begin{pmatrix} 1 & 1 \\ 1 & -1 \end{pmatrix}}_{n}. \end{aligned}$$

In fact, the Kronecker product of any non-singular 2×2 matrices can be used as a transformation matrix and will produce a canonical representation for the function. We call such transformations **Kronecker transformations.** If the entries of the 2×2 matrix are restricted among $\{0, 1, -1\}$, there are six interesting matrices

$$\begin{pmatrix} 1 & 0 \\ 0 & 1 \end{pmatrix}, \begin{pmatrix} 1 & 0 \\ -1 & 1 \end{pmatrix}, \begin{pmatrix} 1 & 0 \\ 1 & 1 \end{pmatrix}, \begin{pmatrix} 0 & 1 \\ -1 & 1 \end{pmatrix},$$
$$\begin{pmatrix} 0 & 1 \\ 1 & 1 \end{pmatrix}, \text{ and} \begin{pmatrix} 1 & 1 \\ -1 & 1 \end{pmatrix}.$$

All other matrices are either singular or would produce BDDs that are isomorphic to one of the six matrices.

We have applied these transformations to the functions discussed in paper[5]. The transformation can be partitioned into two groups of three each. The MTBDDs of the results after applying the transformations in the same group have the same complexity. Let $X = \sum_{i=0}^{n} x_i 2^i, Y = \sum_{j=0}^{m} y_j 2^j, X_j = \sum_{i=0}^{n_j} x_{ij} 2^i,$

the sizes of the results after the Kronecker transformation are shown in Table 4.5.1. The six base matrices can be divided into 2 classes, the first class consists of matrices $\begin{pmatrix} 1 & 0 \\ 0 & 1 \end{pmatrix}$, $\begin{pmatrix} 1 & 0 \\ 1 & 1 \end{pmatrix}$, and $\begin{pmatrix} 0 & 1 \\ 1 & 1 \end{pmatrix}$; the second class consists of matrices $\begin{pmatrix} 1 & 0 \\ -1 & 1 \end{pmatrix}$, $\begin{pmatrix} 0 & 1 \\ -1 & 1 \end{pmatrix}$ and $\begin{pmatrix} 1 & 1 \\ -1 & 1 \end{pmatrix}$.

Table 4.5.1 Size of BDDs after Kronecker transformations.

class	X	X^2	XY	X^k	$\prod_{j=0}^{k} X_j$
1	$O(2^n)$	$O(2^{2n})$	$O(2^{n+m})$	$O(2^{kn})$	$O(\prod_{j=0}^{k} 2^{n_j})$
2	$O(n)$	$O(n^2)$	$O(nm)$	$O(n^k)$	$O(\prod_{j=0}^{k} n_j)$

The possibility of using BMDs to represent binary valued functions is discussed in [5]. In general, the BMD does not appear to be better than the ordinary BDD for representing binary valued functions. In order to see why this is true, consider the binary Reed-Muller transformation [11], in which operations are performed over Z_2 instead of the integers. The DAG representation of this transformation is sometimes called the Functional Decision Diagram or FDD[9]. This transformation can be obtained by applying the modulo 2 operations to all of the terminal nodes of the BMD. Consequently, the size of FDD is always smaller than the size of the BMD. Since the inverse binary Reed-Muller transformation is the same as the binary Reed-Muller transformation, the FDD for the binary Reed-Muller transformation for f is identical to the original BDD for f. Therefore, for every function f such that $|\mathrm{FDD}_f| < |\mathrm{BDD}_f|$, there exists another function f' which is the binary Reed-Muller transform of f such that $|\mathrm{BDD}_{f'}| < |\mathrm{FDD}_{f'}|$. In particular, both the BMD and the FDD representations for the middle bit of a multiplier are still exponential.

4.6 HYBRID DECISION DIAGRAMS

In the previous sections, we have discussed transformations that can be represented as the Kronecker product of a number of identical 2×2 matrices. If the transformation matrix is a Kronecker product of different 2×2 matrices, we still have a canonical representation of the function. We call transformations obtained from such matrices **hybrid transformations**.

A similar strategy has been tried by Becker [8]. However, his technique only works for the binary domain. When using his technique, all of the transformation matrices, the original function and the resulting function must have binary values. Our technique, on the other hand, works over the integers. By allowing integer values, we can handle a wider range of functions. Moreover, we can obtain larger reduction factors since we have more choices for transformation matrices.

We can apply this idea to reduce the size of MTBDD representation of functions. Since there is no known polynomial algorithm to find the hybrid transformation that minimizes MTBDD size, we use a greedy algorithm to reduce the size. If we restrict the entries in the matrix to the set $\{0, 1, -1\}$, then there are six matrices we can try. For each variable, we select the matrix that gives the smallest MTBDD size. The MTBDDs obtained from such transformations are called hybrid decision diagrams (HDDs).

Although a hybrid transformation can be performed by matrix multiplication, there is a more efficient way of computing it. It can be shown that [2]

$$\bigotimes_{i=0}^{k} A_i = \prod_{i=0}^{k} (I_{2^{i-1}} \otimes A_i \otimes I_{2^{k-i}}),$$

where each A_i is a 2×2 matrix and I_k is the identity matrix of size $k \times k$. A transformation of the form $(I_{2^{i-1}} \otimes A_i \otimes I_{2^{k-i}})$ is called a *basic transformation*. Let $A_i = \begin{pmatrix} a_{00} & a_{01} \\ a_{10} & a_{11} \end{pmatrix}$, and let g be a function represented as a MTBDD, then the basic transformation $g' = (I_{2^{i-1}} \otimes A_i \otimes I_{2^{k-i}}) \times g$ can be computed as

$$\begin{aligned} g' = \quad & \text{if } x_i \text{ then } a_{10}\ g|_{x_i=0} + a_{11}\ g|_{x_i=1} \\ & \text{else } a_{00}\ g|_{x_i=0} + a_{01}\ g|_{x_i=1}. \end{aligned}$$

As a result of this observation, the Kronecker transformation can be performed by a series of basic transformations. Moreover, it can be proved that the order of the basic transformations does not effect the final result.

Suppose the transformation matrix for a hybrid decision diagram can be represented as $\bigotimes_{i=0}^{k} \begin{pmatrix} a_{i1} & a_{i2} \\ a_{i3} & a_{i4} \end{pmatrix}$. Then the hybrid decision diagram nodes at level i is shown in Figure 4.6.1.

We have tried to represent the ISCAS85 benchmark circuits using hybrid decision diagrams 4.6.1. In some cases we have been able to reduce the size of

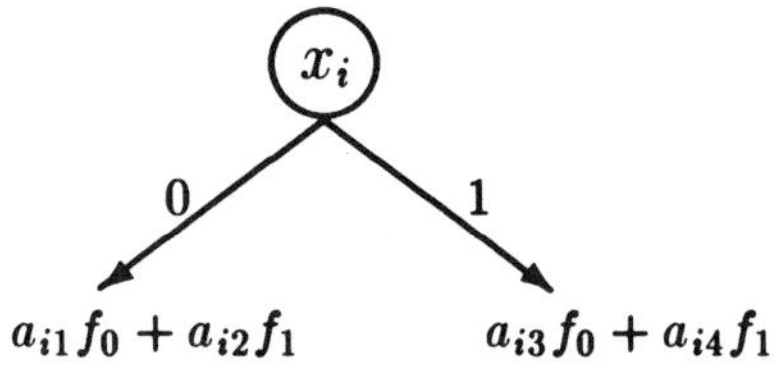

Figure 4.6.1 A hybrid decision diagram node at level i.

Table 4.6.1 Experimental results for hybrid transformations of some ISCAS85 circuits.

circuit	without reordering			with reordering		
circuit	|BDD|	|BMD|	|HDD|	|BDD|	| BMD |	|HDD|
c1355	9419	1217689	2857	4407	478903	1518
c1908	3703	140174	1374	1581	154488	632
c5315	679593	2820	521	108	5106	107

BDD representation by a factor of 1300. However, reductions of this magnitude usually occur when the original function has a bad variable ordering. If dynamic variable ordering is used, then our method gives a much smaller reduction factor.

We have tried several techniques to increase the number of possible matrices. The first technique involves increasing the number of entries in the matrices. This can be accomplished by allowing the entries to take larger values or by using the complex numbers $\{0, 1, -1, i, -i, 1+i, 1-i, i-1, -i-1\}$. Unfortunately, neither extension improved the results significantly.

The second technique involves using transformation matrices that are Kronecker products of larger matrices. For example, we have tried hybrid transformations based on 4×4 matrices instead of 2×2 matrices. Although we have been able to reduce the BDD size even further using this technique, the time it takes to find such transformations is much bigger since the number of possibilities is considerably larger.

Note that our technique can achieve comparable and sometimes better results than dynamic variable reordering. Thus, in some cases, it can serve as an alternative to dynamic variable reordering. We conjecture that the combination of both techniques together may result in reductions that neither technique can achieve alone.

In order to make the techniques described in the previous sections more useful, it is desirable to be able to perform various arithmetic operations and arithmetic relations on hybrid decision diagrams. This problem is discussed in [6] with detail.

4.7 SUMMARY AND DIRECTIONS FOR FUTURE RESEARCH

In this chapter, we have used MTBDDs to represent functions that map vectors with binary values into integers. We have also shown how to represent large integer matrices concisely and perform standard matrices operations such as scalar multiplication, matrix addition and matrix multiplication.

The Walsh and Reed-Muller transforms are given by matrices that have simple recursive definitions. Because of this, the transforms can be computed efficiently using MTBDDs. In fact, we are able to find the transforms of binary valued functions with several hundred variables.

We discuss the relationship between spectral transforms and binary moment diagrams and describe a generalization called the hybrid decision diagram which is often much more concise. We have also discussed a method to generalize the hybrid decision diagrams by using permutations.

In [7], we show how our technique for computing the Walsh transform can be used in technology mapping. Permutation and complementation of input variables does not change the sorted absolute values of the Walsh spectrum of a binary valued function. Thus, by comparing the Walsh spectra of two binary valued functions, we obtain a necessary condition for determining if one can be changed to the other by these operations.

There are other possible applications of the techniques discussed in this chapter. MTBDDs enable us to represent and manipulate very large matrices efficiently. Some potential applications include image compression, numerical solution of partial differential equations and computation of limit state probabilities for Markov Chains. Since hybrid decision diagrams tend to be more concise than multi-terminal BDDs, they may prove even more useful for this type of application.

REFERENCES

[1] R. I. Bahar, E. A. Frohm, C. M. Gaona, G. D. Hachtel, E. Macii, A. Pardo, and F. Somenzi, " Algebraic Decision Diagrams and their applications," *Proceedings of the 1993 IEEE International Conference on Computer Aided Design*, pp. 188-191, IEEE Computer Society Press, November 1993.

[2] R. Bellman, *Introduction to Matrix Analysis*, chapter 12, pp. 231-248, McGraw-Hill, 1970.

[3] J. Bern, C. Meinel, and A. Slobodova, " Efficient OBDD-based Boolean manipulation in CAD beyond current limits," *Proceedings of the 32nd ACM/IEEE Design Automation Conference*, pp. 408-413, IEEE Computer Society Press, June 1995.

[4] R. E. Bryant, " Graph-based algorithms for Boolean function manipulation," *IEEE Transactions on Computers*, C-35(8), pp. 677-691, 1986.

[5] R. E. Bryant and Y. A. Chen, " Verification of arithmetic functions with Binary Moment Diagrams," *Proceedings of the 32nd ACM/IEEE Design Automation Conference*, pp. 535-541, IEEE Computer Society Press, June 1995.

[6] E. M. Clarke, M. Fujita, and X. Zhao, " Hybrid Decision Diagrams – overcoming the limitations of MTBDDs and BMDs," *Proceedings of the 1995 Proceedings of the IEEE International Conference on Computer Aided Design*, pp. 159-163, IEEE Computer Society Press, November 1995.

[7] E. M. Clarke, K. McMillan, X. Zhao, M. Fujita, and J. Yang, " Spectral transforms for large Boolean functions with applications to technology mapping," *Proceedings of the 30th ACM/IEEE Design Automation Conference*, pp. 54-60, IEEE Computer Society Press, June 1993.

[8] R. Drechsler, A. Sarabi, M. Theobald, B. Becker, and M. A. Perkowski, " Efficient representation and manipulation of switching functions based on Ordered Kronecker Functional Decision Diagrams," *Proceedings of the 32nd ACM/IEEE Design Automation Conference*, pp. 415-419, IEEE Computer Society Press, June 1994. (Also Chapter 7 of this book).

[9] R. Drechsler, M. Theobald, and B. Becker, " Fast OFDD based minimization of fixed polarity Reed-Muller expressions," *Proceedings of the 1994 European Design Automation Conference*, pp. 2-7, IEEE Computer Society Press, June 1994.

[10] S. L. Hurst, D. M. Miller, and J. C. Muzio, *Spectral Techniques in Digital Logic*, Academic Press, 1985.

[11] D. E. Muller, " Application of Boolean algebra to switching circuit design and error detection", *IRE Trans.*, 1:6–12, pp. 6-12, 1954.

5

EDGE VALUED BINARY DECISION DIAGRAMS

Sarma B. K. VRUDHULA[1]
Massoud PEDRAM[2]
Yung-Te LAI[3]

[1] *Department of Electrical and Computer Engineering*
University of Arizona
Tuscon, Arizona

[2] *Department of Electrical Engineering - Systems*
University of Southern California
Los Angeles, California

[3] *Hitachi Labs*
179 East Tasman Dr.
San Jose, California

Abstract— We describe a canonical and compact data structure, called Edge Valued Binary Decision Diagrams (EVBDD), for representing and manipulating pseudo Boolean functions (PBF). EVBDDs are particularly useful when both arithmetic and Boolean operations are required. We describe a general algorithm on EVBDDs for performing any binary operation that is closed over the integers. Next, we discuss the relation between the probability expression of a Boolean function and its representation as a pseudo Boolean function. Utilizing this, we present algorithms for computing the probability spectrum and the Reed-Muller spectrum of a Boolean function directly on the EVBDD. Finally, we describe an extension of EVBDDs which associates both an additive and a multiplicative weight with the true edges of the function graph.

5.1 INTRODUCTION

The development of efficient techniques for representing and manipulating Boolean functions lies at the heart of much of the recent advances in logic synthesis, verification and testing. Reduced, ordered binary decision diagrams (ROBDD) [3, 4, 5] and their numerous variants, provide such a representation, and thus have become ubiquitous in many problems that arise in computer-aided design of digital systems.

In many applications it is often convenient and more efficient to represent and manipulate Boolean functions at the *word* level or in the integer domain. For example, in attempting to verify data path components that perform various types of arithmetic operations, it is easier to specify the behavior in the integer domain, and transform the logic implementation into a similar representation. Other applications where integer level representations are preferred are spectral analysis of Boolean functions [16], multiple output decomposition [1, 14, 23, 26], etc. The advantages of integer level representations is not just limited to logic design. A large collection of discrete, or combinatorial optimization problems that involve binary or integer variables, but where the range of the objective function is some finite subset of integers, also require compact representations and algorithms for their manipulation [26].

In this chapter, we describe a canonical and compact representation of *pseudo* Boolean functions, called **Edge Valued Binary Decision Diagrams** (EVBDD) [21, 26]. EVBDDs can be used to efficiently represent and manipulate integer valued functions. Since Boolean functions are a subset of pseudo Boolean functions, EVBDDs can also be used to represent Boolean functions. This allows the construction of general procedures for performing both arithmetic and Boolean operations. Following a description of EVBDDs and their properties, we show how certain *spectral* transformations on Boolean functions can be carried out on EVBDDs. One particular transform, called the *probability spectrum* (PS) [19], will be examined. The PS of a Boolean function has important applications in testing of digital circuits [19]. Additionally, there is a direct relation between the PS and the *Reed-Muller* (RM) coefficients of a Boolean function [19]. We present an algorithm that transforms an EVBDD representation of a Boolean function to another EVBDD that represents its PS. Next, we show how the EVBDD representation of the PS can be transformed into another EVBDD that represents the RM spectrum. Finally, we show that by associating both an additive and a multiplicative weight with the edges of an EVBDD, we can develop a more compact representation of *pseudo* Boolean functions. The new data structure, which was proposed in [39] and is called **Factored Edge Valued Binary Decision Diagram**, reduces complexity of certain arithmetic operations and allows direct implementation of the complement edges.

5.2 PSEUDO BOOLEAN FUNCTIONS

Definition 5.2.1 *A* **pseudo Boolean function** *(PBF) $f(x_0, \ldots, x_{n-1})$, with $x_i \in \{0,1\}$, is a polynomial that is linear in each variable and whose coefficients are elements of the set of integers $\mathcal{I}$. The general representation of a PBF is of the form*

$$f(x_0, \ldots, x_{n-1}) = \sum_{\substack{(i_0,\ldots,i_{n-1}) \\ i_j \in \{0,1\}}} a_{i_0,\ldots,i_{n-1}} x_0^{i_0} x_1^{i_1} \cdots x_{n-1}^{i_{n-1}}, \; x_i \in \{0,1\}, \; a_{i_0,\ldots,i_{n-1}} \in \mathcal{I}. \quad (5.2.1)$$

In (5.2.1), if $a_{0,0,\ldots,0} = 0$, then $f(x_0, \ldots, x_{n-1})$ are said to be **constant-free.**

With respect to a specific ordering $\langle x_0, x_1, \ldots, x_{n-1} \rangle$ of the variables, (5.2.1) can be expressed as

$$f(x_0, \ldots, x_{n-1}) = c + x_0(\alpha + f_\ell(x_1, \ldots, x_{n-1})) + (1 - x_0)(\beta + f_r(x_1, \ldots, x_{n-1})) \quad (5.2.2)$$

where $f_\ell(x_1, \ldots, x_{n-1})$ and $f_r(x_1, \ldots, x_{n-1})$ are constant-free, and $c, \alpha, \beta \in \mathcal{I}$.

For the sake of brevity, the right side of (5.2.2) will be indicated by a 6-tuple, $\langle c,\ x_{n-1},\ f_\ell,\ f_r,\ \alpha,\ \beta \rangle$. The representation given in (5.2.2) is analogous to the Shannon decomposition of a Boolean function. However, (5.2.2) is not in *standard* or **canonical form.** To see what this means, consider the function $f(x_0,\ x_1,\ x_2) = 3 + 2x_0 - 7x_0x_1 - 5x_0x_2 + 6x_0x_1x_2 + 3x_1 - 5x_1x_2$. Below are two different representations of this function in the form of (5.2.2).

$$\begin{aligned} f(x_0, x_1, x_2) &= \\ &1 + x_0(4 - 4x_1 - 5x_2 + x_1x_2) + (1 - x_0)(2 + 3x_1 - 5x_1x_2) = \\ &8 + x_0(-3 - 4x_1 - 5x_2 + x_1x_2) + (1 - x_0)(-5 + 3x_1 - 5x_1x_2) \quad (5.2.3) \end{aligned}$$

In fact there are an infinite number of expressions of the form given in (5.2.2) that denote the same function. Therefore, given two 6-tuples, $\langle c_1,\ x_0,\ f_\ell,\ f_r,\ \alpha_1,\ \beta_1 \rangle$, and $\langle\ c_2,\ x_0,\ g_\ell,\ g_r,\ \alpha_2,\ \beta_2\ \rangle$, one cannot determine whether or not these two representations denote the same function by simply checking the equality of the components. Among all the expressions of the form (5.2.2), we designate that one in which $\beta = 0$ to be the *standard* or *canonical* form. Then two such representations denote the same function if and only if the corresponding components of their canonical forms are equal.

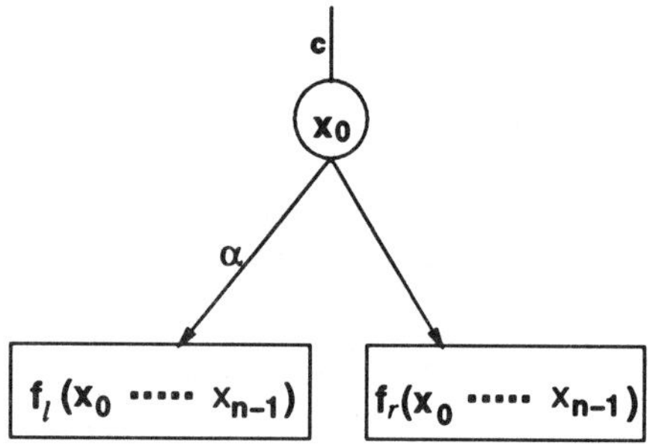

Figure 5.3.1 EVBDD of the pseudo Boolean function shown in (5.2.2) ($\beta = 0$).

5.3 EDGE VALUED BINARY DECISION DIAGRAMS

An Edge Valued Binary Decision Diagram (EVBDD) is a graph representation of a pseudo Boolean function, that is expressed in standard form, e.g., (5.2.2) with $\beta = 0$.

Definition 5.3.1 *An* EVBDD *is a tuple* $\langle c, \mathbf{f} \rangle$*, where c is a constant, and* $\mathbf{f}$ *is a directed acyclic graph consisting of two types of nodes:*

1. *A non-terminal node* $\mathbf{v}$ *described by a 4-tuple* $\langle var(v), child_\ell(\mathbf{v}), child_r(\mathbf{v}), value \rangle$*, where* $var(\mathbf{v}) \in \{x_0, \ldots, x_{n-1}\}$*,* $child_\ell(\mathbf{v})$ *and* $child_r(\mathbf{v})$ *are* EVBDD*s that represent the subexpressions* $f_\ell(x_1, \ldots, x_{n-1})$ *and* $f_r(x_1, \ldots, x_{n-1})$ *in (5.2.2) and* $value = \alpha$ *in (5.2.2).*
2. *The single terminal node, denoted by* $\mathbf{0}$*, which represents the value* 0.

Definition 5.3.2 *An* EVBDD $\langle c, \mathbf{f} \rangle$ *denotes the arithmetic function* $c + f : \{0,1\}^n \rightarrow$ *integer where f is the function f denoted by* $\mathbf{f} = \langle x, \mathbf{f}_\ell, \mathbf{f}_r, value \rangle$*. The terminal node* $\mathbf{0}$ *represents the constant function* $f = 0$*, and* $\langle x, \mathbf{f}_\ell, \mathbf{f}_r, value \rangle$ *denotes the arithmetic function* $f = x \cdot (value + f_\ell) + (1 - x) \cdot f_r$*.*

Fig. 5.3.1 shows a general EVBDD representation of (5.2.2).

5.3.1 Reduced and Ordered EVBDDs

The decomposition of a *PBF* given in (5.2.2) assumes a specific ordering of the variables. This means that there is an index function $index(x) \in \{0, \ldots, n-1\}$, such that for every non-terminal node **v**, either $child_\ell(\mathbf{v})$ is a terminal node or $index(var(\mathbf{v})) < index(var(child_\ell(\mathbf{v})))$, and either $child_r(\mathbf{v})$ is a terminal node or $index(var(\mathbf{v})) < index(var(child_r(\mathbf{v})))$. If **v** is the terminal node, then $index(\mathbf{v}) = n$. Such an EVBDD is said to be **ordered.** An EVBDD is **reduced** if there is no non-terminal node such that $child_\ell(\mathbf{v}) = child_r(\mathbf{v})$ with $value = 0$, and there are no two nodes **u** and **v** such that $\mathbf{u} = \mathbf{v}$. We only consider *reduced, ordered* EVBDDs.

Example 5.3.1 *Consider the function* $f(x_0, x_1, x_2)$ *given in (5.2.3). With respect to the ordering* $\langle x_0,\ x_1,\ x_2\rangle$, *the function can be decomposed as follows:*

$$\begin{aligned} f(x_0,\ x_1,\ x_2) &= 3 + 2x_0 - 7x_0x_1 - 5x_0x_2 + 6x_0x_1x_2 + 3x_1 - 5x_1x_2 \\ &= 3 + x_0(2 - 4x_1 - 5x_2 + x_1x_2) + (1 - x_0)(3x_1 - 5x_1x_2). \end{aligned}$$

$$\begin{aligned} 2 - 4x_1 - 5x_2 + x_1x_2 &= 2 + x_1(-4 - 4x_2) + (1 - x_1)(-5x_2) \\ 3x_1 - 5x_1x_2 &= 0 + x_1(3 - 5x_2) + (1 - x_1)(0) \\ -4 - 4x_2 &= -4 + x_2(-4 + 0) + (1 - x_2)(0) \\ -5x_2 &= 0 + x_2(-5 + 0) + (1 - x_2)(0) \\ 3 - 5x_2 &= 3 + x_2(-5 + 0) + (1 - x_2)(0). \end{aligned}$$

Fig. 5.3.2 shows the EVBDD *representation of* $f(x_0,\ x_1,\ x_2)$. ∎

A path in an EVBDD corresponds to an assignment of values to the variables associated with the path. The value of a *PBF* $f(x_0, \ldots, x_{n-1})$, for a given assignment $(x_0, \ldots, x_{n-1})$ is obtained by summing the values along the path as follows.

Definition 5.3.3 *Given an* EVBDD $\langle c, \mathbf{f}\rangle$ *representing* $f(x_0, \ldots, x_{n-1})$ *and a function* Φ *that for each variable* x *assigns a value* $\Phi(x)$ *equal to either 0 or 1, the function EVeval is defined as*

$$EVeval(\langle c, \mathbf{f}\rangle, \Phi) = \begin{cases} c & \mathbf{f} \textit{ is the terminal node } \mathbf{0} \\ EVeval(\langle c + value, child_\ell(\mathbf{f})\rangle, \Phi) & \Phi(variable(\mathbf{f})) = 1 \\ EVeval(\langle c, child_r(\mathbf{f})\rangle, \Phi) & \Phi(variable(\mathbf{f})) = 0 \end{cases}$$

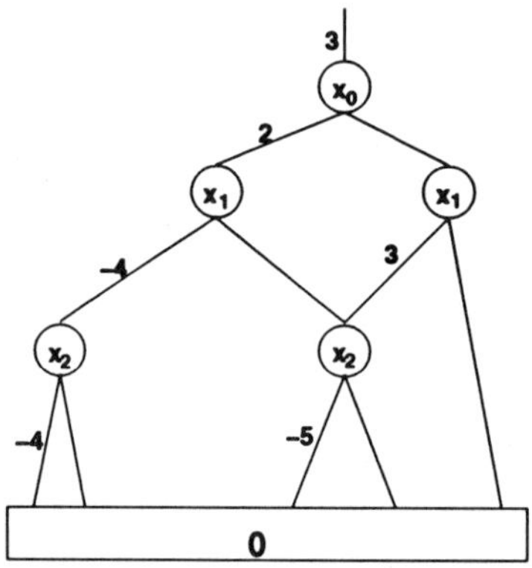

Figure 5.3.2 EVBDD of $f(x_0,\ x_1,\ x_2) = 3 + 2x_0 - 7x_0x_1 - 5x_0x_2 + 6x_0x_1x_2 + 3x_1 - 5x_1x_2$.

An extension of OBDDs, called Multi-valued Decision Diagrams (MDDs), was given in [38]. In an MDD, a nonterminal node can have more than two children and a terminal node assumes integer values. All operations are carried out through the CASE operator, which although works for arbitrary discrete functions, cannot directly perform arithmetic operations. Recently, another extension to OBDDs, called Multi-Terminal Binary Decision Diagram (MTBDD) was presented in [6, 8]. An MTBDD corresponds to a fully *expanded* version of an EVBDD. In general, for functions where the number of distinct terminal values is large, an MTBDD will require larger number of nodes than an EVBDD. However, for functions where the number of distinct terminal values is small, an MTBDD may require less storage space depending on the number of nodes in the corresponding graphs.

An EVBDD requires $n+1$ nodes to represent $2^{n-1}x_0 + \cdots + 2^0x_{n-1}$ (an n-bit integer), while an MTBDD requires $2^{n+1}-1$ nodes to represent the same function. When there are only two different terminal nodes (e.g., 0 and 1), EVBDDs, MTBDDs, and OBDDs are equivalent in terms of the number of nodes and the topology of the graph [26]. In this case, an EVBDD will require more space to represent the the edge-values. The worst case time complexity for performing operations on EVBDDs is the same as that for MTBDDs. However, many operations satisfy certain properties [26] that can be exploited so that EVBDDs are much more efficient than MTBDDs.

In [17] a useful calculus for manipulating pseudo-Boolean functions (referred by authors as **A**-Transforms) has been described. Also, to represent such pseudo-Boolean functions, a graph representation called Semi-Numeric Decision Diagrams (snDD) is introduced. In snDD apart from standard BDD nodes, various operator nodes are also allowed; an operator node corresponds to basic arithmetic operations such as +, -, *. Also, numeric values are allowed to be stored graph terminals. Thus snDD contain MTBDDs or ADDs.

5.3.2 Operations on EVBDDs

Let $\langle c_f, \mathbf{f}\rangle$, with $\mathbf{f} = \langle x, \mathbf{f}_\ell, \mathbf{f}_r, v_f\rangle$, and $\langle c_g, \mathbf{g}\rangle$, with $\mathbf{g} = \langle y, \mathbf{g}_\ell, \mathbf{g}_r, v_g\rangle$, be EVBDDs that represent two *PBF*s. We present a procedure, called *apply*, that computes $\langle c_h, \mathbf{h}\rangle = \langle c_f, \mathbf{f}\rangle$ *op* $\langle c_g, \mathbf{g}\rangle$, where *op* is any binary operator that is closed over the integers. Examples of *op* include the integer arithmetic operators, such as $+$, $-$, and multiplication by a constant; relational operators, minimum and maximum, logical shift, and modulo c, for some constant c. Furthermore, if the EVBDD represents a Boolean function, then *apply* can be used directly to perform Boolean operations, as well.

Let $\mathbf{h} = \langle\ var(\mathbf{h}), \mathbf{h}_\ell, \mathbf{h}_r, v_h\rangle$, and consider the case where $index(x) < index(y)$. This means that y will appear below x in $\mathbf{h}$ and $var(\mathbf{h}) = x$. Now, when $x = 1$, $\langle c_f, \mathbf{f}\rangle_{x=1} = \langle c_f + v_f, \mathbf{f}_\ell\rangle$, and when $x = 0$, $\langle c_f, \mathbf{f}\rangle_{x=0} = \langle c_f, \mathbf{f}_r\rangle$. Therefore, $\langle c_h, \mathbf{h}\rangle_{x=1} = \langle c_{h_\ell}, \mathbf{h}_\ell\rangle = \langle c_f + v_f, \mathbf{f}_\ell\rangle\ op\ \langle c_g, \mathbf{g}\rangle$, and $\langle c_h, \mathbf{h}\rangle_{x=0} = \langle c_{h_r}, \mathbf{h}_r\rangle = \langle c_f, \mathbf{f}_r\rangle\ op\ \langle c_g, \mathbf{g}\rangle$. Once the *left* and *right* children of $\langle c_h, \mathbf{h}\rangle$ have been computed, they have to be combined and the result must be expressed in the canonical form. The step involved in this *standardization* is expressed as follows:

$$x(c_{h_\ell} + h_\ell) + (1 - x)(c_{h_r} + h_r) \\ = c_{h_r} + x(c_{h_\ell} - c_{h_r} + h_\ell) + (1 - x)(h_r).$$

Therefore, the result returned is

$$(c_h, \mathbf{h}) = (c_{h_r}, \langle x, \mathbf{h}_\ell, \mathbf{h}_r, c_{h_\ell} - c_{h_r}\rangle).$$

Fig. 5.3.3 shows this final step. A similar situation exists if $index(x) \geq index(y)$.

The steps involved in the procedure *apply* are shown in Fig. 5.3.4. There are a set of *terminal* or default cases specific to each operator, the results of which can be returned without any further computation. As with ROBDD implementations, the efficient construction and manipulation of EVBDDs is made possible

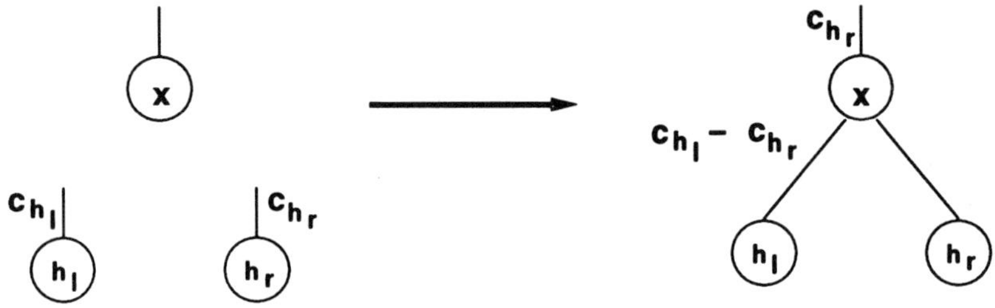

Figure 5.3.3 Combining the left and right children and converting to a canonical form.

through the use of two tables that permit sharing of previously computed subgraphs without having to reconstruct them in the course of a computation. Thus, at any time only one EVBDD for each distinct function is ever maintained. The first table, called the *unique_table*, contains a unique entry for each EVBDD node. The second table, called the *computed_table* maintains the correspondence between a function that has been computed and the node in the *unique_table*. Thus in the course of a computation, *computed_table* is first checked, and if the result is there, it is returned. Otherwise, after the operation is performed, the node associated with the new EVBDD is added to *unique_table* and the function denoted by the node is entered into *computed_table*, along with a pointer to the node in *unique_table*. Additional enhancements such as the use of complement edges [26] are also possible.

Steps 4, 5, 13 and 14 of *apply* may generate new EVBDDs since equality of two EVBDDs $\langle c_f, \mathbf{f}\rangle$ and $\langle c_g, \mathbf{g}\rangle$ requires $c_f = c_g$ and **f** and **g** to be isomorphic. This implies that the time complexity of *apply* is not simply $O(|\langle c_f, \mathbf{f}\rangle| \cdot |\langle c_g, \mathbf{g}\rangle|)$, where $|\langle c_f, \mathbf{f}\rangle|$ and $|\langle c_g, \mathbf{g}\rangle|$ denote the number of nodes in EVBDDs $\langle c_f, \mathbf{f}\rangle$ and $\langle c_g, \mathbf{g}\rangle$. A complete analysis of the time complexity of *apply* is beyond the scope of this paper. Details appear in [26]. However, most of the operators in question satisfy certain properties which can be exploited to reduce the time complexity. As a result, the time complexity of binary arithmetic operations, Boolean operations, and relational operations is $O(|\langle c_f, \mathbf{f}\rangle| \cdot |\langle c_g, \mathbf{g}\rangle|)$. Thus, when Boolean functions are represented by EVBDDs, the time complexity is the same as in the case of ROBDDs.

```
apply(⟨c_f, f⟩, ⟨c_g, g⟩, op)
{
    if (terminal_case(⟨c_f, f⟩, ⟨c_g, g⟩, op)
        return(⟨c_f, f⟩ op ⟨c_g, g⟩));
    if (comp_table_lookup(⟨c_f, f⟩, ⟨c_g, g⟩, op, ans))
        return(ans);
    if (index(f) ≥ index(g)) {
        ⟨c_gl, gl⟩ = ⟨c_g + value(g), child_l(g)⟩;
        ⟨c_gr, gr⟩ = ⟨c_g, child_r(g)⟩;
        var = variable(g);
    }
    else {
        ⟨c_gl, gl⟩ = ⟨c_gr, gr⟩ = ⟨c_g, g⟩;
        var = variable(f);
    }
    if (index(f) ≤ index(g)) {
        ⟨c_fl, fl⟩ = ⟨c_f + value(f), child_l(f)⟩;
        ⟨c_fr, fr⟩ = ⟨c_f, child_r(f)⟩;
    }
    else { ⟨c_fl, fl⟩ = ⟨c_fr, fr⟩ = ⟨c_f, f⟩;}
    ⟨c_hl, hl⟩ = apply(⟨c_fl, fl⟩, ⟨c_gl, gl⟩, op);
    ⟨c_hr, hr⟩ = apply(⟨c_fr, fr⟩, ⟨c_gr, gr⟩, op);
    if ((⟨c_hl, hl⟩ == ⟨c_hr, hr⟩) return ((⟨c_hl, hl⟩));
    h = find_or_add(var, hl, hr, c_hl - c_hr);
    comp_table_insert(⟨c_f, f⟩, ⟨c_g, g⟩, op, ⟨c_hr, h⟩);
    return ((⟨c_hr, h⟩));
}
```

Figure 5.3.4 Procedure apply.

5.3.3 Some Applications of EVBDDs

In this section we present a summary of some important applications of EVBDDs that have been in investigated [26].

Integer Linear Programming: Integer linear programming (ILP) is an NP-hard problem that appears in many applications. In [18] an OBBD based approach for solving 0-1 programming problems is presented. For operations involving integers such as conversion of linear inequality constraints into Boolean functions and optimization of non-binary objective functions, BDDs are not directly applicable. This shortcoming limits the caching of intermediate computations to only Boolean operations. Our approach to solving the ILP is to combine the benefits of EVBDDs (e.g., subgraph sharing and caching of intermediate results) with the state-of-art ILP techniques. In [26], we describe an algorithm using EVBDDs that computes optimal solution to a given linear objective function subject to linear constraints. Even without the use of sophisticated bounding techniques, experimental results show that the EVBDD based ILP solver is as efficient as a commercial ILP package.

Hierarchical Verification: The process of logic verification is to show the equivalence between the specification and the implementation. OBDDs can be used to verify logic circuits, only at the *Boolean level.* For example, to verify a 64-bit adder, one would have to first derive the OBDDs ($obdd_{imp}$) for each of the 65 outputs of the logic circuit and the OBDDs ($obdd_{spec}$) from the 65 Boolean expressions of each output function and then show equivalence. This would only show that each of the 65 outputs does realize the corresponding Boolean expression. However, since EVBDDs can be used to represent discrete integer and Boolean functions, they provide a means to verify the a logic circuit, where the specification is expressed in the integer domain. Thus the specification of a 65-bit adder would simply be $x+y+c$, where c is the carry-in. If $(b_0, b_1, \ldots, b_{n-1})$ represent the outputs of a 65-bit adder, then the verification process would first construct 65 EVBDDs for the b_i's and construct a single EVBDD for $2^0 \times b_0 + \cdots + 2^{64} \times b_{64}$. This EVBDD would be compared with the EVBDD of the specification, thus showing that the adder does indeed perform *addition.* Details of the use of EVBDDs for verification are described in [21].

Decomposition of Multiple Output Boolean Functions: The decomposition of Boolean functions had been the subject of extensive research. Much of the classical work on decomposition based on Karnaugh maps [1, 9] and cubes [34], and more recently using OBDDs [2, 7, 23, 36], are applicable to single Boolean functions. EVBDDs allow decomposition of multiple Boolean functions. In [26] a set of EVBDD based algorithms for decomposition of single and multiple output functions are described. This includes disjunctive and non-disjunctive decomposition, and decomposition of completely and incompletely specified functions.

5.4 THE PROBABILITY TRANSFORM AND ITS SPECTRUM

Many properties of Boolean functions that are difficult to deduce in the Boolean domain are often very easy to establish using an alternate representation. The alternate representation is called the *spectral* domain. A spectral transformation of a Boolean function of n variables is typically represented in the following form:

$$T_n \widetilde{Z}_n = \widetilde{R}_n, \tag{5.4.1}$$

where $\widetilde{Z}_n$ is a $2^n \times 1$ vector representing the truth table of the function, and T_n is a $2^n \times 2^n$ transformation matrix. $\widetilde{R}_n$ is a $2^n \times 1$ vector, and is called the *spectrum* of the Boolean function. Different transformation matrices generate different spectra. Some of the more extensively studied transformations are the Hadamard, Walsh and Reed-Muller [16, 33, 29]. These transforms have found extensive use in function classification, verification, logic synthesis, testing and fault diagnosis.

The key obstacle in using (5.4.1) is that the transformation matrix is of size $2^n \times 2^n$. Thus (5.4.1) can be used in practice only when n is small. This is where EVBDDs play an important role. In situations where T_n has a recursive structure, the transformation given in (5.4.1) can be carried out directly on the EVBDD representation of the Boolean function.

In this section, we examine one particular transform, called the **probability transform**, which is also known as the **arithmetic transform**. The fundamental properties of this transform and its extensive applications to testing were investigated by Kumar [19]. However, many of the results presented in [19] require computations whose complexity is exponential in the number of variables. The probability spectrum of a Boolean function is directly related to the pseudo Boolean representation of the function. Using this relation, and the fact that EVBDDs provide a canonical and compact representation of *PBF*s, we present an algorithm to compute the probability spectrum directly on the EVBDD. The resulting structure is called a **spectral** EVBDD. The only difference between an EVBDD and the corresponding spectral EVBDD is in the interpretation - the sum of the values along the edges of a path in a spectral EVBDD is the spectral coefficient associated with the input assignment. After showing how the probability spectrum can be computed using EVBDDs, we show how the spectral EVBDD can be transformed, so that the resulting structure represents the Reed-Muller spectrum of a Boolean function. The resulting structure is called an RMEVBDD.

Let $f(x_0, \ldots, x_{n-1})$ be a Boolean function. Let $X_i = Pr(x_i = 1)$, with $X_i \in [0, 1]$, and $F(X_0, \ldots, X_{n-1}) = Pr(f(x_0, \ldots, x_{n-1}) = 1)$. $F(X_0, \ldots, X_{n-1})$ is called the **probability expression** of $f(x_0, \ldots, x_{n-1})$. The probability expression $F(X_0, \ldots, X_{n-1})$ is obtained by the repeated application of the following rules [19, 31]:

$$Pr(\overline{x_i} = 1) \quad = \quad 1 - X_i$$

$$\begin{aligned} Pr(x_i \wedge x_i = 1) &= X_i \\ Pr(f_1 \vee f_2 = 1) &= F_1 + F_2 - F_1 F_2, \end{aligned}$$

where $F_i = Pr(f_i = 1)$.

The probability expression $F(X_0, \ldots, X_{n-1})$ represents a real valued function whose domain and range are $[0,1]^n$ and $[0,1]$, respectively. Now, if the variables X_i are restricted to be *integers* that can assume either 0 or 1, (i.e., $X_i \in \{0,1\}$), then the expression $F(X_0, \ldots, X_{n-1})$ is the pseudo Boolean representation of $f(x_0, \ldots, x_{n-1})$.

Example 5.4.1 *Consider* $f(x_0, x_1, x_2) = \overline{x_0} x_1 \vee x_1 x_2$. *Applying the above rules, we obtain*

$$F(X_0, X_1, X_2) = X_0 X_1 X_2 - X_0 X_1 + X_1. \tag{5.4.2}$$

Now the pseudo Boolean representation of $f(x_0, x_1, x_2)$ *is the same as the right side of (5.4.2), with* $X_i \in \{0,1\}$. ∎

Definition 5.4.1 *(from [19]) With respect to the ordering,* $\langle X_0, \ldots, X_{n-1} \rangle$, *the probability* spectrum *of a Boolean function* $f(x_0, \ldots, x_{n-1})$ *is a vector* $\widetilde{S}_f = [s_0, s_1, \ldots, s_{2^n-1}]^T$, *where* s_i *is the coefficient of the term* $X_0^{i_0} X_1^{i_1} \cdots X_{n-1}^{i_{n-1}}$ *in* $F(X_0, \ldots, X_{n-1})$, *and* $\langle i_0 i_1 \cdots i_{n-1} \rangle$ *is the binary representation of the integer* i.

Example 5.4.2 *Consider the probability expression given in (5.4.2). Expanding the expression into a canonical sum of products form, we obtain*

$$\begin{aligned} F(X_0, X_1, X_2) = {} & 0 \cdot X_0^0 X_1^0 X_2^0 + 0 \cdot X_0^0 X_1^0 X_2^1 + 1 \cdot X_0^0 X_1^1 X_2^0 + 0 \cdot X_0^0 X_1^1 X_2^1 \\ & + 0 \cdot X_0^1 X_1^0 X_2^0 + 0 \cdot X_0^1 X_1^0 X_2^1 + (-1) \cdot X_0^1 X_1^1 X_2^0 + 1 \cdot X_0^1 X_1^1 X_2^1. \end{aligned}$$

The probability spectrum $\widetilde{S}_f = [0\ 0\ 1\ 0\ 0\ 0\ -1\ 1]^T$ ∎

Theorem 5.4.1 *(from [19]) Define a* $2^n \times 2^n$ *matrix* P_n *as follows:*

$$P_0 = 1, \qquad P_n = \begin{bmatrix} P_{n-1} & 0 \\ -P_{n-1} & P_{n-1} \end{bmatrix}.$$

Let $\widetilde{Z}$ *be a* $2^n \times 1$ *vector of values of* $f(x_0, \ldots, x_{n-1})$. *Then the probability spectrum* $\widetilde{S}_f$ *is given by*

$$\widetilde{S}_f = P_n \times \widetilde{Z}. \tag{5.4.3}$$

We now show how the transformation given in (5.4.3) can be carried out on an EVBDD representation of f. With respect to the ordering $\langle x_0, \ldots, x_{n-1}\rangle$, $\widetilde{Z}_n$ can be partitioned as $[\widetilde{Z}^0_{n-1}, \widetilde{Z}^1_{n-1}]^T$, where $\widetilde{Z}^0_{n-1}$ ($\widetilde{Z}^1_{n-1}$) corresponds to the min-terms of f with $x_0 = 0$ ($x_0 = 1$). Then (5.4.3) can be expressed as

$$\begin{bmatrix} \widetilde{S}^0_f \\ \widetilde{S}^1_f \end{bmatrix} = \begin{bmatrix} P_{n-1} & 0 \\ -P_{n-1} & P_{n-1} \end{bmatrix} \begin{bmatrix} \widetilde{Z}^0_{n-1} \\ \widetilde{Z}^1_{n-1} \end{bmatrix} = \begin{bmatrix} P_{n-1} \times \widetilde{Z}^0_{n-1} \\ P_{n-1} \times \widetilde{Z}^1_{n-1} - P_{n-1} \times \widetilde{Z}^0_{n-1} \end{bmatrix}. \tag{5.4.4}$$

In (5.4.4), $\widetilde{Z}^0_{n-1}$ and $\widetilde{S}^0_f$ ($\widetilde{Z}^1_{n-1}$ and $\widetilde{S}^1_f$) correspond to the *right* (*left*) children of the EVBDD of $f(x_0, \ldots, x_{n-1})$, and the resulting spectral EVBDD, respectively. Thus, (5.4.4) states that to compute the spectral EVBDD we perform the following steps recursively:

1. replace the *right child* with the spectral EVBDD of the *right child,*
2. replace the *left child* with the spectral EVBDD of the *left child* minus the spectral EVBDD of the *right child.*

Note: Since an EVBDD is reduced, a node of the form $v = \langle x, \mathbf{f}, \mathbf{f}, 0\rangle$ would never appear. However, the spectral transform of the subgraph rooted at v would result in a node $v' = \langle x, \mathbf{0}, \mathbf{f}', 0\rangle$, where $\mathbf{f}'$ is the result of spectral transform when applied to the right child of v. For this reason, the algorithm for computing the spectral transform must keep track of the level of each node. If the index of the variable associated with the node currently being visited is greater than its level, then an a new EVBDD, rooted at a node of the form v' is returned. The basic steps of the procedure that converts an EVBDD of a Boolean function to a spectral EVBDD is shown in Fig. 5.4.1.

Example 5.4.3 *Consider the Boolean function $f(x_0, x_1, x_2) = \overline{x_0}x_1 \vee x_1x_2$ of Example 5.4.1. Its representation as a* PBF *is given in (5.4.2). Using this, its* EVBDD *representation is shown Fig. 5.4.2. The execution of the procedure evbdd_to_spevbdd on the* EVBDD *of Fig. 5.4.2 results in the spectral* EVBDD *shown in Fig. 5.4.3. Summing the values along the edges of the spectral* EVBDD *will result in the spectral coefficients shown in Example 5.4.2.* ■

```
evbdd_to_spevbdd(ev, level, n)
{
    if (level == n) return ev;
    if (ev == 0) return 0;
    if (index(ev) > level) {
        right = evbdd_to_spevbdd(ev, level + 1, n);
        left = 0;
        return new_evbdd(level, left, right);
    }
    right = evbdd_to_spevbdd(child_r(ev), level + 1, n);
    sp_l = evbdd_to_spevbdd(child_l(ev), level + 1, n);
    left = evbdd_sub(sp_l, right);
    return new_evbdd(level, left, right);
}
```

Figure 5.4.1 Procedure to convert an EVBDD to a spectral EVBDD.

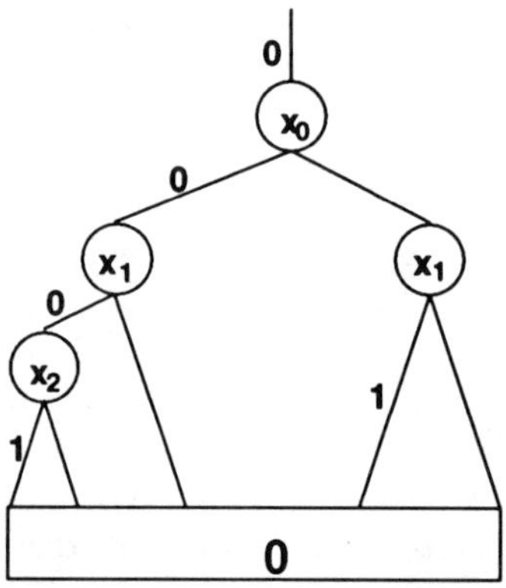

Figure 5.4.2 EVBDD of $f(x_0, x_1, x_2) = \overline{x_0} x_1 \vee x_1 x_2$.

5.5 REED-MULLER COEFFICIENTS

The standard symbolic interpretation of an OBDD results in a sum of products representation of a Boolean function. It is well known [29, 33, 41], that another canonical representation is possible if only *AND* and *XOR* operators are used. The resulting expression is called the Reed-Muller (RM) representation of the function. Specifically, if $f(x_0, \ldots, x_{n-1})$ is a Boolean function, then the RM

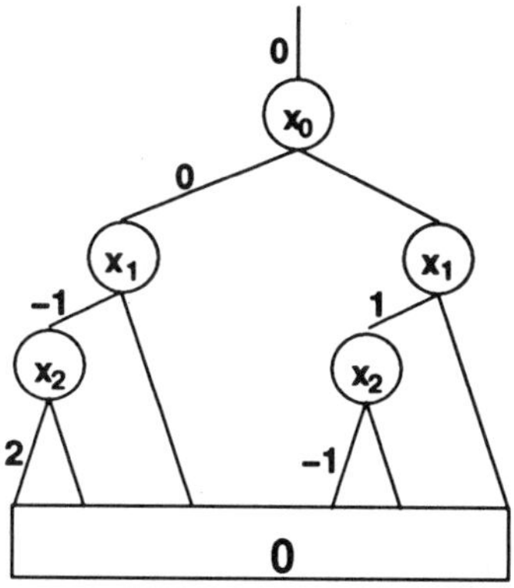

Figure 5.4.3 Spectral EVBDD of the EVBDD shown in Fig. 5.4.2.

representation of f has the form

$$f(x_0, \ldots, x_{n-1}) = a_{-1} \oplus a_0 \tilde{x}_{n-1} \oplus a_1 \tilde{x}_{n-2} \oplus \cdots \oplus a_{2^n-1} \tilde{x}_0 \tilde{x}_1 \cdots \tilde{x}_{n-1}, \quad (5.5.1)$$

where $a_i \in \{0, 1\}$, and $\tilde{x}$ is either complemented or un-complemented variable x_i.

We assume that each variable on the right side of (5.5.1) appears either as complemented or un-complemented, but not both. Digital circuits built using the RM form have a number of useful properties. They belong to a class of easily testable networks since test sets are independent of the function realized by the circuit and depend linearly on the number of inputs [32]. This has motivated interest in minimization of RM forms. Minimizing an RM form requires determining the polarity of each variable so that the resulting RM expression for the function has the least number terms.

The RM coefficients of a Boolean function, represented by an EVBDD can be obtained using a transformation nearly identical to the one given in Theorem 5.4.1. The resulting representation will be called an RMEVBDD. Let $\widetilde{A}_f$ denote the RM coefficients of a Boolean function f, ordered in the same way as $\widetilde{S}_f$ (see Theorem 5.4.1). Define a $2^n \times 2^n$ matrix $\mathcal{R}_n$ as follows:

$$\mathcal{R}_0 = 1, \qquad \mathcal{R}_n = \begin{bmatrix} \mathcal{R}_{n-1} & 0 \\ \mathcal{R}_{n-1} & \mathcal{R}_{n-1} \end{bmatrix}.$$

Then the RM spectrum $\widetilde{A}_f$ is given by

$$\widetilde{A}_f = \mathcal{R}_n \times \widetilde{Z}_n. \tag{5.5.2}$$

The RM spectrum can also be computed from the probability spectrum [19]. The relation is simply

$$\widetilde{A}_f = |\widetilde{S}_f| \; mod \; 2. \tag{5.5.3}$$

Thus, the RMEVBDD can be generated using either (5.5.2) or (5.5.3). Thus, to transform an EVBDD of a Boolean function to an RMEVBDD, line 10 of the procedure *evbdd_to_spevbdd* shown in Fig. 5.4.1 has to be modified as follows:

10 $\quad left = evbdd\_add(sp_l, right);$

Example 5.5.1 *Fig. 5.5.1 shows the* RMEVBDD *of the function* $f(x_0, x_1, x_2) = \overline{x_0}x_1 \vee x_1x_2$ *of Example 5.4.1. It was derived from the* EVBDD *shown in Fig. 5.4.2. In an* RMEVBDD, *a path from the root to the terminal node that involves a* right *edge of a node* v, *such that* $var(v) = x_i$, *corresponds to a product term in which* x_i *is not present in the RM representation. For example, consider the path corresponding to the assignment* $x_0 = 0, x_1 = 1, x_2 = 1$, *in the* RMEVBDD *shown in Fig. 5.5.1. This assignment corresponds to a path that includes the right edge of a node* v, *with* $var(v) = x_0$. *Therefore, this path corresponds to the term* x_1x_2. *The associated RM coefficient is the sum of the edge values along this path, i.e.,* $0 + 0 + 1 + -1 = 0$. ■

There are 2^n RM forms for a function $f(x_0, \ldots, x_{n-1})$, corresponding to the 2^n possible combinations of polarities for each variable. From (5.5.3), we see that finding the minimal RM expression of a function requires identifying the polarity vector that will result in the least number of odd terms [19]. The following theorem shows the effect of changing the polarity of a variable in the RMEVBDD representation of a Boolean function.

Theorem 5.5.1 *Let* E *and* E' *represent the* RMEVBDDs *of* $f(x_0, \ldots, x_{i-1}, x_i, x_{i+1}, \ldots, x_{n-1})$ *and* $f(x_0, \ldots, x_{i-1}, \overline{x}_i, x_{i+1}, \ldots, x_{n-1})$, *respectively. Then* E' *is obtained from* E *by performing the following transformation on every node* u, *such that* $var(u) = x_i$.

$$child_r(u) \leftarrow (child_\ell(u) + child_r(u)) \; mod \; 2.$$

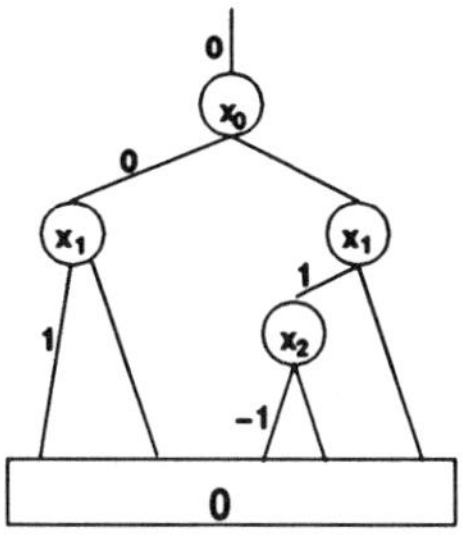

Figure 5.5.1 RMEVBDD of the EVBDD shown in Fig. 5.4.2.

```
/* swith_polarity switches the polarity of
a variable in an RMEVBDD.
ev is an RMEVBDD. i is the index of the
variable whose polarity is to be switched.
/
switch_polarity(ev, level, i, n)
{
    if (level == n) return ev;
    if (ev == 0) return 0;
    if (index(ev) > level) {
        right = 0;
        return new_evbdd(level, child_l(ev), right);
    }
    if (index(ev) == i) {
        sp_r = evbdd_add(child_l(ev), child_r(ev));
        right = mod2(sp_r);
        return new_evbdd(level, child_l(ev), right);
    }
    left = switch_polarity(child_l(ev), level + 1, i, n);
    right = switch_polarity(child_r(ev), level + 1, i, n);
    return new_evbdd(level, left, right);
}
```

Figure 5.5.2 Procedure to switch the polarity of a variable in an RMEVBDD.

Fig. 5.5.2 shows the procedure to switch the polarity of a variable in an RMEVBDD.

5.6 FACTORED EDGE VALUED BINARY DECISION DIAGRAMS

Factored Edge Valued Binary Decision Diagrams (FEVBDD) are an extension to EVBDDs which were presented in [39]. By associating both an additive and a multiplicative weight with the true-edges, FEVBDDs offer a more compact representation of linear functions, since common sub-functions differing only by an *affine* transformation can now be expressed by a single subgraph. Additionally, they allow the notion of complement edges to be transferred from OBDDs to FEVBDDs.

Definition 5.6.1 *An* FEVBDD *is a tuple* $\langle c, w, \mathbf{f} \rangle$, *where* c *and* w *are constant values, and* $\mathbf{f}$ *is a rooted, directed acyclic graph consisting of two types of vertices:* [1]

1. *A non-terminal vertex* $\mathbf{v}$ *is represented by a 5-tuple* $\langle var(\mathbf{v}), child_\ell(\mathbf{v}), child_r(\mathbf{v}), value, factor \rangle$ *where value and factor are rational numbers and* $var(\mathbf{v}) \in \{x_0, \ldots, x_{n-1}\}$.
2. *The single terminal vertex* $\mathbf{v}$ *with value 0 is denoted by* $\mathbf{0}$. *By definition all branches leading to* $\mathbf{0}$ *have an associated weight* $w = 0$.

There is no non-terminal vertex $\mathbf{v}$ *such that* $child_\ell(\mathbf{v}) = child_r(\mathbf{v})$, $value = 0$, *and* $factor = 1$, *and there are no two nonterminal vertices* $\mathbf{v}$ *and* $\mathbf{u}$ *such that* $\mathbf{v} = \mathbf{u}$. *Furthermore, there exists an index function* $index(x) \in \{0, \ldots, n-1\}$ *such that the following holds for every nonterminal vertex. If* $child_\ell(\mathbf{v})$ *is also nonterminal, then we must have* $index(var(\mathbf{v})) < index(var(child_\ell(\mathbf{v})))$. *If* $child_r(\mathbf{v})$ *is nonterminal, then we must have* $index(var(\mathbf{v})) < index(var(child_r(\mathbf{v})))$.

Definition 5.6.2 *A* FEVBDD $\langle c_f, w_f, \mathbf{f} \rangle$ *denotes the arithmetic function* $c_f + w_f \cdot f$ *where* f *is the function* f *denoted by* $\mathbf{f} = \langle x, \mathbf{f}_\ell, \mathbf{f}_r, value, factor \rangle$. *The terminal node* $\mathbf{0}$ *represents the constant function* $f = 0$, *and* $\langle x, \mathbf{f}_\ell, \mathbf{f}_r, value, factor \rangle$ *denotes the arithmetic function* $f = x \cdot (value + factor \cdot f_\ell) + (1 - x) \cdot f_r$.

[1] Here, we will only describe the "rational rule" for weight normalization which requires the use of fractions. See [39] for the "GCD rule" which requires a multiplicative weight to be associated with both edges.

Tabular Form

x	y	z	f
0	0	0	15
0	0	1	6
0	1	0	5
0	1	1	2
1	0	0	13
1	0	1	7
1	1	0	5
1	1	1	2

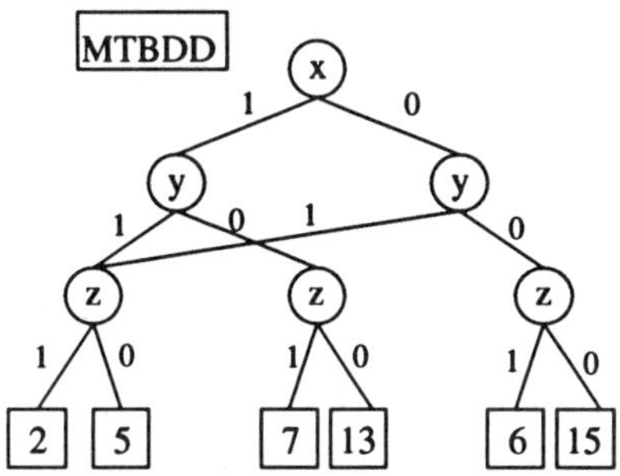

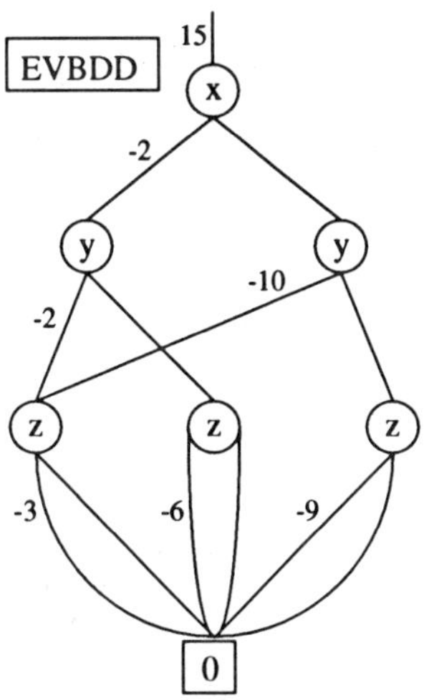

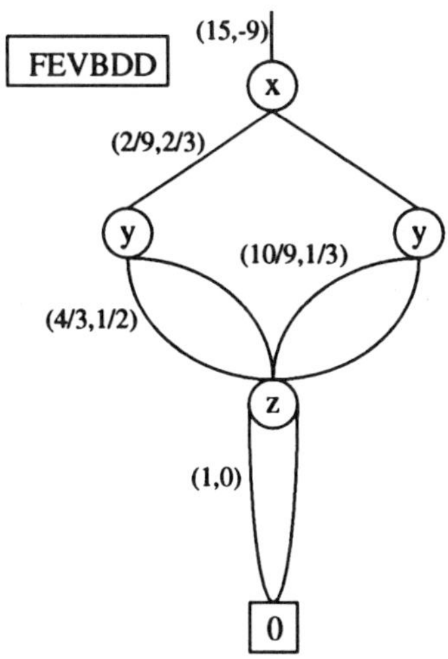

Figure 5.6.1 Example of various function graphs.

As an example, we construct the various function graphs based on the different decompositions of function f given in its tabular form in Fig. 5.6.1.

$$\begin{aligned} f(x,y,z) &= 15(1-x)(1-y)(1-z)+6(1-x)(1-y)z+ & (5.6.1)\\ &\quad 5(1-x)y(1-z)+2(1-x)yz+ \\ &\quad 13x(1-y)(1-z)+7x(1-y)z+ \\ &\quad 5xy(1-z)+2xyz \\ &= 15+x(-2+y(-8+z(-3))+(1-y)(z(-6)))+ & (5.6.2)\\ &\quad (1-x)(y(-10+z(-3))+(1-y)(z(-9))) \\ &= 15-9(x(\frac{2}{9}+\frac{2}{3}(y(\frac{4}{3}+\frac{1}{2}z)+(1-y)z))+ & (5.6.3)\\ &\quad (1-x)(y(\frac{10}{9}+\frac{1}{3}z)+(1-y)z)). \end{aligned}$$

Equation (5.6.1) is in a form that directly corresponds to the function decomposition for MTBDDs or ADDs and the tabular form. Equations (5.6.2) and (5.6.3) reflect the structure of the decomposition rules for EVBDDs and FEVBDDs, respectively. The different function graphs are shown in Fig. 5.6.1.

A path in an FEVBDD corresponds to an assignment of values to the variables associated with the path. The value of a *PBF* $f(x_0,\ldots,x_{n-1})$, for a given assignment $(x_0,\ldots,x_{n-1})$ is obtained as follows.

Definition 5.6.3 *Given a* FEVBDD $\langle c_f, w_f, \mathbf{f}\rangle$ *representing* $f(x_0,\ldots,x_{n-1})$ *and a function* Φ *that for each variable* x *assigns a value* $\Phi(x)$ *equal to either 0 or 1, the function FEVeval is defined as:*

$$FEVeval(\langle c_f, w_f, \mathbf{f}\rangle, \Phi) =$$

$$\begin{cases} c_f & \mathbf{f} \text{ is the terminal node } \mathbf{0}\\ c_f + w_f \cdot FEVeval(\langle value, factor, child_\ell(\mathbf{f})\rangle, \Phi) & \Phi(variable(\mathbf{f})) = 1\\ c_f + w_f \cdot FEVeval(\langle 0, 1, child_r(\mathbf{f})\rangle, \Phi) & \Phi(variable(\mathbf{f})) = 0 \end{cases}$$

An example of a FEVBDD representing a Boolean function with complement edges is given in Fig. 5.6.2. This FEVBDD represents the four output functions of a 3-bit adder. It has the same topology (except for the terminal edges) as the corresponding OBDD depicted in the same figure. As it is shown in this example, FEVBDDs successfully extend the use of EVBDDs to represent Boolean functions as they inherently offer a way to represent complement edges. Furthermore, the Boolean operation '*not*' can now be performed in constant time since it only requires manipulation of the weights of the root node.

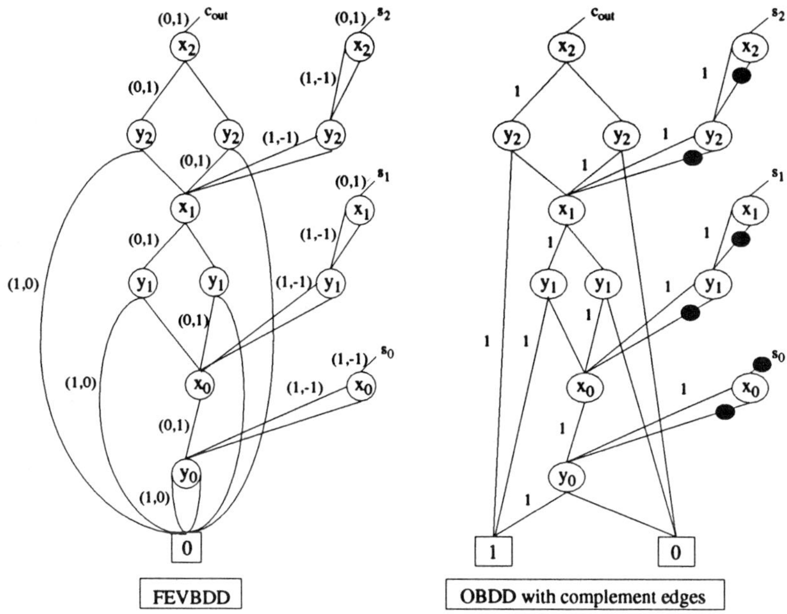

Figure 5.6.2 FEVBDD representation of the four output functions of a 3-bit adder.

A FEVBDD-based matrix package was introduced in [39]. This package was used to solve the Chapman-Kolmogorov equations. Experimental results showed that in the majority of cases FEVBDDs win over the corresponding EVBDD representation of the matrices in terms of number of nodes and overall memory consumption. In general, however, since the memory consumption per node of the FEVBDD is higher than that of the EVBDD, if the number of nodes in the FEVBDD and EVBDD are the same (for example, due to the sparse structure of matrices), then EVBDDs will require less memory.

5.7 SUMMARY

Edge Valued Binary Decision Diagrams (EVBDD) are a novel data structure for representing discrete functions. EVBDDs generalize ROBDDs and are particularly useful for performing both integer and Boolean operations. This extension to the *word* level provides a basis for a variety of applications, including verification (where the specification can be expressed in the arithmetic domain), multiple output decomposition of logic functions, discrete function optimization and others. In this chapter we showed how *evbdds* can be used to efficiently represent and manipulate various types of spectra of Boolean functions. In particular, we showed how the probability spectrum and the Reed-Muller spectrum of a Boolean function can be computed directly on an EVBDD without having to resort to traditional methods that require matrices of exponential size. The relation between the probability spectrum and the Reed-Muller coefficients of a Boolean function was utilized to develop an algorithm for constructing the EVBDD that represents the Reed-Muller form when the polarity of one or more variables is changed. The current direction of this work is to utilize this algorithm to develop efficient heuristics for generating a near minimal Reed-Muller form of a Boolean function. In addition, we showed an extension of EVBDDs which associates both an additive and a multiplicative weight with the true edges of the function graph in order to achieve more compact representations, and in some cases, more efficient operations.

REFERENCES

[1] R. L. Ashenhurst, "The decomposition of switching functions," *Ann. Computation Lab.*, Harvard University, vol. 29, pp. 74-116, 1959.

[2] M. Beardslee, B. Lin and A. Sangiovanni-Vincentelli, "Communication based logic partitioning," *Proc. of the European Design Automation Conference*, pp. 32-37, 1992.

[3] K. S. Brace, R. L. Rudell, and R. E. Bryant, "Efficient implementation of a BDD package," *Proc. of the 27th Design Automation Conference*, pp. 40-45, 1990.

[4] R. E. Bryant, "Graph-based algorithms for Boolean function manipulation," *IEEE Transactions on Computers*, C-35(8): 677-691, August 1986.

[5] R. E. Bryant, "Symbolic Boolean manipulation with ordered binary-decision diagrams," *Computing Surveys*, Vol. 24, No. 3, pp. 293-318, Sept. 1992.

[6] E. M. Clarke, M. Fujita, P. C. McGeer, K. L. McMillan, and J. C.-Y. Yang, "Multi-terminal binary decision diagrams: An efficient data structure for matrix representation," *International Workshop on Logic Synthesis*, pp. 6a:1-15, May 1993. (Also Chapter 4 of this book).

[7] S-C. Chang and M. Marek-Sadowska, "Technology mapping via transformations of function graphs," *Proc. International Conference on Computer Design*, pp. 159-162, 1992.

[8] E. M. Clarke, K. L. McMillan, X. Zhao, M. Fujita, and J. C.-Y. Yang, "Spectral transforms for large Boolean functions with applications to technology mapping," *Proc. of the 30th Design Automation Conference*, pp. 54-60, 1993.

[9] H. A. Curtis, *A New Approach to the Design of Switching Circuits*, Von Nostrand, Princeton, N.J., 1962.

[10] C. R. Edwards, "The application of the Rademacher-Walsh transform to Boolean function classification and threshold-logic synthesis," *IEEE Transactions on Computers*, C-24, pp. 48-62, 1975.

[11] B. J. Falkowski, I. Schafer and M. A. Perkowski, "Effective computer methods for the calculation of Rademacher-Walsh spectrum for completely and incompletely specified Boolean functions," *IEEE Transaction on Computer-Aided Design*, Vol. 11, No. 10, pp. 1207-1226, Oct. 1992.

[12] P. L. Hammer and S. Rudeanu, *Boolean Methods in Operations Research and Related Areas*, Heidelberg, Springer Verlag, 1968.

[13] P.L. Hammer and B. Simeone, "Order relations of variables in 0-1 programming," *Annals of Discrete Mathematics*, 31, North-Holland, pp. 83-112, 1987.

[14] S. He and M. Torkelson, "Disjoint decomposition with partial vertex chart," *International Workshop on Logic Synthesis*, pp. p2a 1-5, 1993.

[15] M. Helliwell and M. Perkowski, "A fast algorithm to minimize multi-output mixed-polarity generalized Reed-Muller forms," *Proc. 25th Design Automation Conf.*, pp. 427-432, 1988.

[16] S. L. Hurst, D. M. Miller and J. C. Muzio, *Spectral Techniques in Digital Logic,* London, U.K. : Academic, 1985.

[17] J. Jain, J. A. Abraham, J. Bitner, and D. S. Fussel, "Probabilistic verification of boolean functions," *Formal Methods in Systems Design*, 1(1), pp. 63-115, July 1992. (Also Chapter 6 of this book).

[18] S-W. Jeong and F. Somenzi, "A new algorithm for 0-1 programming based on binary decision diagrams," Logic Synthesis and Optimization, Sasao ed., Kluwer Academic Publishers, pp. 145-165, 1993.

[19] S. K. Kumar, *Theoretical Aspects of the Behavior of Digital Circuits Under Random Inputs.* Technical Report DISC/81-3, Dept. of Electrical Engineering-Systems, University of Southern California, Los Angeles, CA., Sept. 1981.

[20] Y-T. Lai and S. Sastry, "HINTS: A hardware interpretation system," International Workshop on Formal Methods in VLSI Design 1991.

[21] Y-T. Lai and S. Sastry, "Edge-valued binary decision diagrams for multi-level hierarchical verification," *Proc. of 29th Des. Automation Conf.*, pp.608-613, 1992.

[22] Y-T. Lai, S. Sastry and M. Pedram, "Boolean matching using binary decision diagrams with applications to logic synthesis and verification," *Proc. International Conf. on Computer Design*, pp.452-458, 1992.

[23] Y-T. Lai, M. Pedram and S. Sastry, "BDD based decomposition of logic functions with application to FPGA synthesis," *Proc. of 30th Design Automation Conf*, pp. 642-647, 1993.

[24] Y-T. Lai, K-R. Pan, M. Pedram, and S. Vrudhula, "FGMap: A technology mapping algorithm for Look-Up Table type FPGAs based on function graphs," *International Workshop on Logic Synthesis*, 1993.

[25] Y-T. Lai, M. Pedram and S. B. K. Vrudhula, "FGILP: An integer linear program solver based on function graphs," *Proc. Int'l Conf. on Computer Aided Design,* November 1993, pages 685-689.

[26] Y-T. Lai, M. Pedram, and S. B. Vrudhula, "EVBDD-based Algorithms for linear integer programming, spectral transformation and function decomposition," *IEEE Transactions on CAD*, vol. 13, no. 8, pp. 959-975, Aug. 1994.

[27] Y-T. Lai, M. Pedram, and S. B. Vrudhula, "Formal verification using edge-valued binary decision diagrams," to appear in *IEEE Trans. on Computers,* 1996.

[28] H-T. Liaw and C-S Lin, "On the OBDD-representation of general Boolean functions," *IEEE Trans. on Computers*, C-41(6): 661-664, June 1992.

[29] D. E. Muller, "Application of Boolean algebra to switching circuit design and error detection," *IEEE Trans. on Computers*, vol. C-21, pp. 6-12, 1974.

[30] G. L. Nemhauser and L. A. Wolsey, *Integer and Combinatorial Optimization*, Wiley, New York, 1988.

[31] K. P. Parker and E. J. McCluskey, "Probabilistic treatment of general combinational networks," *IEEE Trans. on Computers*, vol. C-24, pp. 668-670, June 1975.

[32] S. M. Reddy, "Easily testable realizations for logic functions," *IEEE Trans. on Computers*, vol. C-21, pp. 1183-1188, 1972.

[33] I. S. Reed, "A class of multiple-error-correcting codes and their decoding scheme," *IRE Trans.*, vol. IT-4, pp. 38-49, 1954.

[34] J. P. Roth and R. M. Karp, "Minimization over Boolean graphs," *IBM Journal.*, pp. 227-238, April 1962.

[35] A. Sarabi and M. Perkowski, "Fast exact and quasi-minimal minimization of highly testable fixed polarity AND/XOR canonical networks," *Proc. 29th Design Automation Conf.*, pp. 30-35, 1992.

[36] T. Sasao, "FPGA design by generalized functional decomposition," *Logic Synthesis and Optimization*, Sasao ed., Kluwer Academic Publishers, pp. 233-258, 1993.

[37] D. Singh, J. Rabaey, M. Pedram, F. Catthoor, S. Rajgopal, N. Sehgal and T. Mozdzen. "Power-conscious CAD tools and methodologies: a perspective," *Proc. of the IEEE,* April 1995.

[38] A. Srinivasan, T. Kam, S. Malik and R. Brayton, "Algorithms for Discrete Function Manipulation," *Proc. Int. Conf. CAD*, pp. 92-95, 1990.

[39] P. Tafertshofer and M. Pedram, *Factored EVBDDs and Their Application to Matrix Representation and Manipulation,* CENG Tech. Rept. 94-27, Dept. of EE-Systems, Univ. of Southern California, October 1994 (Also to appear in Formal Methods in System Design, Kluwer Academic Publishers, 1996).

[40] J. S. Wallis, "Hadamard Matrices," *Lecture Notes No. 292,* Springer-Verlag, New York, 1972.

[41] X. Wu, X. Chen and S. L. Hurst, "Mapping Reed-Muller coefficients and the minimization of Exclusive-OR switching functions," *Proc. IEE*, vol. E129, pp. 15-20, 1982.

6

ARITHMETIC TRANSFORM OF BOOLEAN FUNCTIONS

Jawahar JAIN

Fujitsu Laboratories of America
Santa Clara CA
USA

Abstract— In many applications where logic functions need to be analyzed it can be useful if we transform Boolean (or switching) functions to arithmetic functions. Such arithmetic transformations can give us new insight into solving some interesting problems. For example, the transformed functions can be easily evaluated (simulated) on integers or real numbers. Through such arithmetic simulation we can probabilistically verify a pair of functions with much more confidence than two-valued Boolean simulation. The arithmetic transform of any Boolean function can be easily computed from its BDD. To help evaluate a Boolean function on non-binary inputs, and to represent multi-variable linear polynomials with integer coefficients, a BDD like data structure *snDD* can be used; for many arithmetic expressions, snDDs are a very compact representation. The error in such probabilistic verification of property of a function is quantifiable and extremely low. Also, the procedures are computationally very efficient. Using a real-valued or integer-valued representation we can derive testability measures for elements of a digital circuit, or conduct the reliability analysis for various networks.

6.1 ARITHMETIC TRANSFORMS: WHY THEY NEED BE STUDIED

Analysis of switching or Boolean functions is well known to be very important in numerous fields ranging from combinatorics to artificial intelligence to computer-aided design (CAD) of digital circuits as well as in applied mathematics. For example, efficiently answering questions about satisfiability, tautology, equivalence etc. on Boolean functions has immense applications in various problems in CAD such as synthesis, verification, testing etc.[1] It is a standard practice to write Boolean functions as expressions or formulas where the operations allowed between two variables are binary operations AND, OR, XOR etc; and the variables take binary values 0 (False), or 1 (True). Interestingly, in many cases, analysis of a Boolean expression can be greatly improved if we "transform" the given expression into one where the familiar arithmetic operations of addition, subtraction and multiplication are allowed. What we obtain will be an arithmetic expression, and clearly we can then evaluate it by, say, assigning integers (instead of binary values) to its variables. Such evaluation of a transform of Boolean functions can be quite useful for various problems as we will see in this chapter.

Consider verification problem. Suppose, we want to check if $f_1 = (x_1 \wedge x_2) \vee (x_2 \wedge x_3)$ is equivalent to $f_2 = x_3 \wedge (x_1 \vee (\overline{x_1} \wedge \overline{x_2}))$. The two functions are not equivalent which is popularly found by random simulation or by building their canonical representations such as OBDDs [5].

In random Boolean simulation, if the first vector we had generated were $x_1 = 1, x_2 = 1, x_3 = 1$, we would have found that both functions f_1, f_2 evaluate to 1. Also, suppose the next two vectors tried were $x_1 = x_2 = x_3 = 0$ and $x_1 = 0, x_2 = 1, x_3 = 0$. In each of these cases, both of our given functions would evaluate to 0. Although simulation is guaranteed to discover the difference between the two functions, it may need to generate an exponential number of vectors before the difference could be proved. (For two equivalent functions, all 2^n vectors need be examined to declare their equivalence.) This is clearly unsatisfactory for large n. In general, random Boolean simulation works well when the difference between a design and its specifications are many. However, for equivalent functions or functions with small differences, it is now increasingly recognized as quite inadequate.

Similarly, note that verification using OBDDs is known to be very expensive and can require exponentially sized graphs for many functions [5, 6]. Since they are canonical representations, when they take exponential time, they also take exponential space, and require memory infeasible on typical computers.

[1] For purpose of this chapter, we will not make distinction between *Boolean functions* and *switching functions* and use the more popular term Boolean functions. Note every switching function $\{0,1\}^n \rightarrow \{0,1\}$ is also a Boolean function. For a clear distinction between Boolean and switching functions, [4] forms an excellent reading. The Boolean algebra of switching functions is called switching algebra. However, in a Boolean Algebra $\{B, 0, 1, +, .\}$, the carrier B may have more than two elements.

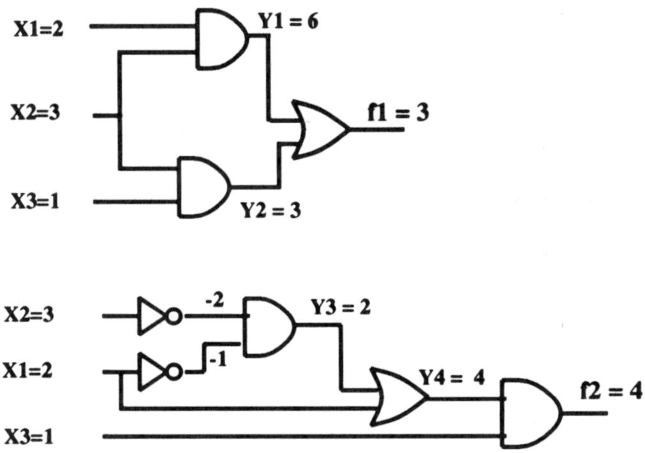

Figure 6.1.1 Verification and Numeric Simulation

In contrast, consider the following approach. As explained in Section 6.2 and 6.3, we can create the arithmetic transforms $\mathbf{A}[f_1]$ and $\mathbf{A}[f_2]$; specifically, $\mathbf{A}[f_1] = x_1 \cdot x_2 + x_2 \cdot x_3 - x_1 \cdot x_2 \cdot x_3$. Similarly, $\mathbf{A}[f_2] = x_3 - x_2 \cdot x_3 + x_1 \cdot x_2 \cdot x_3$. The advantage of making such arithmetic transforms can be now seen in the fact that $\mathbf{A}[f_1]$ and $\mathbf{A}[f_2]$ can now be evaluated on non-Boolean values too. For example, suppose $x_1 = 2, x_2 = 3, x_3 = 1$ was randomly chosen as an assignment. $\mathbf{A}[f_1](2,3,1)$ now evaluates to 3, and $\mathbf{A}[F_2](2,3,1)$ maps to 4. Since these two numeric values, also called as *hash codes*, are different, we can now conclude $f_1 \not\equiv f_2$ (see Section 6.5 for details). Remarkably, such numeric simulations can distinguish any pair of Boolean functions with a quantifiable high probability of success. The probability of the error that two different functions will evaluate to the same hash code is inversely proportional to the size of the set from which the integer assignments are randomly chosen. If we perform several such runs with different random input variable assignments, the resulting *algebraic simulation* has a *probability of error* in verification that decreases exponentially from an initially small value. In contrast to random Boolean simulation, such *numeric* simulation is quite adequate to verify the equivalence of circuits. In contrast to OBDDs, these hash codes can be evaluated requiring only polynomial space [10]. Also, usually such hash codes can be evaluated much more quickly than constructing OBDDs. For many functions where OBDDs require exponential space and time, these hash codes can be evaluated using only polynomial resources.

Notice one can simulate a circuit on any *Boolean* vector in a time linear in the size of the circuit. Now consider the contrasting scenario where in circuit for f_1 we simply propagate the *numeric* values by a unit-time evaluation at each gate according to the probability expressions $x\mathsf{AND}y = x \cdot y$, $\mathsf{NOT}(x) = 1 - x$, $x\mathsf{OR}y = x + y - x \cdot y$; such expressions were first discussed in [3]. According to the above expressions, a straightforward numeric simulation for inputs $y_1 = 6$, $y_2 = 3$ will produce at gate f_1, $6 + 3 - 6 \cdot 3 = -9$, which is incorrect. The reason, as discussed in detail in Section 6.2 and 6.3, is that between input signals y_1 and y_2 at gate f_1, we ignored the law of *idempotence of multiplication* on variable x_2, namely $x_2 \wedge x_2 = x_2$. Without accounting for idempotence we may simulate circuits for some F_1 and F_2, $F_1 \not\equiv F_2$, and may still come up with a different hash code for each circuit. Unfortunately, this may produce different hash code from two circuits even if they are functionally equivalent, and that is unacceptable. Accounting for such idempotence is central to our numeric simulation method. It is also the computational bottleneck. Accounting for idempotence makes the computational resources intractable in the worst case, and thus requires elaborate analysis of arithmetic transforms to determine when such multiplicative idempotence can be ignored, or calculated more efficiently. Such a study, along with formal definition of our transforms, is undertaken in Section 6.2 and 6.3. In fact, we never explicitly compute the **A**-transform of a function f, but using the laws of such transforms incrementally compute the correct hash code; such computational procedures are explained in detail in [10].

Our **A**-transform based theories are central to verifying Boolean functions probabilistically.[2] They also provide useful theoretical insight into properties of Boolean functions. Or, we can use them to evaluate testability measures of different gates in some logic circuit [16]. Similarly, we can conduct the reliability analysis of a given network [17]. The potential of our theories in computing exact or approximate testability measure or reliability computation appears promising considering that they have been remarkably successful in (probabilistically) verifying properties of some intractable circuits for which Boolean algebra based deterministic schemes require prohibitive space and time.

The concepts and theories explained here have been often excerpted from an earlier publication [10]. In the following we focus on describing an integer-valued arithmetic (algebraic) transform of Boolean and field functions (Section 6.2 and 6.3). BDDs are now frequently used in solving problems where an analysis of the arithmetic aspect of Boolean functions is central. In this context

[2]We must point out that one can also probabilistically verify equivalences of Boolean functions using spectral techniques in a variety of ways. For example, one can choose to compare the two functions on a subset of their spectra only. However, such techniques are not a focus of this chapter.

note that the transforms we discuss are at the heart of integer-valued interpretation of any BDD. To represent arithmetic expressions we discuss in Section 6.4 a suitable graph representation called *Semi-Numeric Decision Diagram* (snDD). We will briefly discuss applications of our theory in probabilistically verifying properties of Boolean functions (Section 6.5).

6.2 AN INTEGER-VALUED ARITHMETIC TRANSFORM

To generate a hash code for a Boolean function $\beta(x_1, \ldots, x_n)$ of Boolean input variables x_1 through x_n, a random integer value is selected for each x_i and a transformed version of β is evaluated using these values. More generally, the values can be selected from any *field,* which we will denote as $\mathcal{F}$. Computationally, we wish to deal with a finite field, so instead of using unbounded integers, we will use Z_p, the field of integers *modulo* p, for some prime p.

Transforms similar to ours have been studied in [8, 13, 14, 15] etc. However, simply defining the transform does not by itself yield an efficient computational procedure. We define algebraic transforms for Boolean as well as field functions. We will also discuss new properties of such transforms which lead to efficient computational strategies for hashing a given function.

We now show how to transform β into an "equivalent" *field function,* defined over $\mathcal{F}$, so that we can compute its value given the randomly selected input values. We do this by defining a *functional transformation,* $\mathbf{A}[\beta]$, which transforms β into a field polynomial. To denote its n parameters explicitly, we write $\mathbf{A}[\beta](x_1, \ldots, x_n)$. Evaluating $\mathbf{A}[\beta]$ with the chosen integer values provides the desired hash code. $\mathbf{A}[\beta]$ is a canonical transformation; that is, given equivalent Boolean functions β_1, β_2, their field polynomials $\mathbf{A}[\beta_1]$ and $\mathbf{A}[\beta_2]$ will be identical.

6.2.1 Definition of the A-Transform and Fundamental Properties

To define the $\mathbf{A}$-transform, we associate a "key polynomial" (defined below) with each of the 2^n input assignments to a Boolean function $\beta(x_1, \ldots, x_n)$. We then sum the polynomials associated with all such assignments that produce the output value *true,* and interpret the result as an integer-valued function to obtain the required algebraic transform. Similarly, we also obtain a transform of a field function, allowing for their probabilistic verification as well. Thus our techniques can have an application beyond the familiar problems in Boolean function analysis.

Key Polynomials

The key polynomial W_n for a given row of the truth table is a product of terms: $\prod_{i=1}^{n} \mathbf{w}(b_i, x_i)$, where each term $\mathbf{w}(b_i, x_i)$ is associated with a particular input variable x_i. Here b_i represents the truth value assigned to x_i in a given row of the truth table, and $\mathbf{w}(b_i, x_i)$ is x_i if $b_i = 1$ else $1 - x_i$. Thus, b_i acts as a selector.

Definition 6.2.1 *For any $n \geq 0$, let $\mathbf{W}_n : \mathcal{F}^{2n} \to \mathcal{F}$ be defined as*

$$\mathbf{W}_n(b_1, \ldots, b_n, x_1, \ldots, x_n) = \prod_{i=1}^{n} \mathbf{w}(b_i, x_i).$$

For convenience, we will write $\mathbf{W}_n(b_1, \ldots, b_n, x_1, \ldots, x_n)$ as $\mathbf{W}_n(\vec{b}, \vec{x})$, where $\vec{b}$ and $\vec{x}$ are vectors of length n, and $f(\vec{b})$ as an abbreviation for $f(b_1, \ldots, b_n)$.

Example 6.2.1 $\mathbf{W}_3(0, 0, 0, x_1, x_2, x_3)$, *the key polynomial for that row in the truth table of $\beta(x_1, x_2, x_3)$ in which all variables are assigned 0 is simply $(1 - x_1)(1 - x_2)(1 - x_3)$. Similarly, $\mathbf{W}_3(1, 1, 0, x_1, x_2, x_3) = x_1 x_2 (1 - x_3)$.* ■

A Constructive Definition of A-Transform

$\mathbf{A}[f]$ can be now defined as the sum of these key polynomials, each multiplied by the Boolean result of its corresponding assignment, thus summing only those polynomials for "true" assignments.

Definition 6.2.2 *Given any function $f : \{0,1\}^n \to \mathcal{F}$, the polynomial $\mathbf{A}[f] : \mathcal{F}^n \to \mathcal{F}$ is defined as*

$$\mathbf{A}[f](\vec{x}) = \sum_{\text{all } \vec{b} \in \{0,1\}^n} f(\vec{b}) \cdot \mathbf{W}(\vec{b}, \vec{x}). \tag{6.2.1}$$

The following example explains the derivation of polynomial $\mathbf{A}[f]$.

Example 6.2.2 *In the following we find arithmetic polynomial for $\beta = x_1 \oplus x_2$.*

x_1	x_2	W_2	β
0	0	$(1-x_1)\cdot(1-x_2)$	0
0	1	$(1-x_1)\cdot x_2$	1
1	0	$x_1\cdot(1-x_2)$	1
1	1	$x_1\cdot x_2$	0.

Since $\beta = x_1 \oplus x_2$ is 1 for the second and third rows of the truth table, we add the corresponding key polynomials to obtain: $\mathbf{A}[\beta] = x_1(1-x_2) + (1-x_1)x_2 = x_1 + x_2 - 2x_1x_2$. ∎

Note that equation (1) can be applied to any field function and not just to Boolean function. Therefore, we can also take the algebraic transform of a field function of n-variables. Note, an arbitrary field function need not have an **A**-transform equivalent to the **A**-transform of any of the 2^{2^n} possible Boolean functions. In the next section we will more formally define the notion of transform equivalences.

Example 6.2.3 *Let* $F_1(x,y) = x + 3y^2$. *Then* $\mathbf{A}[F_2](x,y) =$

$$(0+0)\cdot(1-x)(1-y) + (0+3)\cdot(1-x)y + (1+0)\cdot x(1-y) + (1+3)\cdot xy,$$

which simplifies to $x + 3y$. ∎

6.2.2 On Equivalences Between A-Transforms

Clearly, a field function f *is* defined for inputs $x_i \notin \{0,1\}$. But to define **A**-transform, we only consider the values f produces for assignments $x_i \in \{0,1\}$. Therefore, if two field functions f_1 and f_2 are equal for every Boolean vector, they are treated identically by the **A**-transform. Therefore we say:

Definition 6.2.3 f_1 *is* 0,1-equivalent *to* f_2, *written* $f_1 \overset{0,1}{=} f_2$, *if and only if* f_1 *and* f_2 *are two functions such that* $f_1(\vec{b}) = f_2(\vec{b})$ *for any Boolean vector* $\vec{b}$.

Given two field functions such that $f_1 \overset{0,1}{=} f_2$, $\mathbf{A}[f_1] = \mathbf{A}[f_2]$. If f_1 and f_2 are not 0,1-equivalent, i.e., there is some Boolean vector that distinguishes them, can we conclude $\mathbf{A}[f_1] \neq \mathbf{A}[f_2]$? To answer this question, let us examine the behavior of $\mathbf{A}[f]$ when it is evaluated at a Boolean vector. When an **A**-transform is evaluated on a Boolean vector, all **W**'s *vanish,* except the one for the particular vector, which evaluates to 1. Hence the term *key* polynomial. Using these facts, we can prove the following theorem [10]:

Theorem 6.2.1 *For any given* $\vec{b}_0 \in \{0,1\}^n$, $\mathbf{A}[f](\vec{b}_0) = f(\vec{b}_0)$, *i.e.,* $\mathbf{A}[f] \stackrel{0,1}{=} f$.

This theorem, and the preceding discussions, have many important implications:

Observations: (1) Though applying the **A**-transform to a Boolean function f increases its domain from $\{0,1\}^n$ to $\mathcal{F}^n$, $\mathbf{A}[f]$ still yields the same values as f when evaluated at a Boolean vector. Therefore,

(2) The algebraic representations for two distinct Boolean functions differ on all Boolean vectors that distinguish the two Boolean functions. Thus, the representations themselves will differ, i.e., if $\beta_1 \neq \beta_2$, then $\mathbf{A}[\beta_1] \neq \mathbf{A}[\beta_2]$. However, this is a rather weak observation. It can be shown that the algebraic representations differ for nearly all vectors where the inputs can assume integer values [10].

(3) Because of points (1) and (2), when we wish to develop a data structure for algebraic representations, any data structure for representing *Boolean functions* can be used, *as long as* we provide a theory to allow its interpretation as an algebraic representation.

Example 6.2.4 *Consider expressions:* $B_1 = (x_1 \wedge x_2) \vee (\overline{x_1} \wedge x_3)$; $B_2 = (x_1 \wedge x_2) \vee (\overline{x_1} \wedge x_3) \vee (x_2 \wedge x_3)$.

Using the above given discussion, the reader can easily work out transforms for these functions, namely:

$$\mathbf{A}[B_1] = x_3 + x_1 \cdot x_2 - x_1 \cdot x_3,$$

$$\mathbf{A}[B_2] = x_3 + x_1 \cdot x_2 - x_1 \cdot x_3.$$

The two functions represented by B_1 and B_2 are indeed equivalent, and hence their algebraic transforms are also identical, though derived from different expressions. ∎

6.2.3 Computing the A-Transform of a Boolean Expression

Computing **A**-transform of Boolean expressions using key polynomials is not convenient as it usually needs exponential number of terms. We would prefer to compute the **A**-transform of a complex expression *incrementally* by computing and then combining the **A**-transforms of its component terms.

The first step in computing $\mathbf{A}[\beta]$ is to replace Boolean function β by a field function f, such that $\mathbf{A}[\beta] = \mathbf{A}[f]$. We need merely replace Boolean function β with a 0,1-equivalent field function f because $\beta \overset{0,1}{=} f$ implies $\mathbf{A}[\beta] = \mathbf{A}[f]$. This is a special case of a very powerful rule for manipulating **A**-transforms: If $f_1 \overset{0,1}{=} f_2$, we can *replace* the f_1 in $\mathbf{A}[f_1]$ with f_2 without changing the algebraic representation. Similarly, we can replace a subexpression of f_1 with a 0,1-equivalent subexpression without changing the value of $\mathbf{A}[f_1]$:

0,1-Substitution Rule: If ψ_1 is a subexpression of ψ and $\psi_2 \overset{0,1}{=} \psi_1$, then ψ_1 can be replaced by ψ_2 in $\mathbf{A}[\psi]$.

To find an f that is 0,1-equivalent to a given Boolean expression, we can use the 0,1-equivalences shown in the second column of Table 6.2.1. We need only replace the Boolean operation $\odot$ by the *corresponding* field operation $\odot_{\mathcal{F}}$, as defined in Table 6.2.1. We will call these field operations *extended Boolean operations.*

On Idempotence of Multiplication

Though the above lets us manipulate functions inside the brackets, it does not suffice for incremental computation. To compute $\mathbf{A}[\beta_1 \wedge \beta_2]$ for instance, we cannot, in general, simply compute $\mathbf{A}[\beta_1] \cdot \mathbf{A}[\beta_2]$. For $\beta_1 = \beta_2 = x$, this would result in $\mathbf{A}[x] \cdot \mathbf{A}[x] = x \cdot x = x^2$, while the correct answer is $\mathbf{A}[x \wedge x] = x$, from equation (1). Though x and x^2 agree for Boolean values, i.e., $x \overset{0,1}{=} x^2$, they do not agree for other values from $\mathcal{F}$.

The Boolean equivalence $x \wedge x = x$ can be used to make a correct calculation: $\mathbf{A}[x \wedge x] = \mathbf{A}[x] = x$. Can we do the same using field functions? After replacing $\wedge$ by $\wedge_{\mathcal{F}}$, we have $\mathbf{A}[x \wedge x] = \mathbf{A}[x \cdot x] = \mathbf{A}[x^2]$. Now we can use the 0,1-equivalence $x^2 \overset{0,1}{=} x$ to *replace* x^2 by x *inside* the **A**-transform, resulting in $\mathbf{A}[x]$, and the correct result.

This replacement *enforces idempotence,* i.e., the removal of the exponent of x to obtain the correct result. As illustrated above, exponents can be removed *within* an **A**-transform using the 0,1-equivalence $x^i \stackrel{0,1}{=} x^j$, for $i, j > 0$. For example, $\mathbf{A}[x^2 \cdot y^2] = \mathbf{A}[x^2 \cdot y] = \mathbf{A}[x \cdot y]$. That is, the **A**-transform is *exponent insensitive.*

Exponent Reduction Rule: In $\mathbf{A}[\psi]$, any occurrence of x_i^k ($k > 1$) can be replaced by x_i.

Therefore, our method for enforcing idempotence is simply to apply the **A**-transform after combining the **A**-transforms of the sub-functions, i.e., we will compute $\mathbf{A}[\beta_1 \wedge \beta_2]$ by $\mathbf{A}[\mathbf{A}[\beta_1] \cdot \mathbf{A}[\beta_2]]$. To derive a formal rule, recall $\mathbf{A}[f] \stackrel{0,1}{=} f$ (Theorem 6.2.1). Given a field expression ψ, this allows the substitution of $\mathbf{A}[f]$ for occurrence(s) of f in $\mathbf{A}[\psi]$. Thus, $\mathbf{A}[\beta_1 \wedge \beta_2] = \mathbf{A}[\beta_1 \cdot \beta_2] = \mathbf{A}[\mathbf{A}[\beta_1] \cdot \mathbf{A}[\beta_2]]$.

We can perform this final step for any extended Boolean operation. Since $x \odot_{\mathcal{F}} y$ must be a polynomial in x and y, we have the 0,1-equivalence $x \odot_{\mathcal{F}} y \stackrel{0,1}{=} c_1 + c_2 \cdot x + c_3 \cdot y + c_4 \cdot x \cdot y$ for some $c_1, \ldots, c_4 \in \mathcal{F}$, and therefore, any extended Boolean operation can be assumed to have this bilinear form.[3] Replacing each occurrence of x and y in $\mathbf{A}[x \odot_{\mathcal{F}} y]$ by $\mathbf{A}[x]$ and $\mathbf{A}[y]$ respectively results in $\mathbf{A}[\mathbf{A}[x] \odot_{\mathcal{F}} \mathbf{A}[y]]$, and we have:

Theorem 6.2.2 $\mathbf{A}[f_1 \odot_{\mathcal{F}} f_2] = \mathbf{A}[\mathbf{A}[f_1] \odot_{\mathcal{F}} \mathbf{A}[f_2]]$, *for any extended Boolean operation* $\odot_{\mathcal{F}}$.

The need to enforce idempotence occurs only when f_1 and f_2 have a common variable in their *support* set.[4]

Thus, if f_1 and f_2 have a disjoint support, their **A**-transforms can be combined directly ($\mathbf{A}[f_1] \odot_{\mathcal{F}} \mathbf{A}[f_2]$), without applying the **A**-transform to the result, as in Theorem 6.2.2. This gives the following very important theorem:

Theorem 6.2.3 *(Disjointness Theorem) If* f_1 *and* f_2 *have disjoint support,* $\mathbf{A}[f_1 \odot_{\mathcal{F}} f_2] = \mathbf{A}[f_1] \odot_{\mathcal{F}} \mathbf{A}[f_2]$ *for any extended Boolean operation,* $\odot_{\mathcal{F}}$.

[3] After all, an extended Boolean operation is chosen to be 0,1-equivalent to a Boolean operation. This bilinear form is also 0,1-equivalent.

[4] The support of any function f is the set of variables on which f depends. Informally, two functions f_1, f_2 have a *disjoint support* if they do not depend on a variable common to both. See Definition 6.3.1 for a precise meaning of "disjoint support".

Table 6.2.1 Extended Boolean Operations.

Boolean $\odot$	0,1 equivalence	Extension $\odot_{\mathcal{F}}$	$=$ *Definition*
$\neg\beta$	$\neg\beta \overset{0,1}{=} 1-\beta$	$\neg_{\mathcal{F}}(\beta)$	$=1-\beta$
$\beta_1 \wedge \beta_2$	$\beta_1 \wedge \beta_2 \overset{0,1}{=} \beta_1 \cdot \beta_2$	$\beta_1 \wedge_{\mathcal{F}} \beta_2$	$=\beta_1 \cdot \beta_2$
$\beta_1 \vee \beta_2$	$\beta_1 \vee \beta_2 \overset{0,1}{=} \beta_1 + \beta_2 - \beta_1 \cdot \beta_2$	$\beta_1 \vee_{\mathcal{F}} \beta_2$	$=\beta_1 + \beta_2 - \beta_1 \cdot \beta_2$
$\beta_1 \oplus \beta_2$	$\beta_1 \oplus \beta_2 \overset{0,1}{=} \beta_1 + \beta_2 - 2 \cdot \beta_1 \cdot \beta_2$	$\beta_1 \oplus_{\mathcal{F}} \beta_2$	$=\beta_1 + \beta_2 - 2 \cdot \beta_1 \cdot \beta_2$

β, β_1 and β_2 are Boolean functions. For each Boolean operation, $\odot$, a 0,1-equivalent field function is given. This defines a corresponding extended Boolean operation, $\odot_{\mathcal{F}}$. Extensions for other Boolean operations can be derived using the 0,1-equivalences for $\neg$ and $\wedge$.

Theorems 6.2.2 and 6.2.3 allow the incremental evaluation of the algebraic transform of a complex Boolean function. We simply evaluate the transforms of its sub-functions, combine them using the appropriate extended Boolean operation, and then apply the **A**-transform.

Example 6.2.5 *Consider* $B_1 = x_1 \wedge x_2 \wedge \overline{x_3}$, *with algebraic transform* $\mathbf{A}[B_1] = x_1 x_2(1-x_3)$, *and* $B_2 = x_1 \wedge x_2$, *with* $\mathbf{A}[B_2] = x_1 x_2$. *To compute the* **A**-*transform of* $B_3 = B_1 \vee B_2$, *we evaluate* $\mathbf{A}[\mathbf{A}[B_1] \vee_{\mathcal{F}} \mathbf{A}[B_2]]$, *i.e.,* $\mathbf{A}[\mathbf{A}[B_1] + \mathbf{A}[B_2] - \mathbf{A}[B_1] \cdot \mathbf{A}[B_2]]$. *Thus* $\mathbf{A}[B_3] = \mathbf{A}[x_1x_2(1-x_3) + x_1x_2 - (x_1x_2(1-x_3) \cdot x_1x_2)] = \mathbf{A}[2x_1x_2 - x_1x_2x_3 - x_1^2x_2^2 + x_1^2x_2^2x_3] = \mathbf{A}[x_1x_2]$.

To derive the **A**-*transform of* $B_4 = B_1 \wedge B_2$, *we compute* $\mathbf{A}[\mathbf{A}[B_1] \wedge_{\mathcal{F}} \mathbf{A}[B_2]]$ *as* $\mathbf{A}[\mathbf{A}[B_1] \cdot \mathbf{A}[B_2]] = \mathbf{A}[x_1x_2(1-x_3) \cdot x_1x_2] = \mathbf{A}[x_1^2x_2^2(1-x_3)] = \mathbf{A}[x_1x_2(1-x_3)]$. ∎

We have discussed how to produce $\mathbf{A}[\beta]$ from a given Boolean function β. We will now discuss how to determine the actual polynomial from the **A**-transform.

6.2.4 Computing the Symbolic Polynomial from an A-Transform

It is difficult to compute the actual polynomial $\mathbf{A}[f]$ using the cumbersome definition of **A**. Therefore, this subsection will develop a calculus for determining the actual polynomial $\mathbf{A}[\psi]$ for a given field expression ψ. However, the utility of the tools we present in this subsection is not at all limited to this rather simple application.

One method for determining this symbolic polynomial is to proceed variable by variable, using an analog of Boole's expansion of a Boolean function. The form of equation (1) (see Definition 6.2.2) makes it easy to see that our algebraic transforms can be expanded in a way analogous to Boole's expansion of a Boolean function for a particular variable x, by collecting all terms containing the factor "x" and all terms containing "$1 - x$." Let β be a field function of two variables, say, x_1 and x_2. We know that $\mathbf{A}[\beta] = \beta(0,0) \cdot (1 - x_1)(1 - x_2) + \beta(0,1) \cdot (1 - x_1)x_2 + \beta(1,0) \cdot x_1(1 - x_2) + \beta(1,1) \cdot x_1 x_2$.

The above expression can be factored as

$$(1 - x_1) \cdot [\beta(0,0) \cdot (1 - x_2) + \beta(0,1) \cdot x_2] + x_1 \cdot [\beta(1,0) \cdot (1 - x_2) + \beta(1,1) \cdot x_2].$$

Note that the two bracketed expressions are simply the algebraic transforms of two *restrictions* of β, where x_1 has been replaced by 0 and 1 respectively. We use the notation $f_{x=R}$ to denote the function which is the restriction of f where x is replaced by R ($R \in \mathcal{F}$). Now we can state a very useful theorem (we omit all proofs through out this article; for details, please refer to [10]).

Theorem 6.2.4 *(Linear Expansion Theorem) [10] Let variable x occur in the parameter list of $\mathbf{A}[f]$, i.e., $\mathbf{A}[f](\ldots, x, \ldots)$, then $\mathbf{A}[f] = (1 - x) \cdot \mathbf{A}[f_{x=0}] + x \cdot \mathbf{A}[f_{x=1}]$.*

A second method is to develop a sufficient set of rules for obtaining the polynomial $\mathbf{A}[\psi]$. The following theorem [10], along with the exponent reduction rule, can be used to simplify the expression.

Theorem 6.2.5 *For any field functions f_1 and f_2 and any constant $c \in \mathcal{F}$,*

(a) $\mathbf{A}[c \cdot f_1] = c \cdot \mathbf{A}[f_1]$, *(d)* $\mathbf{A}[x_i] = x_i$,
(b) $\mathbf{A}[f_1 + f_2] = \mathbf{A}[f_1] + \mathbf{A}[f_2]$, *(e)* $\mathbf{A}[c] = c$,
(c) $\mathbf{A}[f_1 - f_2] = \mathbf{A}[f_1] - \mathbf{A}[f_2]$,
(f) $\mathbf{A}[f_1 \cdot f_2] = \mathbf{A}[f_1] \cdot \mathbf{A}[f_2]$, *if f_1, f_2 have disjoint support.*

These rules are sufficient to calculate the actual field expression for $\mathbf{A}[\psi]$ for any field expression ψ. After the application of rules (a)–(d) and the exponent reduction rule, rules (e) and (f) can be used to yield the actual polynomial.

Example 6.2.6 *Consider* $f = (f_1 \wedge_{\mathcal{F}} f_2) + c_1 \cdot (f_3 \wedge_{\mathcal{F}} f_4) - f_5$,

where, $f_1 = x_1 \vee_{\mathcal{F}} x_2$, $f_2 = x_3 \wedge_{\mathcal{F}} x_4$, $f_3 = x_1 \wedge_{\mathcal{F}} x_3$, $f_4 = x_3$, $f_5 = x_5, c_1 = 2$.

Applying the above rules, we can immediately write $\mathbf{A}[f]$ *as* $\mathbf{A}[f_1 \wedge_{\mathcal{F}} f_2] + c_1 \cdot \mathbf{A}[f_3 \wedge_{\mathcal{F}} f_4] - \mathbf{A}[f_5]$.

Now by Disjointness Theorem, $\mathbf{A}[f_1 \wedge_{\mathcal{F}} f_2] = \mathbf{A}[f_1] \cdot \mathbf{A}[f_2]$
$= (x_1 + x_2 - x_1 \cdot x_2) \cdot (x_3 \cdot x_4) = x_1 \cdot x_3 \cdot x_4 + x_2 \cdot x_3 \cdot x_4 - x_1 \cdot x_2 \cdot x_3 \cdot x_4$.

Also, using the Exponent Reduction rule, the transform for $\mathbf{A}[f_3 \wedge_{\mathcal{F}} f_4]$ *can be written as* $\mathbf{A}[x_1 \cdot x_3^2] = \mathbf{A}[x_1 \cdot x_3] = x_1 \cdot x_3$. *Finally,* $\mathbf{A}[f_5] = \mathbf{A}[x_5] = x_5$.

Hence, $\mathbf{A}[f] = -x_5 + 2 \cdot x_1 \cdot x_3 + x_1 \cdot x_3 \cdot x_4 + x_2 \cdot x_3 \cdot x_4 - x_1 \cdot x_2 \cdot x_3 \cdot x_4$. ∎

6.3 MORE ON A-TRANSFORMS: INTRODUCING NUMERIC VALUES

It is simple to introduce a numeric value into an algebraic representation by substituting the given value for the equivalent symbol and then performing any resulting simplifications. This produces a representation which we refer to as a *semi-numeric polynomial.* This can be defined mathematically using the concept of *functional restriction.*

6.3.1 Functional Restrictions in Evaluating A-Transforms

Since every variable has an associated random value, we will denote a semi-numeric polynomial by specifying a *restriction set* on a function f (denoted by $\mathbf{A}_{\{\rho\}}[f]$), which gives the set of all variables upon which restrictions are to be performed. For example, consider the following:

Example 6.3.1 *Let $B_1 = x_1 \wedge x_2 \wedge \overline{x_3}$, with random values $x_1 = 2$, $x_2 = 4$ and $x_3 = 3$. Now, $\mathbf{A}[B_1] = x_1x_2(1 - x_3)$, so $\mathbf{A}_{\{x_1\}}[B_1] = 2x_2(1 - x_3)$, $\mathbf{A}_{\{x_1,x_2\}}[B_1] = 8(1 - x_3)$, and $\mathbf{A}_{\{x_1,x_2,x_3\}}[B_1] = -16$ is the hash code for B_1. Also, consider field function f described in example 6.2.6. Let $x_1 = 2, x_2 = 1, x_3 = 4, x_4 = 0, x_5 = 4$ be the random integer assignment. Assume we are working in a finite field Z_{19}. We can now work out $\mathbf{A}_{\{x_1,x_2,x_3,x_4,x_5\}}[f]$ as 12.*

Care must be taken when semi-numeric polynomials are combined using Theorem 6.2.2. Consider $B_5(x) = x \wedge x$, with x assigned the value 2. Since $\mathbf{A}[B_5] = x$, the desired hash code is 2. However, combining the numeric values for the sub-functions yields $\mathbf{A}[\mathbf{A}_{\{x\}}[x] \cdot \mathbf{A}_{\{x\}}[x]] = \mathbf{A}[2 \cdot 2] = 4$, which is incorrect. ∎

Intuitively, two semi-numeric polynomials cannot be combined if they have been restricted on a common variable. If this pitfall is avoided, the combination will yield the correct result. Therefore we will only restrict one of the polynomials on a variable, x, when the other *does not depend* on x. Formally, a function f does not depend on variable x if and only if, for all $R_1, R_2 \in \mathcal{F}$, $f_{x=R_1} = f_{x=R_2}$. We will write this as $f \perp x$, that is, f does not depend on x. This gives the generalization of Theorem 6.2.2 to semi-numeric polynomials:

Theorem 6.3.1 *(Semi-Numeric Combination Theorem) Let f_1 and f_2 be functions and ρ a restriction set. Let $\rho_1 = \{x \in \rho \mid f_2 \perp x\}$ and $\rho_2 = \{x \in \rho \mid f_1 \perp x\}$. Then*

$$\mathbf{A}_{\{\rho\}}[\mathbf{A}[f_1] \odot_{\mathcal{F}} \mathbf{A}[f_2]] = \mathbf{A}_{\{\rho-\rho_1-\rho_2\}}[\mathbf{A}_{\{\rho_1\}}[f_1] \odot_{\mathcal{F}} \mathbf{A}_{\{\rho_2\}}[f_2]],$$

where "−" denotes set difference, and $\odot_{\mathcal{F}}$ is any extended Boolean operation.

We will wait to illustrate the use of this theorem until Section 6.4, where a convenient data structure is developed for representing semi-numeric polynomials. Intuitively, when we have to hash the function obtained from combining functions f_1 and f_2, we will obtain the same hash code as if we

1. Substitute numeric values of all such variables in f_1 that are not present in f_2; and vice-versa. This process is also called *partial hashing*.
2. Combine the partially hashed polynomials $\mathbf{A}_{\{\rho_1\}}[f_1]$ and $\mathbf{A}_{\{\rho_2\}}[f_2]$ using laws of $\mathbf{A}$-transforms.

3. Hash the resulting data structure, after accounting for idempotence, on the variables that are common to both f_1 and f_2.

Theorem 6.3.1 can be generalized from binary operations to n-ary operations. The n-ary operations we consider are simply **A**-transforms with n arguments.[5]

Theorem 6.3.2 $\mathbf{A}_{\{\rho\}}[F](\mathbf{A}[f_1],\ldots,\mathbf{A}[f_k]) = \mathbf{A}_{\{\rho_0\}}[F](\mathbf{A}_{\{\rho_1\}}[f_1],\ldots,\mathbf{A}_{\{\rho_k\}}[f_k])$, *where, for* $1 \le i \le k$, $\rho_i = \{x \mid \textit{for all } j \ne i,\ f_j \perp x\}$, *and* $\rho_0 = \rho - \rho_1 - \cdots - \rho_k$.

We will illustrate the use of these theorems in the next section through an example.

6.3.2 Combining Numeric Values

We noted in the previous subsection that the combination of semi-numeric polynomials yields the correct result only when certain precautions are taken as to which variables are to be numeric by enforcing multiplicative idempotence. Computationally speaking, in an ideal case we can avoid all explicit (and expensive) checks for idempotence, which typically requires symbolic manipulation, and can thus hash by just appropriately combining numeric values.[6] We now discuss the circumstances in which we can hash a function f by hashing its sub-functions, and then directly combining their hash codes.

First, the numeric analog of Theorem 6.2.3 allows the hash codes for two sub-functions $\mathbf{A}[f_1]$ and $\mathbf{A}[f_2]$ to be combined when f_1 and f_2 have no variables in common. Formally,

Definition 6.3.1 *Functions f_1 and f_2 have a* disjoint *support if and only if for any variable x, $f_1 \perp x$ or $f_2 \perp x$.*

If f_1 and f_2 have a disjoint support, $\mathbf{A}[f_1 \odot_{\mathcal{F}} f_2] = \mathbf{A}[f_1] \odot_{\mathcal{F}} \mathbf{A}[f_2]$, and any restriction set can be applied to both sides of the equation. This gives:

[5] Just as we had to restrict our attention to bilinear operations in the Semi-Numeric Combination Theorem, here we must restrict ourselves to **A**-transforms, and not consider arbitrary field functions.

[6] A tree circuit, without any reconvergent fanout, is such an example; a tree circuit of size N is easily hashed in time $O(N)$.

Theorem 6.3.3 *(Disjoint Combination Theorem) If f_1 and f_2 have a disjoint support then $\mathbf{A}_{\{\rho\}}[f_1 \odot_{\mathcal{F}} f_2] = \mathbf{A}_{\{\rho\}}[f_1] \odot_{\mathcal{F}} \mathbf{A}_{\{\rho\}}[f_2]$ for any ρ.*

If f_1 and f_2 do not have a disjoint support, we can only write $\mathbf{A}[f_1 \odot_{\mathcal{F}} f_2] = c_1 + c_2 \cdot \mathbf{A}[f_1] + c_3 \cdot \mathbf{A}[f_2] + c_4 \cdot \mathbf{A}[f_1 \cdot f_2]$, where $\odot_{\mathcal{F}}$ denotes any extended Boolean operation in the bilinear form, i.e.,

$$x \odot_{\mathcal{F}} y = c_1 + c_2 \cdot x + c_3 \cdot y + c_4 \cdot xy. \tag{6.3.1}$$

The $\mathbf{A}[f_1 \cdot f_2]$ term prevents the direct combination of the hash codes for an arbitrary f_1 and f_2. Therefore, let us consider conditions on f_1 and f_2 which eliminate this term. For example, the simplest is $\mathbf{A}[f_1 \cdot f_2] = 0$. More generally, we have:

Lemma 6.3.1 *If $\mathbf{A}[f_1 \cdot f_2] = a_1 + a_2 \cdot \mathbf{A}[f_1] + a_3 \cdot \mathbf{A}[f_2]$ then*

$$\mathbf{A}[f_1 \odot_{\mathcal{F}} f_2] = (c_1 + c_4 a_1) + (c_2 + c_4 a_2) \cdot \mathbf{A}[f_1] + (c_3 + c_4 a_3) \cdot \mathbf{A}[f_2].$$

We now consider two applications of Lemma 6.3.1. The first is for the condition $\mathbf{A}[f_1 \cdot f_2] = 0$. Note that this corresponds to requiring that Boolean functions β_1 and β_2 are *orthogonal,* i.e., $\beta_1 \wedge \beta_2 = 0$. When this is the case, we can show $\mathbf{A}[\beta_1 \vee \beta_2] = \mathbf{A}[\beta_1] + \mathbf{A}[\beta_2]$, which can be extended by induction to:

Theorem 6.3.4 *(The Orthogonality Theorem) If $\beta_1, \ldots, \beta_k$ are mutually orthogonal Boolean functions, $\mathbf{A}[\beta_1 \vee \ldots \vee \beta_k] = \mathbf{A}[\beta_1] + \ldots + \mathbf{A}[\beta_k]$.*

Therefore, $\mathbf{A}_{\{\rho\}}[\beta_1 \vee \ldots \vee \beta_k] = \mathbf{A}_{\{\rho\}}[\beta_1] + \ldots + \mathbf{A}_{\{\rho\}}[\beta_k]$ for any restriction set ρ. Applying Lemma 6.3.1 to this case allows the numeric combination for any extended Boolean operation.

Theorem 6.3.5 *(Orthogonal Partitioning Theorem) If f_1 and f_2 are field functions such that $\mathbf{A}[f_1 \cdot f_2] = 0$, then $\mathbf{A}[f_1 \odot_{\mathcal{F}} f_2] = c_1 + c_2 \cdot \mathbf{A}[f_1] + c_3 \cdot \mathbf{A}[f_2]$.*

Second, let us consider the dual condition of orthogonality: $\beta_1 \vee \beta_2 = 1$. Since $\mathbf{A}[\beta_1 \vee \beta_2] = \mathbf{A}[\beta_1] + \mathbf{A}[\beta_2] - \mathbf{A}[\beta_1 \wedge \beta_2]$, the dual of the Orthogonality Theorem is $\mathbf{A}[\beta_1 \wedge \beta_2] = \mathbf{A}[\beta_1] + \mathbf{A}[\beta_2] - 1$. The following theorem allows the combination of such a pair of field functions.

Theorem 6.3.6 *(Disjunctive Partitioning Theorem) If f_1 and f_2 are field functions such that $\mathbf{A}[f_1 \vee_{\mathcal{F}} f_2] = 1$, i.e., $\mathbf{A}[f_1] + \mathbf{A}[f_2] - \mathbf{A}[f_1 \cdot f_2] = 1$, then*

$$\mathbf{A}[f_1 \odot_{\mathcal{F}} f_2] = (c_1 - c_4) + (c_2 + c_4) \cdot \mathbf{A}[f_1] + (c_3 + c_4) \cdot \mathbf{A}[f_2].$$

Example 6.3.2 *Let $f = (\beta_1 \vee \beta_2) \wedge \beta_3$, where $\beta_1 = x_1 \wedge x_2$, $\beta_2 = \overline{x_1} \wedge x_3$, $\beta_3 = x_4 \wedge x_5$.*

Let $x_1 = 4, x_2 = 3, x_3 = 8, x_4 = 5, x_5 = 2$ be some random integer assignment to input variables. As in the previous example, assume we are working in the field of integer Z_{19}.

By Theorem 6.3.4 (The Orthogonality Theorem), $\mathbf{A}[\beta_1 \vee \beta_2] = \mathbf{A}[\beta_1] + \mathbf{A}[\beta_2]$. Hence, $\mathbf{A}[f] = \mathbf{A}[(\mathbf{A}[\beta_1] + \mathbf{A}[\beta_2]) \wedge_{\mathcal{F}} \mathbf{A}[\beta_3]]$.

Since function β_3 has disjoint support with function $\beta_1 \vee \beta_2$, hence using Theorem 6.2.3 (Disjoint Combination Theorem), we can separately hash both disjoint functions to write

$\mathbf{A}_{\{\rho\}}[f] = (\mathbf{A}_{\{\rho\}}[\beta_1] + \mathbf{A}_{\{\rho\}}[\beta_2]) \cdot \mathbf{A}_{\{\rho\}}[\beta_3] = (4 \cdot 3 + (1-4) \cdot 8) \cdot (5 \cdot 2) \mathsf{mod} 19 = 13.$

Suppose we had to hash function $g = f \vee \beta_4$, where $\beta_4 = x_4 \wedge x_6 \wedge x_7$ and $x_6 = 3, x_7 = 2$. Here, we could have hashed $\mathbf{A}[g]$ using Theorem 6.3.1 (Semi-Numeric Combination Theorem), as

$\mathbf{A}_{\{\rho - x_4\}}[g] = \mathbf{A}[(4 \cdot 3 + (1-4) \cdot 8) \cdot (x_4 \cdot 2) \vee_{\mathcal{F}} (x_4 \cdot 3 \cdot 2)].$

Or, $\mathbf{A}_{\{\rho - x_4\}}[g] = \mathbf{A}[(-24 \cdot x_4) \vee_{\mathcal{F}} (6 \cdot x_4)]$. Using the Exponent Reduction Rule, we can write $\mathbf{A}_{\{\rho - x_4\}}[g] = 126 \cdot x_4$. Calculated mod19, $\mathbf{A}_{\{\rho\}}[g] = 3$. ■

6.3.3 How to Efficiently Hash Boolean Functions

As described in the following section, we can interpret a BDD as an algebraic representation. Hence, we can compute a hash code by building a BDD (or a similar decision diagram) from the input function and then applying some simple procedure to reducing this structure to the appropriate integer. Of course, the efficiency of such a scheme would be limited by the need to create and evaluate a BDD representation of the entire function. A more efficient approach is to first symbolically decompose the function, and then hash it incrementally. This is done by first hashing some of its parts using the properties of **A**-transforms (such as disjointness, orthogonality etc.) that we discussed in previous section. Of course, now we need to represent these intermediate forms efficiently so that our representation may become ***smaller*** than original BDD. Assuming this can be done, and our resulting representations are not only smaller but also just as easily manipulable as BDDs, we can use these more compact intermediate forms to complete the hashing of the entire function. This is done by applying the *Semi-numeric combination* Theorem. We discussed one such method, a new algorithmic equivalent to the *compose* algorithm for OBDDs [5], called *collapse-with-compose,* in [10] which allows us to do this efficiently.

On Using Semi-Numeric Combination Theorem

Consider the function $f = (\neg v_1 \wedge x) \wedge (v_2 \wedge x)$, with random values 2, 3, and 4 for v_1, v_2, and x respectively. Obviously, the two terms of f are not disjoint, so we cannot apply the Disjoint Combination Theorem to evaluate f. Instead we can use the Semi-Numeric Combination Theorem if we represent x symbolically throughout the computation and perform the restriction on x *after* the representations for the sub-functions have been combined. In the left term, we can numerically evaluate $\neg v_1 = -1$ directly, but we cannot use the numeric value of x; thus we represent x symbolically as a BDD. (Similarly, x must be symbolic in the evaluation of the right term.) When we combine this BDD with a numerically represented function, we obtain a new data structure which we will call as a *Semi-Numeric Decision Diagram* (snDD).

Figure 6.3.1 shows the sequence of operations on snDDs that would be done to compute a numeric value for f. Algorithms similar to *apply* and *reduce* of [5] can be used to manipulate snDDs, and we can evaluate them in a bottom up fashion, using the Linear Expansion Theorem, to obtain numeric representations of functions.

On Using Partitioning Theorem

To illustrate how a partitioning Theorem may be used in hashing, consider Orthogonal Partitioning Theorem: If we can represent a function f using two (simpler) orthogonal functions f_1, f_2, then $\mathbf{A}[f](\vec{v})$ can be determined even if we just separately compute $\mathbf{A}[f_1](\vec{v})$ and $\mathbf{A}[f_2](\vec{v})$. (Note, to check if $f \equiv g$, we can partition and evaluate g independently from f.) Such partitioning not only reduces the space requirements but can reduce time required too since both f_1 and f_2 can be ordered independently, and thus more optimally, and can be even further partitioned. In such recursive hashing scenario, during evaluation of

```
1. ¬v1                       ¬F(2) = 1 − 2          =  −1

                                        x              x
2. ¬v1 ∧ x                   −1  ∧F    / \         =  / \
                                      0   1          0  -1

                                        x              x
3. v2 ∧ x                     3  ∧F    / \         =  / \
                                      0   1          0   3

                                x       x              x
4. (¬v1 ∧ x) ∧ (v2 ∧ x)        / \  ∧F / \         =  / \
                              0  -1   0   3          0  -3

5. Evaluation of (4)      0·(1 − 4) + (−3)·4       =  −12
```

$\neg_{\mathcal{F}}$ and $\wedge_{\mathcal{F}}$ are extended Boolean operations, as defined in Table 6.2.1. For clarity, we have omitted the fact that negative values must be taken modulo p.

Figure 6.3.1 Incremental Evaluation of $f = (\neg v_1 \wedge x) \wedge (v_2 \wedge x)$.

each partition, we can obviously apply other evaluation schemes as well, such as *collapse-with-compose.* By the above discussion, it is clear that apart from the laws of **A**-transforms, a critical step in realizing efficient computational procedure for probabilistic verification is to be able to efficiently represent and manipulate semi-numeric polynomials.

6.4 FIELD EXPRESSIONS AND BDDS: SEMI-NUMERIC DECISION DIAGRAMS

During hashing, we may often need to analyze *semi-numeric* polynomials derived from a Boolean function or some other arithmetic expressions. For example, in function $f = (x_1 \wedge x_2) \vee (x_3 \wedge x_4)$ when we substitute $x_2 = 3, x_4 = 2$, the resulting arithmetic transform is a semi-numeric expression $\mathbf{A}_{\{x_2,x_4\}}[f] = 3 \cdot x_1 + 2 \cdot x_3 - 6 \cdot x_1 \cdot x_3$. Such a flattened form is usually exponentially sized, where even disjunction of n variables requires $O(2^n)$ terms. Also, during hashing, we frequently need to *manipulate* semi-numeric polynomials, such as adding or multiplying two given expressions. Thus, it is very important to investigate how to obtain compact and efficiently manipulable representations for semi-numeric polynomials.

As mentioned after Theorem 6.2.1, $\mathbf{A}[f]$ yields the *same values* as f when evaluated on a Boolean vector. In addition, $\mathbf{A}[f]$ is defined for non-Boolean vectors. Thus, $\mathbf{A}[f]$ contains the information in f, along with some additional information. Therefore, it is reasonable to try to extend data structures for representing Boolean functions to represent semi-numeric algebraic representations (polynomials). We can do so, if we provide a method for *interpreting* it as an algebraic representation. Our data structure, Semi-Numeric Decision Diagrams (snDDs), for representing semi-numeric polynomials is modeled after BDDs since they are the most generally successful data structure for representing Boolean functions.

In the example shown in Figure 6.4.1, semi-numeric polynomials were represented by storing elements from $\mathcal{F}$ in the terminal nodes, not just 0 and 1. An examination of the *apply* algorithm [5] for OBDDs, along with the properties of the **A**-transform discloses other possible sites (besides the terminals) for incorporating numeric values into the OBDD-structure and guides in developing algorithms for manipulating snDDs [10]. For such a possibility, we need to define how the numeric values *are to be interpreted* (i.e., what semi-numeric polynomial the snDD is supposed to represent).

6.4.1 How to Interpret an OBDD as an Algebraic Representation

For a node v in a BDD, let $v\!\uparrow\!.low$ denote v's 0-child, and $v\!\uparrow\!.high$ its 1-child. Further, if v is a non-terminal, let X_v be the symbolic variable associated with node v. The Boolean function represented by an entire BDD can be defined recursively in terms of $F(v)$, the Boolean function represented by the subgraph rooted at non-terminal node v, as follows

$$F(v) = \left[\overline{X}_v \wedge F(v\!\uparrow\!.low)\right] \vee \left[X_v \wedge F(v\!\uparrow\!.high)\right], \tag{6.4.1}$$

and if v is a terminal, $F(v)$ merely yields *true* or *false*. Hence, by the Orthogonality Theorem,

$$\mathbf{A}[F(v)] = \mathbf{A}[(1 - X_v)\cdot F(v\!\uparrow\!.low)] + \mathbf{A}[X_v\cdot F(v\!\uparrow\!.high)]. \tag{6.4.2}$$

Equation (6.4.2) is simplified if the BDD is *free:* a BDD is called to be *free* if and only if each variable occurs *at most once* along any path. Since this

```
      v1:x            v3:x
       / \             / \
   v2:y   1           6   1
     / \
    0   1
```

Figure 6.4.1 An OBDD for $x \vee y$ and the snDD after the replacement $y := 6$.

property is guaranteed for OBDDs, they form a special case of free BDDs. In a free BDD, variable X_v will not occur in v's sub-BDDs, and we can apply the Disjointness Theorem to equation (6.4.2) to obtain

$$\mathbf{A}[F(v)] = (1 - X_v)\cdot\mathbf{A}[F(v\uparrow.low)] + X_v\cdot\mathbf{A}[F(v\uparrow.high)]. \tag{6.4.3}$$

Using (6.4.3), a free BDD can be hashed in a time linear in the size of such graph by a single top-down pass. Note, through such evaluations one can also obtain the size of the satisfying set (testability measure) of a given Boolean function (circuit) from its free BDD. The following example illustrates use of the above equation to determine the algebraic representation of a free BDD.

Example 6.4.1 *Consider Figure 6.4.1. For the OBDD shown on the left,* $F(v_2) = y$ *and* $F(v_1) = x \vee y$. *Applying equation (6.4.3) to node* v_2 *gives* $\mathbf{A}[F(v_2)] = y$, *and* $\mathbf{A}[F(v_1)]$ *is calculated as* $(1 - x)\cdot y + x\cdot 1 = x + y - xy$, *which is simply* $\mathbf{A}[x \vee y]$. *The snDD shown on the right results from the replacement of the sub-BDD rooted at* v_2 *by the numeric value for its function, i.e., 6. The interpretation for node* v_3 *is simply* $(1 - x)\cdot 6 + x \cdot 1 = 6 - 5x$, *i.e.,* $\mathbf{A}_{\{y\}}[x \vee y]$. *Thus, a suitable interpretation of a terminal is simply its numeric value.* ■

6.4.2 snDDs: A Scheme to Represent Semi-Numeric Expressions

For sake of computational efficiency, it should be clear that snDDs must have the following characteristics.

Requirement 1: $\mathbf{A}[f]$ must be represented compactly.

Requirement 2: The representation of $\mathbf{A}[f]$ must be easy to manipulate. In other words, algorithms to manipulate two snDDs should require time at most quadratic in the size of the two snDDs (as is the case with OBDDs).

Now, snDDs must store numeric values apart from Boolean (field) variables. Hence we must ask at which sites in a BDD can numeric values be easily stored so that the above requirements are well satisfied. The possible sites for storing numeric values are (1) Terminals; (2) Edges; (3) Standard (non-terminal) Nodes. In [10] it was proved that storing numeric values in nodes *along* with a variable leads to inefficient AND operation between resulting graphs. If numeric values are placed on each edge then though Boolean AND may be carried out, complementation operation becomes difficult. However, a desired representation scheme obeying above requirements exists if it stores numeric values at following sites:

1. **At Terminals.** In this respect they contain features of *Multi-Terminal BDDs* (MTBDDs) [7], and *Algebraic Decision Diagrams* (ADDs)[1]. Much research needs be done to determine if the applications in which MTBDDs and ADDs are applied will benefit by the features of snDDs.[7]
2. **At special nodes.** These nodes, Σ, Δ, Π are special operator nodes, and can respectively represent operations of *addition*, *subtraction* and *multiplication.*

Understanding snDD Manipulation

When *apply* [5] encounters a pair of nodes labeled x, representing functions $G = \overline{x}G_0 \vee xG_1$ and $H = \overline{x}H_0 \vee xH_1$, two recursive calls are made to construct x's sub-BDDs, using the equation

$$G \odot H = \overline{x}[G_0 \odot H_0] \vee x[G_1 \odot H_1]. \tag{6.4.4}$$

These equations are readily adapted for use in snDDs. For example, we can apply the $\mathbf{A}$-transform to equation (6.4.4) to obtain

$$\mathbf{A}\Big[\mathbf{A}[G] \odot_{\mathcal{F}} \mathbf{A}[H]\Big] = (1-x) \cdot \mathbf{A}\Big[\mathbf{A}[G_0] \odot_{\mathcal{F}} \mathbf{A}[H_0]\Big] + x \cdot \mathbf{A}\Big[\mathbf{A}[G_1] \odot_{\mathcal{F}} \mathbf{A}[H_1]\Big],$$

[7] Note, in an MTBDD, a node v has two outgoing edges: 0 and 1; in an snDD, the outgoing edges are labeled $1 - v$ and v. When an snDD only represents the numerical values in the terminal *only*, then the differences between the two schemes are only semantical. An MTBDD is not used to represent arbitrary field polynomials but represents functions $B \rightarrow \mathcal{F}$.

where $\odot_{\mathcal{F}}$ is the extended Boolean operation corresponding to $\odot$. Therefore, if all non-terminals in the snDD are "standard nodes," having the interpretation given in equation (6.4.3), the resulting snDDs can be combined in the same way as OBDDs: a standard node labeled x is created with children $\mathbf{A}[G_0] \odot_{\mathcal{F}} \mathbf{A}[H_0]$ and $\mathbf{A}[G_1] \odot_{\mathcal{F}} \mathbf{A}[H_1]$. The only difference is that $\odot_{\mathcal{F}}$ is used to combine terminal values instead of $\odot$.

Non-Standard Nodes and Edges

Summation (Σ) node do not have a variable attached, merely weights. Thus the interpretation for such a node v is

$$\mathbf{A}[F(v)] = l_v \cdot \mathbf{A}[F(v\uparrow.low)] + h_v \cdot \mathbf{A}[F(v\uparrow.high)], \qquad (6.4.5)$$

where l_v and h_v denote the multipliers on the "low" and "high" edges leaving node v.

Binary operations can be easily performed: When *apply* encounters a Σ-node paired with another node, it need not combine them (as it does when it encounters two nodes labeled with the same variable), since Σ-nodes have no variable attached. If the children of the Σ-node represent functions F_0 and F_1, and the other node represents function G, a Σ-node can simply be created in the result graph, with children $F_0 \odot_{\mathcal{F}} G$ and $F_1 \odot_{\mathcal{F}} G$.

Having introduced a type of node that corresponds to a summation, it is natural to consider two other kinds of nodes which correspond to other field operations: Difference nodes and Multiplicative nodes. Instead of being used in routine *apply* operations, these nodes are more useful for performing high-level functional transformations on an snDD to put it into a more computationally useful form. We will not discuss these nodes in detail but merely give a brief description.

A *difference node* (Δ-node) represents the subtraction of algebraic representations. It may have more than two edges. One of the edges is designated as the *leading edge,* and the others are referred to as *trailing edges.* If L is the function rooted at the leading edge and $T_1, \ldots, T_k$ are the function(s) rooted at the trailing edges, the interpretation of the node is $\mathbf{A}[L - T_1 - \cdots - T_k]$. The interpretation of a *multiplicative node* (Π-node) is simply the $\mathbf{A}$-transform of the product of the interpretations of its children, multiplied by a weight stored in the node. Thus, a Π-node with weight w and three edges respectively leading to sub-functions f_1, f_2 and f_3 has the interpretation $w \cdot \mathbf{A}[f_1 \cdot f_2 \cdot f_3]$. The weight in a Π-node is used to absorb the numeric values of the sub-functions as they are hashed, if it can be guaranteed that they are disjoint. It can similarly be used to absorb the numeric value of any function leading to the Π-node. Finally, *dotted edges* [12] also fit naturally into our scheme. A dotted edge from v to w simply modifies the interpretation of v to use $1 - F_w$ instead of F_w.

snDDs and EVBDDs

It was also proved in [10] that it is *not* possible to have weights on edges (as constant multipliers) or numeric weights on internal nodes and still have a polynomial time manipulation algorithm between arbitrary snDDs. In this respect, however, note that it was proved by Lai and Sastry in [11] that if weights are placed on only the *non-zero* edge, with the polynomial represented by the graph recursively interpreted in an "additive" fashion, the resulting data structure *Edge-Valued BDD* (EVBDD) can be easily manipulated. We can use such features to augment our snDDs also. Also, note that storing semi-numeric polynomials as EVBDDs can be much more compact than storing them as snDDs in many cases.

6.5 APPLICATION IN PROBABILISTIC EQUIVALENCE VERIFICATION

The preceding discussion helps us to evaluate (the **A**-transform of) a Boolean function on random numeric values. To probabilistically check if two functions B_1 and B_2 are equivalent, we compare whether $\mathbf{A}[B_1](\vec{v})$ and $\mathbf{A}[B_2](\vec{v})$ map to same hash code under a given random assignment $\vec{v}$. Let $\mathbf{A}[B_1](\vec{v}) = H_1$ and $\mathbf{A}[B_2](\vec{v}) = H_2$.

1. If $H_1 \neq H_2$, we guaranteedly know $B_1 \not\equiv B_2$.
2. If $H_1 = H_2$, $B_1 \equiv B_2$ (probabilistically).

What is the probability that for $H_1 = H_2$, $B_1 \not\equiv B_2$? Let us denote the field by $\mathcal{F}$, and assume its size is p. We will assume the random integers are selected from the entire field.

Theorem 6.5.1 *Let f_1 and f_2 be any two field functions of n variables, such that $\mathbf{A}[f_1] \neq \mathbf{A}[f_2]$, and let $\vec{v}$ be a vector of length n whose elements are chosen* *randomly from $\mathcal{F}$. Then*

$$\epsilon = Prob(\mathbf{A}[f_1](\vec{v}) = \mathbf{A}[f_2](\vec{v})) \leq 1 - \left(\frac{p-1}{p}\right)^n.$$

The theorem handles two *Boolean functions* as a special case. If ϵ denotes the upper bound on the error probability, then if $p \gg n$, a reasonable assumption, $\epsilon \approx n/p$. Otherwise $\epsilon \approx 1 - e^{-n/p}$. The probability of erroneously deciding that functions are equivalent decreases *exponentially* with the number of runs: after k runs, the error probability is ϵ^k. Identical bounds exist for deciding a function's satisfiability (or tautology) as well. Note, any two inequivalent functions of n-variables differ in at least $(p-1)^n$ number of assignments. Thus, the error probability will hold for *any* pair of 2^{2^n} Boolean functions of n-variables. Interestingly, the most difficult error to detect by simulation (difference in a single minterm), is found by our method with probability 1 if we disallow assignments of 0 and 1 to any variable.

Note that during simulation, for each variable, we can choose to use (1) a symbol, (2) a random integer, or (3) a random bit. The computational cost can generally be expected to decrease from (1) to (3), while the probability of error increases. Thus, consider *partial simulation,* where v variables are assigned random integer values, and the rest are kept symbolic. Now, two functions are declared equivalent if and only if identical semi-numeric expressions are obtained for each. Note, *partial* simulation describes a range of strategies from completely arithmetic simulation ($v=n$) to deterministic verification ($v=0$).

Theorem 6.5.2 *In partial simulation, $\epsilon \leq v/p$.*

Another verification scenario is *mixed mode simulation.* In this case, v variables receive random integer values, and the rest receive random Boolean values. *Mixed mode* simulation describes a spectrum of strategies between completely arithmetic simulation ($v=n$) and random Boolean simulation ($v=0$).

Theorem 6.5.3 *In mixed mode simulation, $\epsilon \leq 1 - (1/2^{n-v}) * (1 - v/p)$.*

Some Observations on Size of p

Assuming a 32-bit word, integer codes can be multiplied or added without overflow if $(p-1)^2 < 2^{31}$, and then reduced modulo p. However, by using multiple-word arithmetic operations, we can very easily use a larger prime; with double word arithmetic, any prime $p \approx 2^{31}$ can be used without a great deal of difficulty. Our verification experiments show that the corresponding time overhead in using double-word arithmetic is only minimal (majority of time is spent in graph manipulation etc.). A double-word arithmetic using 64 bit words will let one use primes as large as $p \approx 2^{63.5}$, leading to a vanishing error even in a single run for any practical circuit. For a circuit with even 2^{20} (> a million) inputs, the error is at most $\epsilon = n/p = 2^{20}/2^{63.5} = 2^{-43.5}$. Now consider simulating a single binary vector. Here the above error probability expression reduces to $\epsilon = 1 - (1/2)^n$. Thus, after simulating k number of vectors where $1 \ll k \ll 2^n$, the error is $\epsilon = 1 - k/2^n$ and is thus still almost 1 (even though, practically speaking, k may be a very large number).

6.5.1 Contrasting Integer-Transform Based Approach with a Free BDD Approach

One approach, though limited in theoretical as well as in practical utility, to probabilistically compare equivalence of two free BDDs was developed in [2] who provide an *error bound* of 1/2 on a single run. Note that **A**-transforms is crucial for hashing, and thus probabilistic verification of Boolean logic, multi-valued logic, or of arbitrary field functions since evaluating such transforms is an intermediate step in reducing a function to an integer hash code. Apart from capturing this wide phenomenon, the use of **A**-transforms effect the procedure (the mechanics) as well as the performance of probabilistic verification. For example, the *Semi-numeric Combination* Theorem allows us to incrementally reduce portions of a function representation to numbers without risking violation of idempotence. This is critical to get an efficient probabilistic verification scheme. Also, **A**-transform based theory allows function representations, and different forms of circuit simulation, which mix the use of Boolean, integer, and symbolic forms including free BDDs [9, 10]. Thus our approach, on theoretical as well as practical grounds, properly contains the free BDD approach.

6.5.2 Results on Probabilistic Verification

For all difficult circuits in ISCAS-85 benchmark circuits, such as c2670, c3540, c6288, c7552 etc. it was shown in the [10] that the probabilistic method significantly outperforms monolithic OBDD methods in both time and memory requirements; for example, it is relatively easy to generate the hash code for the 16th output of c6288, a 16×16-bit multiplier, allowing its probabilistic verification. We can also compute the hash code from the circuit representation of the *Hidden-Weighted Bit* (HWB) function [6] function in only $\Theta(n^2 \log n)$ space and $\Theta(n^3 \log n)$ time. For a 127 input HWB function, probabilistic methods require around 188,000 nodes as compared to an estimated 239 billion nodes that monolithic OBDDs may require! Similarly, difficult functions such as Fortune-Hopcroft-Schmidt (FHS) function or modified Achilles-heel functions which are known to require exponential resources using OBDDs, can be verified in polynomial time [10]. Also, an n-bit ALU requiring $\Theta(n^2)$ time deterministically with OBDDs requires only linear resources with the probabilistic method [10].

Since one is deriving the integer representation of a function rather than its Boolean representation, one can often obtain the hash code by exploiting the algebraic properties of a higher level representation, circumventing its conversion to a Boolean representation. This was illustrated in [10] where hash code for an n-input HWB function was computed from its abstract specification in $\Theta(n^3)$ time using only $\Theta(n)$ space. Similarly, the hash code for integer multiplier could be obtained 5 times more efficiently from the arithmetic specification of the multiplication function than from its circuit description using only a minimal of space.

6.6 CONCLUSION

We have discussed **A**-transforms, an integer-valued arithmetic representation of Boolean and field functions. We have defined new and computationally useful properties of such transforms. Transforming Boolean functions to arithmetic functions provides useful solutions, and new insights, into important practical problems of verifying properties of Boolean functions such as satisfiability, tautology or functional equivalence. By analyzing an integer-valued representation of Boolean functions, we can efficiently verify equivalence of Boolean functions (circuits) probabilistically with a vanishingly small probability of error.

Our work can be directly used to derive testability measures for elements of a digital circuit, or conduct the reliability analysis for various networks. It is well known that in such applications the central task is to evaluate arithmetic expressions similar to **A**-transforms. We can easily compute arithmetic transforms for any Boolean function from its BDD. Accordingly, wherever such transforms need be analyzed, one can utilize snDDs, a BDD like data structure. Our techniques can also be used to verify arbitrary field functions.

We also propose a modification to snDDs which is *EVsnDDs*. They have features of EVBDDs as well as snDDs. For example, apart from allowing weights on edges (say, in a part of graph) as is allowed in EVBDDs, they have an added capability to store weights at terminals, or at nodes such as Σ nodes or Π-nodes, as is typical in snDDs. We believe such modifications can lead to a highly potent data structure for the applications in which EVBDD, snDDs as well as MTBDDs and ADDs are applied such as computing spectral coefficients, timing analysis, integer linear programming, matrix manipulation, probabilistic verification, etc.

Acknowledgment

The author wishes to acknowledge Dr. Jim Bitner, Professor Jacob A. Abraham, and Professor Donald S. Fussell of University of Texas at Austin, with whom he developed, during his Ph.D., most of the work explained here [9, 10]. The author also wishes to acknowledge Dr. M. Fujita and Professor T. Sasao for constructive comments which greatly improved the quality of this presentation.

REFERENCES

[1] R. I. Bahar, E. A. Frohm, C. M. Gaona, G. D. Hachtel, E. Macii, A. Pardo, and F. Somenzi, "Algebraic decision diagrams and their applications," *in Proc. ICCAD*, pp. 188-191, 1993.

[2] M. Blum, A. K. Chandra, and M. N. Wegman, "Equivalence of free Boolean graphs can be decided probabilistically in polynomial time," *Information Processing Letters*, 10:80-82, March 1980.

[3] G. Boole, *An Investigation of the Laws of Thought*, London, Walton, 1854 (Reprinted by Dover Books, New York, 1954).

[4] M. F. Brown. *Boolean Reasoning: The Logic of Boolean Equations*, Kluwer Academic Publishers, 1990.

[5] R. E. Bryant, "Graph based algorithms for Boolean function representation," *IEEE Trans. Comput.*, C-35, pp. 677-690, August 1986.

[6] R. E. Bryant, "On the complexity of VLSI implementations and graph representations of Boolean functions with application to integer multiplication," *IEEE Trans. Comput.*, Vol. C-40, pp. 206-213, February 1991.

[7] E. M. Clarke, K. L. McMillan, X. Zhao, M. Fujita, and J. Yang, "Spectral transforms for large boolean functions with applications to technology mapping," *in Proc. 30th DAC*, pp. 54-60, 1993. (Also Chapter 4 of this book).

[8] K. D. Heidtmann, "Arithmetic spectrum applied to fault detection for combinational networks," *IEEE Trans. Comput.*, Vol. C-40, pp. 320-324, March 1991.

[9] J. Jain, "On analysis of Boolean functions," *Ph.D. Dissertation, Elec. and Comput. Eng. Dep., The University of Texas at Austin*, May 1993.

[10] J. Jain, J. Bitner, D. S. Fussell, and J. A. Abraham, "Probabilistic verification of Boolean functions," *Formal Methods in System Design*, Vol. 1, pp. 61-115, July 1992.

[11] Y. T. Lai and S. Sastry, "Edge-valued binary decision diagrams for multi-level hierarchical verification," *in Proc. 29th DAC*, pp. 608-613, 1992. (Also Chapter 5 in this book).

[12] J. C. Madre and J. P. Billon, "Proving circuit correctness using formal comparison between expected and extracted behavior," *in Proc. 25th DAC*, pp. 205-210, 1988.

[13] K. P. Parker and E. J. McCluskey, "Correspondence: Probabilistic treatment of general combinational networks," *IEEE Trans. Comput.*, Vol. C-24, pp. 668-670, June 1975.

[14] D. E. Rumelhart, J. L. McClelland, et al, Parallel Distributed Processing, *The MIT Press*, Vol. 1, pp. 423-443, 1986.

[15] S. K. Kumar and M. A. Breuer, "Probabilistic aspects of Boolean switching functions via a new transform," *J. ACM*, Vol. 28, pp. 502-520, July 1981.

[16] S. C. Seth, L. Pan, and V. D. Agarwal, "Predict - probabilistic estimation of digital circuit testability," *in Proc. FTCS*, pp. 220-225, 1985.

[17] K. S. Trivedi, *Probability and Statistics with Reliability, Queuing, and Computer Science Applications.* Prentice-Hall, 1982.

7

OKFDDS - ALGORITHMS, APPLICATIONS AND EXTENSIONS

Rolf DRECHSLER
Bernd BECKER

Institute of Computer Science
Albert-Ludwigs-University
79110 Freiburg im Breisgau, Germany

Abstract— We present Ordered Kronecker Functional Decision Diagrams (OKFDDs), a graph-based data structure for the representation and manipulation of Boolean functions. OKFDDs are a generalization of Ordered Binary Decision Diagrams and Ordered Functional Decision Diagrams and as such provide a more compact representation of the functions than either of the two decision diagrams. We review basic properties of OKFDDs and study methods for their efficient representation and manipulation. These algorithms are integrated in our OKFDD package PUMA whose implementation is briefly discussed. Finally we point out some applications of OKFDDs, demonstrate the efficiency of our approach by some experiments and discuss a promising extension of the concept to also allow representation and manipulation of word-level functions.

7.1 INTRODUCTION

Decision Diagrams (DDs) are often used in *Computer Aided Design* (CAD) systems for the efficient representation and manipulation of Boolean functions. The most popular data structure in this context are *Ordered Binary Decision Diagrams (OBDDs)* [8] that are used in many applications [9]. Nevertheless, some relevant classes of Boolean functions cannot be represented efficiently by OBDDs [5, 38]. As one alternative *Ordered Functional Decision Diagrams (OFDDs)* [31] have been introduced and in the meantime they are used in various applications of XOR-based logic synthesis (see e.g. [17]). If ease of manipulation and canonicity are not main concerns, still other types of DDs, like Ternary Decision Diagrams [39] and Kronecker Functional Decision Diagrams

[30], have proven to be useful especially in the area of technology mapping for multi-level XOR-based circuitry.

Recently, *Ordered Kronecker Functional Decision Diagrams (OKFDDs)* have been introduced as a means for the efficient representation and manipulation of Boolean functions [24]. OKFDDs are a generalization of OBDDs and OFDDs as well and try to combine the advantages of both representations by allowing the use of Shannon decompositions and (positive and negative) Davio decompositions.

From a (more) theoretical point of view it has been shown that there exist certain classes of Boolean functions whose OFDD size is exponentially smaller than the OBDD representation of the same function, and vice versa [4]. Thus, it is useful to consider a representation like OKFDDs, that integrate both, OBDDs and OFDDs. Furthermore, it has been proved that a "restriction" of the OKFDD concept results in families of functions that lose their efficient representations. It follows that OKFDDs in full generality should be considered. On the other hand, based on a formalization of the concept *decomposition type* it has been shown in [1], that OKFDDs are the most general type of *Ordered Decision Diagram* (ODDs). In this sense it is interesting and important to also device effective practical algorithms for representing and manipulating Boolean functions with OKFDDs.

In this chapter we review basic algorithmic properties of OKFDDs and study methods for their efficient representation and manipulation. The data structure allows us to dynamically adapt the representation of a Boolean function to a given problem. But, as well-known for OBDDs [8] and OFDDs [5], OKFDDs are also very sensitive to the variable ordering [24]. In addition to the position of a variable in the ordering so-called decomposition types have to be chosen for OKFDDs. Thus, there is a need for heuristics to choose a suitable variable ordering and decomposition type list for OKFDDs.

In [21] first topology-based heuristics have been presented. These heuristics allow a fast construction of OKFDDs from given circuit descriptions. But, in some cases these heuristics fail to determine small graphs. In [24, 17] it has been shown that dynamic variable ordering methods for OBDDs [29, 38] can also be applied to OKFDDs. These algorithms together with other synthesis algorithms are integrated in our OKFDD package PUMA the implementation of which is discussed below.

Finally we point out some applications of the OKFDD concept to *logic synthesis* and *synthesis for testability.* We demonstrate the efficiency of our approach by some experiments and discuss a promising extension of the concept to also allow for the representation and manipulation of word-level functions.

7.2 ORDERED KRONECKER FUNCTIONAL DECISION DIAGRAMS

In the following, basic definitions of *Decision Diagrams* (DDs) and OKFDDs in particular are presented.

Definition 7.2.1 A *DD* over the variable set $X_n = \{x_1, x_2, \ldots, x_n\}$ is a rooted directed acyclic graph $G = (V, E)$ with edge set E and vertex set V containing two types of vertices, ***non-terminal*** and ***terminal*** vertices. A non-terminal vertex v is labeled with a variable from X_n, called the ***decision variable*** for v, and has exactly two successors denoted by $low(v), high(v) \in V$. All nodes with label x_i are denoted as *level i*. A terminal vertex v is labeled with a 0 or 1 and has no successors. The size of a DD G, denoted by $|G|$, is given by its number of non-terminal nodes.

Further structural restrictions turn out to be important.

Definition 7.2.2 A DD is *complete* if each variable is encountered exactly once on each path in the DD from the root to a terminal vertex. A DD is *free* if each variable is encountered at most once on each path in the DD from the root to a terminal vertex. A DD is *ordered* if it is free and the variables are encountered in the same order on each path in the DD from the root to a terminal vertex.

In the following, the letter O will be used to denote ordered DDs.

DDs can be related to Boolean functions by using the following well-known decomposition types (given here for an arbitrary Boolean function $f : \mathbf{B}^n \to \mathbf{B}$ over the variable set X_n):

$$
\begin{aligned}
f &= \overline{x}_i f_i^0 \vee x_i f_i^1 && \textit{Shannon } (S) && (7.2.1)\\
f &= f_i^0 \oplus x_i f_i^2 && \textit{positive Davio } (pD) && (7.2.2)\\
f &= f_i^1 \oplus \overline{x}_i f_i^2 && \textit{negative Davio } (nD) && (7.2.3)
\end{aligned}
$$

where f_i^0 (f_i^1) denotes the *cofactor* of f with respect to $x_i = 0$ ($x_i = 1$) and f_i^2 is defined as $f_i^2 = f_i^0 \oplus f_i^1$, $\oplus$ being the XOR operation.

Decomposition types are associated with the variables in X_n with the help of a *Decomposition Type List* (DTL) $\mathbf{d} = (d_1, \ldots, d_n)$ where $d_i \in \{S, pD, nD\}$.

Now, the Kronecker Functional Decision Diagrams can formally be defined as follows:

Definition 7.2.3 A *KFDD over* X_n is given by a DD over X_n together with a fixed DTL, $\mathbf{d} = (d_1, \ldots, d_n)$. The function $f_G^{\mathbf{d}} : \mathbf{B}^n \to \mathbf{B}$ represented by G with DTL $\mathbf{d}$ is defined as:

1. If G consists of a single node labeled with 0 (1), then G is a KFDD for the constant 0 (1) function.
2. If G has a root v with label x_i, then G is a KFDD for

$$\begin{cases} \overline{x}_i f_{low(v)} \vee x_i f_{high(v)} & : \quad d_i = S \\ f_{low(v)} \oplus x_i f_{high(v)} & : \quad d_i = pD \\ f_{low(v)} \oplus \overline{x}_i f_{high(v)} & : \quad d_i = nD \end{cases}$$

   where $f_{low(v)}$ ($f_{high(v)}$) is the function represented by the OKFDD rooted at $low(v)$ ($high(v)$).

Of course, a KFDD is an OKFDD iff the underlying DD is ordered.

Example 7.2.1 An OKFDD G with DTL $\mathbf{d} = (nD, pD, S)$ is shown in Figure 7.2.1. The left (right) outgoing edge at each node v denotes $f_{low(v)}$ ($f_{high(v)}$). G represents the function $f : \mathbf{B}^3 \to \mathbf{B}$ with

$$f(x_1, x_2, x_3) = x_3 \oplus \overline{x}_1 x_2 \overline{x}_3$$

A node in a KFDD is called an *S-node* if it is expanded by Shannon decomposition - Equation (7.2.1). It is called a *D-node* if it is expanded by Davio decompositions - Equations (7.2.2) or (7.2.3), the latter being an *nD-node* and the former a *pD-node*. According to the DTL, at every node of a fixed level i in the KFDD, the same decomposition type d_i is applied. (In Figure 7.2.1 the node labeled with x_1 is an nD-node, the nodes labeled with x_2 are pD-nodes

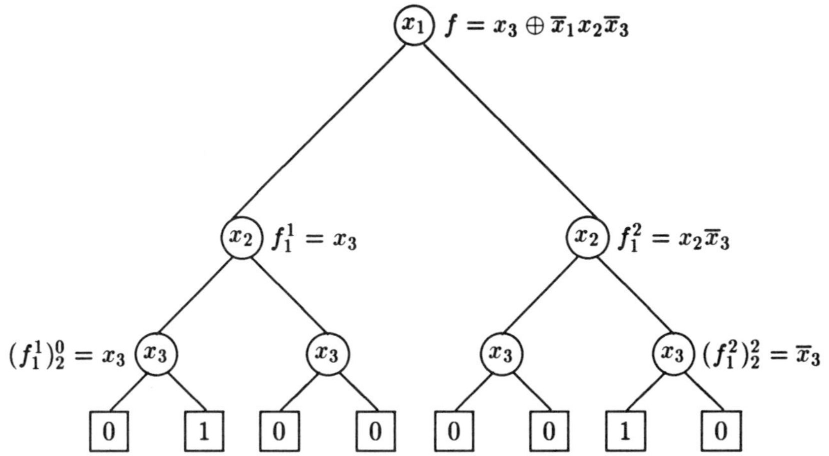

Figure 7.2.1 Example for (non-reduced) OKFDD G with DTL $\mathbf{d} = (nD, pD, S)$ for $f(x_1, x_2, x_3) = x_3 \oplus \overline{x}_1 x_2 \overline{x}_3$.

and the nodes labeled with x_3 are S-nodes.) We directly conclude that KFDDs are a generalization of BDDs and FDDs: If in all levels only Shannon decompositions are applied, the KFDD will be a BDD. If only Davio decompositions are applied, i.e., $d_i \in \{pD, nD\}$ for all i, the KFDD will be an FDD. Analogously positive FDDs (pFDDs) and negative FDDs (nFDDs) are defined.

Reductions of three different types are used to reduce the size of KFDDs:

Type I: Delete a node v' whose label is identical to the label of a node v and whose successors are identical to the successors of v and redirect the edges pointing to v' to point to v.

Type S: Delete a node v whose two outgoing edges point to the same node v' and connect the incoming edges of v to v'.

Type D: Delete a node v whose successor $high(v)$ points to the terminal 0 and connect the incoming edges of the deleted node to $low(v)$.

In Figure 7.2.2, graphical representations of reductions of type S and D are shown. While each node in a KFDD is a candidate for the application of

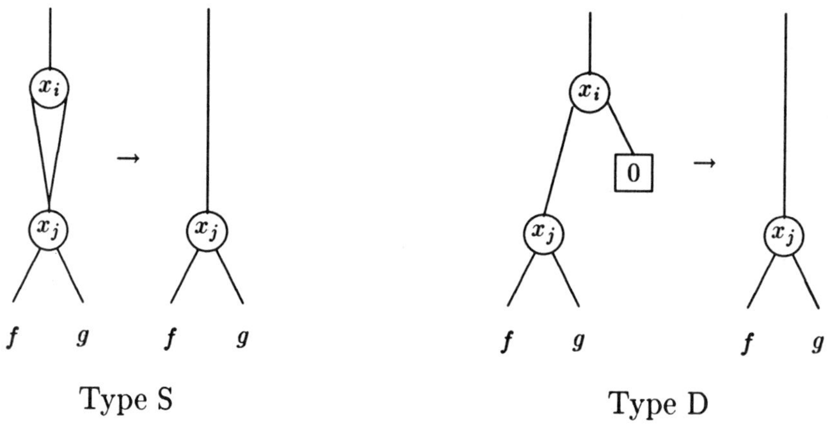

Figure 7.2.2 Type S and type D reduction.

reduction type I, only S-nodes (D-nodes) are candidates for the application of the reduction type S (D). A KFDD is *reduced* iff no reductions can be applied to the KFDD. Two KFDDs G_1 and G_2 (with the same variable ordering and DTL) are *equivalent* iff G_2 results from G_1 by repeated applications of reductions and inverse reductions. A KFDD G_2 is called the *reduction* of a KFDD G_1 iff G_1 and G_2 are equivalent and G_2 itself is reduced.

From [24], it is known that reductions can be used to obtain canonical representations for not only OBDDs and OFDDs, but also for OKFDDs.

Theorem 7.2.1 Reduced OKFDDs are canonical representations for Boolean functions if the variable ordering and the DTL is fixed.

The reduced OKFDD for the example given in Figure 7.2.1 is shown on the lefthand side of Figure 7.2.3.

The size of an OKFDD can be further reduced if *Complement Edges* (CEs) are used analogously to OBDDs [7]. Then a node represents a function and its complement at the same time. If the representation is chosen in a way such that a *low*-edge never is a CE and that a terminal node is always labeled with

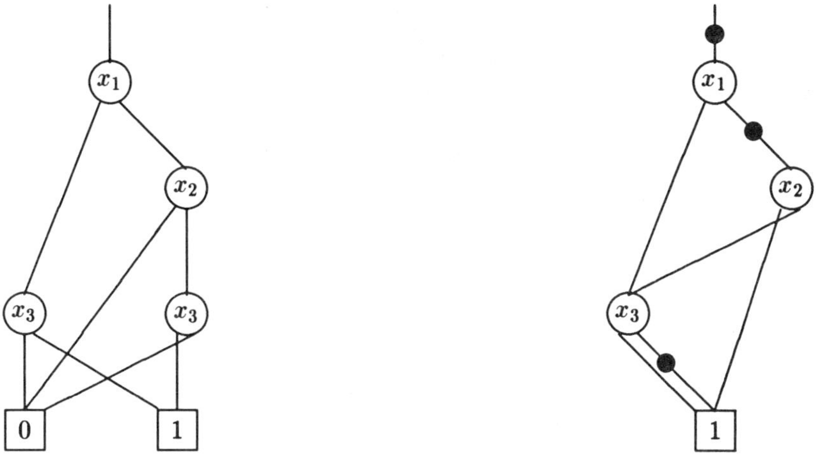

Figure 7.2.3 Reduced OKFDDs with DTL **d** $= (nD, pD, S)$ for $f(x_1, x_2, x_3) = x_3 \oplus \overline{x}_1 x_2 \overline{x}_3$: (1) without complement edges and (2) with complement edges.

1, the representation remains unique [24]. (For illustration see Figure 7.2.4.) It follows:

Corollary 7.2.1 For given variable ordering and DTL, reduced OKFDDs with CEs are canonical representations for Boolean functions.

In Figure 7.2.3 (righthand side) the canonical reduced OKFDD with DTL **d** $=$ (nD, pD, S) and CEs is given for the function $f(x_1, x_2, x_3) = x_3 \oplus \overline{x}_1 x_2 \overline{x}_3$. If not explicitly stated otherwise, all OKFDDs considered in the following are reduced and have CEs.

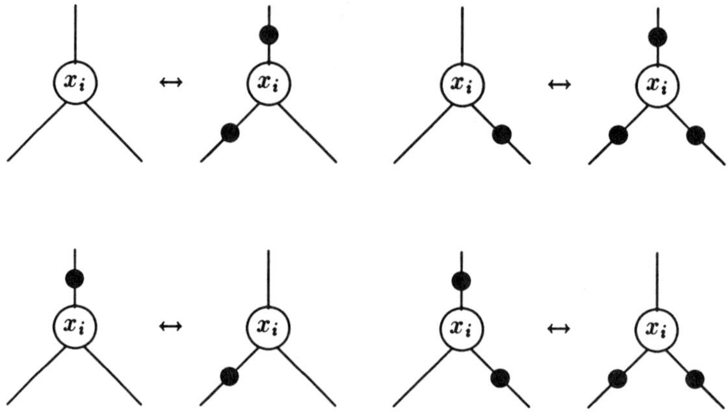

Figure 7.2.4 Complement edges: equivalent pairs.

7.3 BASIC ALGORITHMS ON OKFDDS

7.3.1 Evaluation and Satisfiability

Evaluation of a Boolean function for a given assigment is one of the basic algorithms that should be supported by any data structure for the representation of Boolean functions.

Assume that a Boolean function f is given by an OKFDD G with DTL $\mathbf{d} = (d_1, d_2, \ldots, d_n)$. The value $f(\mathbf{a})$ for an assignment $\mathbf{a} = (a_1, a_2, \ldots, a_n)$ of the variables $x_1, \ldots, x_n$ can be determined on G as follows:

We call edges and paths *(**d**)-active* for **a** if they are considered for the computation of $f(\mathbf{a})$. Depending on the decomposition type of a node one or both outgoing edges are (**d**)-active: Let v be any non-terminal node of G labeled with a variable x_i. The edge $(v, low(v))$ is called *(**d**)-active* for **a**, iff $[d_i \in \{pD, nD\}]$ or $[d_i = S$ and $a_i = 0]$. The edge $(v, high(v))$ is *(**d**)-active* for **a**, iff $[d_i \in \{S, pD\}$ and $a_i = 1]$ or $[d_i = nD$ and $a_i = 0]$. For illustration see Table 7.3.1.

A path in G is called *(**d**)-active* for **a** iff it leads from the root to a terminal node and only contains (**d**)-active edges. The value of a (**d**)-active path is given

Table 7.3.1 (**d**)-active edges at node v for assignment **a**.

node type d_i	(**d**)-active edges for $a_i = 0$	(**d**)-active edges for $a_i = 1$
S	*low*-edge	*high*-edge
pD	*low*-edge	both edges
nD	both edges	*low*-edge

by the value of the corresponding terminal node. (In the case of OKFDDs with CEs the value has to be complemented if the number of CEs on the path is odd.)

It follows easily by induction on the number of variables that $f(\mathbf{a})$ is given by XOR-ing the values of all (**d**)-active paths. Furthermore, it can be shown, that $f(\mathbf{a})$ is computable by a linear time traversal of G.

Vice versa, given a Boolean function f, one may be interested in knowing a satisfying assignment of f, i.e., there should exist an algorithm, that decides whether f is non-zero and in the positive case computes an assignment **a** with $f(\mathbf{a}) = 1$.

For OKFDDs, using the canonicity of the representation it follows directly that the function represented by an OKFDD G is non-zero, iff G is not the terminal 0. A satisfying assignment can be easily computed in time linear in the number of variables by following the "leftmost" path in G that leads to a non-zero subfunction. For some applications it is also interesting to compute the number of satisfying assignments (SATCOUNT). For OBDDs this task can be solved again by a linear time traversal of the OBDD. On the contrary, it can be shown, that already for OpFDDs SATCOUNT is #P-complete [42]. (#P is the "equivalent" of NP for counting problems.) We summarize:

Theorem 7.3.1 Let G be a reduced OKFDD with DTL **d** for the function f. Then it holds:

1. Evaluation of G for a given assignment $\mathbf{a} = (a_1, \ldots, a_n)$ can be performed in $O(|G|)$.

2. A satisfying assignment for f can be computed in time $O(n)$, if $f \neq 0$. The computation of the number of satisfying assignments is #P-complete.

```
kfdd_xor_kfdd (F,G) {
 if (terminal case) {
    return result;
 } else if (computed-table has entry (F,G)) {
    return result;
 } else {
    let v be the top node of (F,G);
    low(v) = kfdd_xor_kfdd (F_low(v), G_low(v));
    high(v) = kfdd_xor_kfdd (F_high(v), G_high(v));
    if Shannon(v) {
      if (high(v) == low(v)) return low(v);
    } else {
      if (high(v) == 0) return low(v);
    }
    R = find_or_add_unique_table (v,low(v),high(v));
    insert_computed_table (F,G,R);
    return R;
  }
}
```

Figure 7.3.1 Algorithm for XOR-operation.

7.3.2 Synthesis Operations

We now present synthesis algorithms for OKFDDs with fixed variable ordering and a fixed DTL.

First, the XOR-operation is presented as it provides the basis for certain other operations. Notice that for two functions, f and g, decomposed by positive Davio expansion, one has:

$$\begin{aligned} f \oplus g &= (f_0 \oplus x_i f_2) \oplus (g_0 \oplus x_i g_2) \\ &= (f_0 \oplus g_0) \oplus x_i (f_2 \oplus g_2). \end{aligned}$$

This equation makes it possible to recursively split up a positive Davio-node into its left and right subgraphs. The algorithm for negative Davio-nodes is performed analogously. This provides an efficient algorithm for Davio-nodes while the basic XOR-operation for Shannon-nodes is based on the following equation:

$$f \oplus g \;=\; \overline{x}_i(f_0 \oplus g_0) \vee x_i(f_1 \oplus g_1).$$

The resulting algorithm for XOR-operation on two OKFDDs (without CEs) is sketched in Figure 7.3.1. Analogously to the presentation in [8] it follows that the algorithm has polynomial worst case behavior.

The realization of the AND-operation turns out to be more complicated for Davio-nodes in comparison to the XOR-operation. The following recursive equation holds for positive Davio-nodes:

$$\begin{aligned} f \cdot g &= (f_0 \oplus x_i f_2) \cdot (g_0 \oplus x_i g_2) \\ &= (f_0 \cdot g_0) \oplus x_i((f_2 \cdot g_2) \oplus (f_0 \cdot g_2) \oplus (g_0 \cdot f_2)). \end{aligned}$$

This equation again defines a recursive algorithm similar to the one from Figure 7.3.1, but this time with exponential worst case running time: In [5] an example has been constructed, where the AND-synthesis of two polynomial OpFDDs results in an OpFDD of exponential size. Since OKFDDs are a superset of OpFDDs this implies the result also for OKFDDs. By application of DeMorgan's rule an analogous result follows for the OR-synthesis, since the computation of the complement can be carried out efficiently (see below).

However, for OKFDDs with a constant number of levels, where the Davio expansion is performed, the AND-operation remains polynomial. A simple method to show this is as follows: The OKFDD is transformed to an OBDD. (Due to the fact that only a constant number of operations have to be performed and each transformation requires only polynomial time (see next subsection) this can be done efficiently.) The AND-operation on OBDDs has polynomial worst case behavior and after the AND-operation the OBDD is transformed back to an OKFDD (again in polynomial time).

The negation of a function f can be computed by observing that $\overline{f} = 1 \oplus f$. Thus, the operation requires an XOR-operation with the constant 1. If CEs are used, negation can be performed even more efficiently, simply by setting a CE.

Now, using the algorithms for the XOR-, AND-, and NOT-operations, any binary operation can be realized.

For an OKFDD G the restriction $G|_{x_i=c}$ for variable x_i and constant c can be computed by traversing the graph and performing the corresponding substitutions. The case for Shannon-nodes is given in [7]. For the case of positive Davio-nodes, if $x_i = 0$, edges from nodes v with label x_i to $high(v)$ have to be deleted and all edges from predecessors of v to v itself are redirected to the low-successor of v. If nodes with indegree 0 result, they and their outgoing edges are also deleted. Clearly, all this can be done in linear time. If $x_i = 1$, then at each node v with label x_i and subfunctions $f_{low(v)}$ and $f_{high(v)}$, the following has to be done. All edges from predecessors of v to v itself have to be redirected to a node being the root node of an OKFDD for $f_{low(v)} \oplus f_{high(v)}$. We therefore perform an XOR-operation on the successors of v and, as before, if nodes with indegree 0 result, they and their outgoing edges are deleted. Due to the complexity of the XOR-operation, we obtain an algorithm with quadratic complexity in this case. For negative Davio-nodes, a similar procedure is required.

All in all, we obtain for the complexity of the synthesis operations:

Theorem 7.3.2 Let G_1 and G_2 be two OKFDDs with the same variable ordering and the same DTL **d** for the functions f_1 and f_2. Then it holds:

1. The negation of $f_i (i = 1, 2)$ can be performed in constant time.
2. An OKFDD for $f_1 \oplus f_2$ can be computed in time and space $O(|G_1| \cdot |G_2|)$.
3. An algorithm for the computation of $f_1 \cdot f_2$ and $f_1 \vee f_2$, respectively, has exponential worst case runtime (independant of the implementation). If the number of Davio-levels is constant the runtime of the algorithm becomes polynomial.
4. The restriction $G|_{x_i=c}$ for a variable x_i and a constant c can be performed efficiently for OKFDDs:
   (a) For Shannon-nodes the algorithm has complexity $O(|G|)$.
   (b) For positive (negative) Davio-nodes and $c = 0$ $(c = 1)$ the algorithm has complexity $O(|G|)$.
   (c) For positive (negative) Davio-nodes and $c = 1$ $(c = 0)$ the algorithm has complexity $O(|G|^2)$.

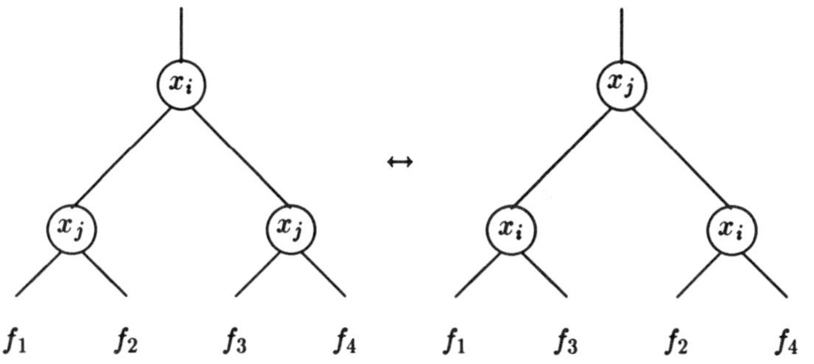

Figure 7.3.2 Exchange of i-th and adjacent variable.

7.3.3 Basic Operations for Optimization of Variable Ordering and DTL

While the variable ordering plays a dominant role in the identification of the minimal OBDD representation of the functions, in OKFDDs both the ordering and the decomposition type are important. Depending on the order of the variables and the particular decomposition type among the possible three, the size of the OKFDD can vary from linear to exponential [5, 4].

Exchange of Neighbouring Variables

It is well-known that in the case of OFDDs and OBDDs, the size of the DD can be minimized by the exchange of adjacent variables. This idea can be extended to OKFDDs: The general case for the exchange of a variable i and an adjacent variable j is shown in Figure 7.3.2. Notice that the exchange pattern is independent of the decomposition types of the nodes. The exchange is performed very quickly since only edges must be redirected. CEs can also be used. Thus, it is possible to use all techniques based on the exchange of adjacent variables also for OKFDDs.

Change of Decomposition Type

It is easy to see that any change of the decomposition type of a single variable can be performed by an XOR-operation on the successors. Due to the results of the previous subsection this implies an algorithm with time and space complexity $O(|G|^2)$.

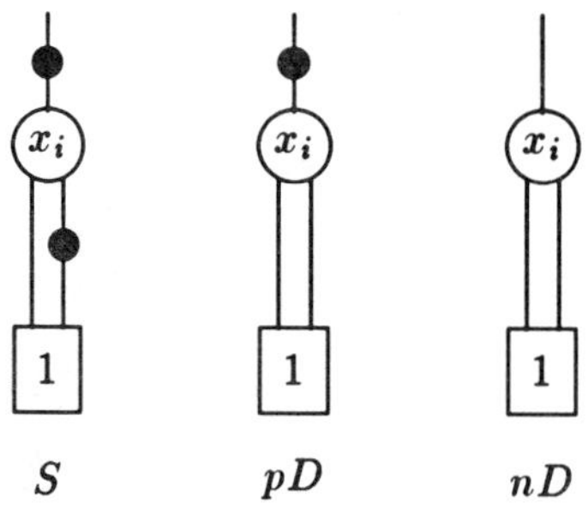

Figure 7.3.3 Different cases for bottom level.

In the following, we briefly discuss another method that is based on reordering [15]: Variable x_i is moved to the bottom level of the OKFDD by exchange of neighbouring variables. In the bottom level there exists exactly one node (due to CEs). The different cases for Shannon, positive and negative Davio nodes are shown in Figure 7.3.3 for the function x_i. The decomposition type of this single node can easily be changed. In some cases an additional DFS-run must be used to restore the canonicity of the OKFDD, i.e., the labels at the edges have to be changed. (Notice, that the DFS-run is only needed if a transformation from (or to) negative Davio is performed.) Experiments have shown that this method is superior to the simple method based on XOR-operations of the successors. On average the new method is two times faster.

7.4 IMPLEMENTATION OF AN OKFDD PACKAGE

The methods described above have been implemented as the OKFDD package PUMA[1] which has been used in all applications and experiments discussed in the following sections.

[1]PUMA is available from the authors by *ftp*. For more details contact the authors under <name>@informatik.uni-freiburg.de.

7.4.1 Technical Details

At first, some programming techniques and methods of implementation used to speed up the package are described. The methods are similar to other packages used for representation and manipulation of OBDDs and OFDDs (see e.g. [7, 37, 3]). Hence, these techniques are only briefly reviewed.

For fast reference to functions, a *hash-based unique table* is used to store the nodes. A *computed table* is implemented for the optimization of the synthesis algorithms.

Memory management is done by *garbage collection.* The nodes are only deleted if the storage place is needed for other nodes. Thus, it is not necessary to recompute the results each time if they were used earlier on. By the unique table, different OKFDDs can *share* the same sub-OKFDDs. Therefore, several functions can efficiently be represented at the same time.

For the realization of CEs the storage of an additional bit is necessary. As suggested in [7, 15] the overhead can be saved if the lowest bit of each pointer is used on machines where this is allowed.

7.4.2 Features of the Package PUMA

The most important features of the package are:

1. The package supports BLIF as standard input format.
2. Several methods for finding good variable orderings and decomposition types can be used:

   Exact minimization: The algorithm for exact minimization of OKFDDs computes an OKFDD with a minimum number of non-terminal nodes (including CEs). The algorithm can be restricted to a single fixed DTL. (For only Shannon-nodes this results in the exact algorithm for OBDD-minimization presented in [27].)

   Heuristic minimization: For the construction of OKFDDs from standard BLIF-files, the package uses *variable interleaving* [28] combined with heuristics for decomposition type choice. Interleaving has been developed for OBDDs but has proven to also work well for OKFDDs [21].

Several methods for dynamic variable ordering based on the basic algorithms of Subsection 7.3.3 are supported, like *sifting* and *window permutation.* Additionally, operations to set an arbitrary variable ordering and/or decomposition type can be used. This allows the user to integrate problem specific ordering methods.

The package not only allows dynamic variable ordering with an upper limit on the absolute number of nodes in the package but also with respect to a growing factor between to subsequent steps.

3. *Zero-suppressed OBDDs* (ZBDDs) [36] are integrated.

4. The package supports an interactive interface.

7.5 APPLICATIONS AND EXTENSIONS

In this section we present some applications of OKFDDs. We performed numerous experiments to demonstrate the efficiency of our approach. All experiments were carried out on the package PUMA. All runtimes are given in CPU seconds on an *HP 9000/710* workstation.

7.5.1 Minimization of Representation Size by Dynamic Variable Ordering

Application of DDs, most notably of OBDDs, in combinational (and sequential) logic verification has long been known [35, 41]. As DDs provide a canonical representation of the functions, the isomorphism checking operation is trivial. On the other hand, a compact representation of the considered functions is essential for the applicability of the concept to real-world examples [32].

To demonstrate the advantages of OKFDDs, we developed several algorithms to minimize the representation size, among them being dynamic variable reordering by sifting.

In the last few years several methods for dynamic variable ordering for OBDDs have been presented and intensively been studied [29, 38]. The most promising approach is the *sifting algorithm* [38]. A comparison between sifting for OBDDs and OKFDDs is given in Table 7.5.1. *in* (*out*) denotes the number of inputs (outputs) of circuit *name*. In column OBDD the OBDD sizes from [38] are

Table 7.5.1 Comparison of sifted OKFDDs and OBDDs.

name	in	out	OBDD	OKFDD
C432	36	7	1.2	1.1
C499	41	32	44.8	13.4
C880	60	26	9.1	4.0
C1355	41	32	36.2	13.4
C1908	33	25	12.4	3.8
C2670	233	140	6.6	1.4
C3540	50	22	27.2	22.4
C5315	178	123	3.1	1.3
C7552	207	108	8.2	3.0
s1423	91	79	5.7	1.3
pair	173	137	4.5	1.9
rot	137	107	5.0	3.2
total			164.0	70.2

given. The column OKFDD denotes the sizes of the OKFDDs using the method described in [15]. (All sizes are given in units of 1000 nodes.) Overall, the representation size could be reduced by more than 50 % on average.

7.5.2 OKFDD-Circuits - FPGA Technology Mapping and Synthesis for Testability

In the following we demonstrate how OKFDDs can be transformed into multi-level circuits. These circuits can be viewed as starting point for FPGA technology mapping and furthermore have good testability properties.

In general, a *Combinational Logic Circuit* (CLC) is defined over a fixed library and modeled as a directed acyclic graph $C = (V, E)$ with some additional properties. Very often a standard library consisting of *Primary Input* (PI) and *Primary Output* (PO) ports, the 2-input, 1-output *AND, OR* gate and the 1-input, 1-output inverter *NOT* is used. Additionally, we consider the 2-input, 1-output *XOR* gate and at first construct three cells realizing the Shannon, positive Davio and negative Davio decomposition, respectively. For an illustration see Figure 7.5.1.

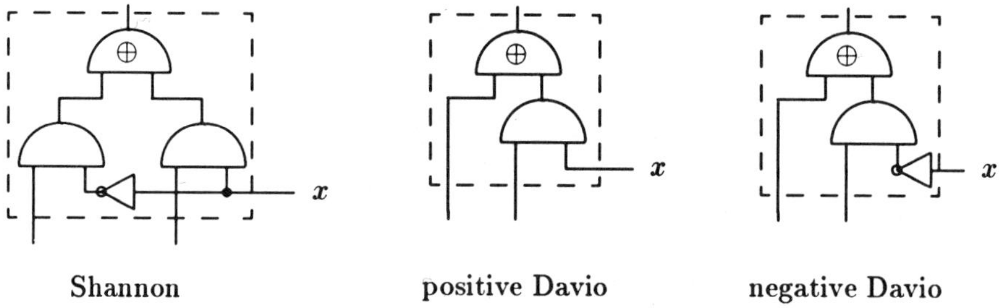

Figure 7.5.1 Circuit realizations of the S, pD and nD-nodes.

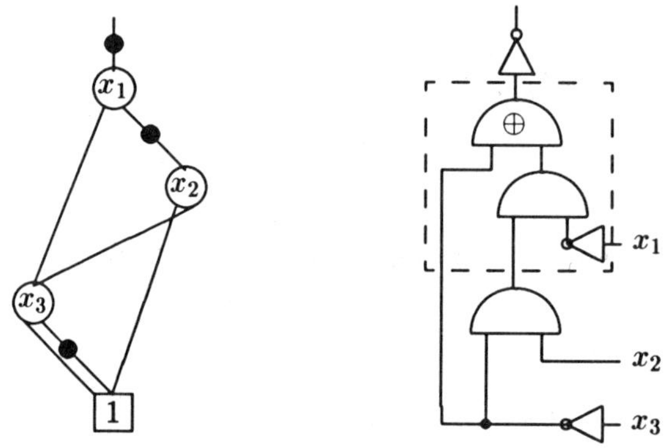

Figure 7.5.2 OKFDD and corresponding OKFDD-circuit.

The edge corresponding to the *low*-successor (*high*-successor) is denoted as the 0-input (1-input). The 0-input and the 1-input are also called data inputs. If one (or both) of the data inputs have constant values 0 or 1, or if the data inputs compute identical (complementary) values for all primary input combinations, the cells in Figure 7.5.1 can easily be simplified.

An OKFDD-circuit can be obtained from an OKFDD by traversing the corresponding graph in topological order and replacing all nodes by the corresponding cells from the library. For an example see Figure 7.5.2.

Table 7.5.2 Results for FPGA design.

name	in	out	SPW	dynamic	time
cc	21	20	117	38	0.1
c8	28	18	237	62	0.2
frg1	28	3	186	77	1.2
pdc	16	40	1118	564	3.5
spla	16	46	724	550	3.2
vg2	25	8	204	76	1.0
total			2586	1367	9.2

Using circuits derived from DDs becomes also more and more popular in the area of technology mapping for *Field Programmable Gate Arrays* (FPGAs). There DD based approaches have also been integrated in commercial FPGA tools [33].

In [40] a method for FPGA design using OKFDDs has been presented. We compare our results for larger circuits with the most powerful heuristic from [40]. The results are given in Table 7.5.2. The results from [40] measured in number of nodes are given in column SPW. The results obtained by dynamic variable ordering and the corresponding execution times are given in the last two columns. As can easily be seen our approach obtains much better results in all considered cases in negligible time. For some benchmarks more than a 50% improvement is obtained. Obviously this simplifies the mapping process tremendously.

Furthermore, testability properties of circuits derived from OKFDDs have been studied with respect to the *Stuck-At Fault Model* and the *Cellular Fault Model* [16, 2]. The computation of complete test sets and of all occuring redundancies can be done easily and efficiently due to the synthesis operations on OKFDDs. Circuits with high testability can be obtained.

7.5.3 Two-level AND/XOR Forms Computed from OKFDDs

OKFDDs can be viewed as a compressed form of two-level AND/XOR expressions: Let G be a (not necessarily reduced) OKFDD with DTL $\mathbf{d}$ [19]. The corresponding two-level AND/XOR expression *2okfdd*$(G, \mathbf{d})$ is determined by recursively traversing the OKFDD and enumerating all 1-paths as follows:

> A uniquely defined product of literals over X_n, called the *term* t_P *of* P is associated. to each 1-path P of G (= a path from the root of G to the the terminal 1). If e is an edge on P originating in a node v with label x_i, then x_i is in the product t_P, iff v is either a Shannon node or a positive Davio node and the endpoint of e is $high(v)$. $\bar{x}_i$ is in t_P, iff either v is a Shannon node and the endpoint of v is $low(v)$ or v is a negative Davio node and the endpoint of e is $high(v)$. In all other cases the edge e contributes 1 to t_P.

2okfdd$(G, \mathbf{d})$ then is the two-level expression obtained by XOR-ing all terms t_P, where P is a 1-path in G. It follows by induction, that *2okfdd*$(G, \mathbf{d})$ is a two-level expression for $f_G^{\mathbf{d}}$, the function represented by the OKFDD G and DTL $\mathbf{d}$.

The number of terms in *2okfdd*$(G, \mathbf{d})$ is called the *size* of *2okfdd*$(G, \mathbf{d})$. We immediately conclude, that the number of 1-paths in G is equal to the size (= number of terms) of $2okfdd(G, \mathbf{d})$ and that the number of nodes in G is less or equal to the number of literals in $2okfdd(G, \mathbf{d})$. Obviously, the number of 1-paths in an OKFDD and thus the number of terms in the corresponding two-level AND/XOR form can be computed in time linear in the size of the OKFDD by a simple DFS algorithm.

The choice of the DTL (and in general also the variable ordering) in the OKFDD largely influences the number of terms of the corresponding two-level expression, i.e., the number may vary from constant to exponential, as is shown by the following example [39]:

Example 7.5.1 Let $f : \mathbf{B}^n \to \mathbf{B}$, given by $f = \overline{x}_1\overline{x}_2 \ldots \overline{x}_n \vee x_1x_2 \ldots x_n$. If all nodes in the reduced OKFDD are Shannon nodes the corresponding two-level expression has only two terms. But if the OKFDD has only Davio nodes (positive or negative) the corresponding two-level expression has an exponential number of terms.

We briefly summarize two-level AND/XOR classes resulting from different classes of OKFDDs. In most applications, OKFDDs are used in reduced form. Nevertheless, also OKFDDs that are not fully reduced, i.e., OKFDDs where the reductions of type S and D are not always performed can be considered. It is also possible to restrict the use of decomposition types.

Also for these classes of OKFDDs the corresponding two-level AND/XOR expression can be defined resulting in the following classes:

GKROs: *Generalized Kronecker Expressions* correspond to general (not necessarily reduced) OKFDDs.

KROs: *Kronecker Expressions*, a class of two-level AND/XOR expressions already considered in [14, 39], result from complete OKFDDs.

RKROs: *Reduced Kronecker Expressions* are given by reduced OKFDDs.

FPRMs: The well-known *Fixed Polarity Reed Muller Expressions* result from OKFDDs with only Davio nodes.

Using these definitions we directly conclude

$$FPRMs \subset RKROs \subset GKROs,$$

$$FPRMs \subset KROs \subset GKROs.$$

We finish this subsection with some comments on the minimization of two-level AND/XOR forms based on OKFDDs. The size of all forms depends on the choice of the DTL and for RKROs and GKROs on the choice of the variable ordering as well. Basic algorithms for changing variable ordering and DTLs have been introduced in previous sections. They have been very successfully used to device minimization algorithms for AND/XOR forms based on OKFDDs [25, 18, 20].

7.5.4 K∗BMDs - An Extension of OKFDDs to the Word-Level

Several types of DDs have been proposed in the CAD area, among them being OBDDs, OFDDs and OKFDDs as they have been defined in the previous sections.

While the aforementioned types of DDs are suitable for representing Boolean functions at the bit-level and have proved useful for many applications in CAD, recently, DDs to represent functions from the Boolean domain into the integers attract more and more interest. Representatives of this type of DDs are *Multi Terminal BDDs* (MTBDDs) [12], *Edge-Valued BDDs* (EVBDDs) [34], *Multiplicative Binary Moment Diagrams* (*BMDs) [10] and *Hybrid DDs* (HDDs) [11].

All these word-level DDs are also graph-based representations where the underlying graph is ordered. MTBDDs [12] and EVBDDs [34] make use of the (integer-valued) Shannon decomposition and allow terminal nodes labeled with integer values. Additionally, in the case of EVBDDs a constant a is added to the function being represented. Thus, in the EVBDD an edge with weight a to a node v labeled with variable x represents the function $< a, f >= a+(1-x)f_{low(v)}+xf_{high(v)}$. (As before f is the function represented at node v; $f_{low(v)}$ ($f_{high(v)}$) denotes the function represented by the *low*-edge (*high*-edge) of v. $-$ and $+$ denote the usual operations in the ring $\mathbf{Z}$.)

BMDs and *BMDs [10] make use of the (integer-valued) positive Davio decomposition and allow terminal nodes labeled with integer values. Additionally, in the case of *BMDs the values at the edges are multiplied with the functions represented. Thus, an edge with weight m to a node v in a *BMD represents the function $< m, f >= m(f_{low(v)} + xf_{high(v)})$.

Using *BMDs it was for the first time possible to verify multipliers of bit length up to $n = 256$. Unfortunately, *BMDs fail for the representation of Boolean functions that can easily be represented using OBDDs [26]. Thus, extensions of *BMDs are of interest. A first step in this direction has been proposed in [11, 13]. There BMDs are combined with MTBDDs. The resulting DDs, called HDDs in [11], are based on integer-valued generalizations of Shannon and Davio decompositions. They combine the advantages of MTBDDs and BMDs in a similar way as OKFDDs combine OBDDs and OFDDs. We therefore prefer to call them Kronecker BMDs (KBMDs) instead of HDDs. With KBMDs it is possible, for example, to handle functions related to data bits as BMDs and those related to control signals as MTBDDs.

*Kronecker *BMDs* (K*BMDs) [22] also use integer-valued generalizations of Shannon and Davio decompositions and in addition allow edge-values as possible with *BMDs and EVBDDs.

Analogously to OKFDDs at the bit-level different decomposition types per variable can be used. Since we consider integer-valued functions, many differing decomposition types are possible. They can be defined by the set $\mathbf{Z}_{2,2}$ of non singular 2×2 matrices over $\mathbf{Z}$ [11]. As for OKFDDs decomposition types are associated to the n Boolean variables with the help of a *Decomposition Type List* (DTL) $\mathbf{d} = (d_1, \ldots, d_n)$ where $d_i \in \mathbf{Z}_{2,2}$, i.e., for each variable one fixed decomposition is chosen. Following [11] it can be shown that the matrices corresponding to Shannon (S), positive Davio (pD) and negative Davio (nD), respectively, are

$$\begin{pmatrix} 1 & 0 \\ 0 & 1 \end{pmatrix} \begin{pmatrix} 1 & 0 \\ -1 & 1 \end{pmatrix} \begin{pmatrix} 0 & 1 \\ 1 & -1 \end{pmatrix}.$$

K*BMDs differ from KBMDs in the fact that they allow the use of additive **and** multiplicative weights in parallel, i.e., we make use of the following type of representation:

$$< (a, m), f >= a + mf.$$

It follows that in the case of Shannon decomposition, positive and negative Davio decomposition the function represented at a node v is given by

$$\begin{aligned} < (a,m), f > &= a + m((1-x)f_{low(v)} + xf_{high(v)}) \\ < (a,m), f > &= a + m(f_{low(v)} + xf_{high(v)}) \\ < (a,m), f > &= a + m(f_{low(v)} + (1-x)f_{high(v)}). \end{aligned}$$

To make K*BMDs a canonical representation, we need some further restrictions on the graph with respect to the edge values. Basically we require that there exists only one terminal and this terminal is labeled 0, and that the *low*-edge of a node has additive weight 0 and that the *greatest common divisor* of the remaining weights is 1.

We obtain a data structure, that is a generalization of *BMDs and OBDDs as well. Thus, on the one hand it is possible to represent functions efficiently, that have a good word-level description, on the the other hand K*BMDs are also applicable to verification problems at the bit-level.

In [23] experimental results for bit-level and word-level representation with K*BMDs in comparison to other types of DDs are given. We review some of these experiments in Tables 7.5.3 and 7.5.4.

In Table 7.5.3 sizes of OBDDs, *BMDs and K*BMDs for bit-level representation of some benchmarks from ISCAS85 and LGSynth93 are given. The results were obtained using dynamic variable reordering during the construction. The best results are given in bold. A dash symbolizes that the corresponding graph

Table 7.5.3 DD sizes for bit-level representation of some benchmarks.

name	OBDD	*BMD	K*BMD
C432	**1209**	-	**1209**
C499	**29561**	-	**29561**
C880	7718	-	**4048**
C1355	**29561**	-	**29561**
C1908	**5028**	-	5944
C2670	4093	-	**3939**
C5315	**2270**	-	2504
des	**3022**	4425	**3022**
pair	3210	4833	**2911**
rot	7702	11191	**5808**

Table 7.5.4 DD sizes for word-level representation of multipliers.

name	OBDD	*BMD	K*BMD
mult-4	134	**8**	**8**
mult-8	10283	**16**	**16**
mult-16	-	**32**	**32**
mult-30	-	**60**	**60**

could not be constructed within a node limit of 100,000 nodes. As can easily be seen the K*BMDs in Table 7.5.3 are nearly **always** smaller. (For only two examples the sizes are slightly larger than the OBDD size.) For none of the ISCAS85 benchmarks the *BMD could be constructed, since too many nodes were needed during the synthesis procedures.

In Table 7.5.4 results for word-level representation of multipliers are given. For OBDDs only circuits with a small number of inputs could be constructed while the word-level representation as (K)*BMDs only requires $2n$ nodes where n is the bit length of the mulitplier.

All in all, seen over the whole set of benchmarks considered, K*BMDs are the only data structure that can represent **all** functions efficiently.

7.6 CONCLUSIONS

We introduced *Ordered Kronecker Functional Decision Diagrams* (OKFDDs), a graph-based data structure for the representation and manipulation of Boolean functions. OKFDDs are a generalization of OBDDs and OFDDs as well and as such provide a more compact representation of the functions than either of the two DDs.

We reviewed basic properties of OKFDDs and in particular discussed methods for dynamic variable ordering. Using these methods the compactness of OKFDDs versus OBDDs (and OFDDs) was confirmed through experimental results. All experiments were performed with the OKFDD package PUMA whose implementation was briefly discussed.

Then we pointed out some applications of OKFDDs in the area of *technology mapping for FPGAs, synthesis for testability* and *synthesis of two-level AND/XOR expressions.*

The canonicity of the OKFDDs and efficient construction and manipulation techniques presented here make them a prime candidate for utilization in applications where OBDDs have been the main construct.

Finally we discussed K*BMDs, a promising extension of the concept to also allow the representation and manipulation of word-level functions. Applications in verification of combinational and sequential circuits are among those where K*BMDs should be useful. In combination with OKFDDs, they incorporate the features of many other ordered DD types and allow the dynamic use of the appropriate DD depending on the specific application.

Acknowledgement

The authors like to thank Andreas Hett, Konrad Nowak and Stefan Ruppertz for their help with the realization of the methods discussed in this chapter. We also acknowledge the helpful discussions with Michael Theobald, Harry Hengster, Andisheh Sarabi and Marek A. Perkowski.

This work was supported by DFG Grant Be 1176/4.

REFERENCES

[1] B. Becker and R. Drechsler, "How many decomposition types do we need?," *Proc. European Design & Test Conf.*, pp. 438–443, Mar. 1995.

[2] B. Becker and R. Drechsler, "Synthesis for testability: Circuits derived from ordered kronecker functional decision diagrams," *Proc. European Design & Test Conf.*, page 592, Mar. 1995.

[3] B. Becker, R. Drechsler, and M. Theobald, "On the implementation of a package for efficient representation and manipulation of functional decision diagrams," *IFIP WG 10.5 Workshop on Applications of the Reed-Muller Expansion in Circuit Design*, pp. 162–169, Sept. 1993.

[4] B. Becker, R. Drechsler, and M. Theobald, "OKFDDs versus OBDDs and OFDDs," *ICALP, LNCS 944*, pp. 475–486, July 1995.

[5] B. Becker, R. Drechsler, and R. Werchner, "On the relation between BDDs and FDDs," *LATIN, LNCS 911*, pp. 72–83, Apr. 1995.

[6] B. Bollig, P. Savicky, and I. Wegener, "On the improvement of variable orderings for OBDDs," *IFIP Workshop on Logic and Architecture Synthesis*, pp. 71–80, Dec. 1994.

[7] K.S. Brace, R.L. Rudell, and R.E. Bryant, "Efficient implementation of a BDD package," *Proc. Design Automation Conf.*, pp. 40–45, June 1990.

[8] R.E. Bryant, "Graph - based algorithms for Boolean function manipulation," *IEEE Trans. Comput.*, vol. C-35, pp. 677–691, Aug. 1986.

[9] R.E. Bryant, "Symbolic boolean manipulation with ordered binary decision diagrams," *ACM, Comp. Surveys*, 24:293–318, 1992.

[10] R.E. Bryant and Y.-A. Chen, "Verification of arithmetic functions with binary moment diagrams," *Proc. Design Automation Conf.*, pp. 535–541, June 1995.

[11] E.M. Clarke, M. Fujita, and X. Zaho, "Application of multi-terminal binary decision diagrams," *IFIP WG 10.5 Workshop on Applications of the Reed-Muller Expansion in Circuit Design*, pp. 21–27, Aug. 1995. (Also Chapter 4 of this book).

[12] E.M. Clarke, K. McMillan, X. Zhao, M. Fujita, and J. Yang, "Spectral transforms for large boolean functions with application to technology mapping," *Proc. Design Automation Conf.*, pp. 54–60, June 1993.

[13] E.M. Clarke and X. Zaho, "Word level symbolic model checking - a new approach for verifying arithmetic circuits," Technical report, CMU-CS-95-161, May 1995.

[14] M. Davio, J.P. Deschamps, and A. Thayse, *Discrete and Switching Functions*. McGraw-Hill, 1978.

[15] R. Drechsler and B.Becker, "Dynamic minimization of OKFDDs," *Proc. Int'l Conf. on Comp. Design*, pp. 602-607, Oct. 1995.

[16] R. Drechsler and B. Becker, "Rapid prototyping of fully testable multi-level AND/EXOR networks," *IFIP WG 10.5 Workshop on Applications of the Reed-Muller Expansion in Circuit Design*, pp. 126–133, Sept. 1993.

[17] R. Drechsler and B. Becker, "Sym*pathy*: Fast exact minimization of Fixed Polarity Reed-Muller expressions for symmetric functions," *Proc. European Design & Test Conf.*, pp. 91–97, Mar. 1995.

[18] R. Drechsler, B. Becker, and N. Göckel, "A genetic algorithm for minimization of Fixed Polarity Reed-Muller expressions," *Proc. Int'l Conf. on Artificial Neural Networks and Genetic Algorithms*, pp. 392–395, Feb. 1995.

[19] R. Drechsler, B. Becker, and N. Göckel, "A genetic algorithm for 2-Level AND/EXOR minimization," *Proc. SASIMI*, pp. 49–56, Aug. 1995.

[20] R. Drechsler, B. Becker, and N. Göckel, "A genetic algorithm for RKRO minimization," *Proc. Int'l Symp. on Artificial Intelligence*, pp. 266-275, Oct. 1995.

[21] R. Drechsler, B. Becker, and A. Jahnke, "On variable ordering and decomposition type choice in OKFDDs," *Proc. IFIP International Conference on VLSI'95*, pp. 805–810, Aug. 1995.

[22] R. Drechsler, B. Becker, and S. Ruppertz, "K*BMDs: a new data structure for verification," *Proc. European Design & Test Conf.*, Mar. 1996.

[23] R. Drechsler, B. Becker, and S. Ruppertz, "Dynamic Minimization of K*BMDs," submitted, 1995.

[24] R. Drechsler, A. Sarabi, M. Theobald, B. Becker, and M.A. Perkowski, "Efficient representation and manipulation of switching functions based on Ordered Kronecker Functional Decision Diagrams," *Proc. Design Automation Conf.*, pp. 415–419, June 1994.

[25] R. Drechsler, M. Theobald, and B. Becker, "Fast OFDD based minimization of Fixed Polarity Reed-Muller expressions," to be published in *IEEE Trans. Comput.*, 1996.

[26] R. Enders, "Note on the complexity of binary moment diagram representations," *IFIP WG 10.5 Workshop on Applications of the Reed-Muller Expansion in Circuit Design*, pp. 191–197, Aug. 1995.

[27] S.J. Friedman and K.J. Supowit, "Finding the optimal variable ordering for binary decision diagrams," *Proc. Design Automation Conf.*, pp. 348–356, June 1987.

[28] H. Fujii, G. Ootomo, and C. Hori, "Interleaving based variable ordering methods for ordered binary decision diagrams," *Proc. Int'l Conf. on CAD*, pp. 38–41, Nov. 1993.

[29] M. Fujita, Y. Matsunga, and T. Kakuda, "On variable ordering of binary decision diagrams for the application of multi-level synthesis," *Proc. European Conf. on Design Automation*, pp. 50–54, Mar. 1991.

[30] P. Ho and M.A. Perkowski, "Free kronecker decision diagrams and their application to Atmel 6000 FPGA mapping," *Proc. European Design Automation Conf.*, pp. 8–13, Sept. 1994.

[31] U. Kebschull, E. Schubert, and W. Rosenstiel, "Multilevel logic synthesis based on functional decision diagrams," *Proc. European Conf. on Design Automation*, pp. 43–47, Mar. 1992.

[32] D.P. Appenzeller and A. Kuehlmann, "Formal verification of a PowerPc microprocessor," *Proc. Int'l Conf. on Comp. Design*, pp. 79-84, Oct. 1995.

[33] V.V. Le, T. Besson, A. Abbara, D. Brasen, H. Bogushevitsh, G. Saucier, and M. Crastes, "ASIC prototyping with area oriented mapping for ALTERA/FLEX devices," *Proc. SASIMI*, pp. 176–183, Aug. 1995.

[34] Y.-T. Lai and S. Sastry, "Edge-valued binary decision diagrams for multi-level hierarchical verification," *Proc. Design Automation Conf.*, pp. 608–613, June 1992. (Also Chapter 5 of this book).

[35] S. Malik, A.R. Wang, R.K. Brayton, and A.L. Sangiovanni-Vincentelli, "Logic verification using binary decision diagrams in a logic synthesis environment," *Proc. Int'l Conf. on CAD*, pp. 6–9, Nov. 1988.

[36] S. Minato, "Zero-suppressed BDDs for set manipulation in combinational problems," *Proc. Design Automation Conf.*, pp. 272–277, June 1993. (Also Chapter 1 of this book).

[37] S. Minato, N. Ishiura, and S. Yajima, "Shared binary decision diagrams with attributed edges for efficient boolean function manipulation," *Proc. Design Automation Conf.*, pp. 52–57, June 1990.

[38] Richard Rudell, "Dynamic variable ordering for ordered binary decision diagrams," *Proc. Int'l Conf. on CAD*, pp. 42–47, Nov. 1993.

[39] T. Sasao, *Logic Synthesis and Optimization.* Kluwer Academic Publisher, 1993. (Also Chapter 2 of this book).

[40] I. Schaefer, M.A. Perkowski, and H.Wu, "Multilevel logic synthesis for cellular FPGAs based on orthogonal expansions," *IFIP WG 10.5 Workshop on Applications of the Reed-Muller Expansion in Circuit Design*, pp. 42–51, Sept. 1993.

[41] H. Touati, H. Savoj, B. Lin, R.K. Brayton, and A.L. Sangiovanni-Vincentelli, "Implicit enumeration of finite state machines using BDDs," *Proc. Int'l Conf. on CAD*, pp. 130–133, Nov. 1990.

[42] R. Werchner, T. Harich, R. Drechsler, and B. Becker, "Satisfiability problems for ordered functional decision diagrams," *IFIP WG 10.5 Workshop on Applications of the Reed-Muller Expansion in Circuit Design*, pp. 206–212, Aug. 1995.

8

EXACT MINIMIZATION OF FPRMS USING MULTI-TERMINAL EXOR TDDS

Tsutomu SASAO
Fumitaka IZUHARA

Department of Computer Science and Electronics,
Kyushu Institute of Technology
Iizuka, Japan

Abstract— This chapter presents methods to derive a fixed polarity Reed-Muller expression (FPRM) and a Kronecker expression (KRO) having the minimum number of products for a given logic function. The minimization methods use EXOR ternary decision diagrams (ETDDs) and multi-terminal binary decision diagrams (MTBDDs) to represent extended truth vectors and weight vectors, respectively. Various techniques to reduce computation time and memory storage are developed. Experimental results up to 94 inputs are shown. The presented method outperforms existing methods.

8.1 INTRODUCTION

In the recent years, EXOR-based logic synthesis received gained more and more interest, since EXOR-based realizations are very efficient for arithmetic functions [24]. Various classes of AND-EXOR two-level logical expressions exist[24]. Among them, exclusive-or sum-of-products expressions (ESOPs), which require both the positive (x_i) and the negative ($\bar{x}_i$) literals for each variable x_i. ESOPs require the fewest products among AND-EXOR expressions. Fixed polarity Reed-Muller expression (FPRMs), on the other hand, require either a positive or a negative literal for each variable. FPRMs have easily testable realizations [18, 20, 9]. For an n-variable function, 2^n different polarities exist for FPRMs. A minimum FPRM for f is one with the minimum number of products. A minimum FPRM can be chosen from all FPRMs with 2^n different polarities. Various minimization methods for FPRMs have been developed. They are classified into two categories:

1) Gray code [10]: This method generates sequentially 2^n different FPRMs from a given FPRM. Memory requirement is $O(2^n)$, but the computation time is $O(4^n)$.

2) Extended truth vector [7, 12, 11]: This method uses an extended truth vector and a weight vector to obtain the number of products for 2^n different expansions simultaneously. Memory requirement and computation time are both $O(3^n)$.

The second method is faster, and we can easily obtain minimum FPRMs up to $n = 15$. However, since the size of extended truth table is 3^n, we cannot minimize large FPRMs. Thus, various heuristic minimization algorithms have been developed [2, 21, 26]. Among these, [26] and [21] can only handle single output functions. The logic minimizer in this chapter can deal with multiple output functions, which is much more complicated than single-output case [8].

In this chapter, we use an EXOR ternary decision diagram (ETDD) to represent an extended truth vector, and a multi-terminal binary decision diagram (MTBDD) to represent a weight vector. An ETDD is used to represent a binary vector of 3^n elements, and an MTBDD is used to represent an integer vector of 2^n elements.

By using this data structure, we successfully obtained exact minimum FPRMs with more than 20 inputs. Also, by using similar technique, we have obtained the exact minimum KROs (Kronecker expressions) for a function with 94 inputs.

8.2 DEFINITION AND BASIC PROPERTIES

In this section, we define various classes of AND-EXOR expressions. Also, we show the relations among them.

Theorem 8.2.1 *(Expansion Theorem) An arbitrary function f can be represented as*

$$f = f_0 \oplus x f_2, \tag{8.2.1}$$

$$f = \bar{x} f_2 \oplus f_1, \tag{8.2.2}$$

$$f = \bar{x} f_0 \oplus x f_1, \text{ or} \tag{8.2.3}$$

where $f_0(f_1)$ is f with x replaced by 0(1), and $f_2 = f_0 \oplus f_1$. (8.2.1) is the **positive Davio expansion**; *(8.2.2) is the* **negative Davio expansion**; *and (8.2.3) is the* **Shannon expansion.**

Definition 8.2.1 *If we expand the given function f by using only the positive Davio expansion, we have the following expression:*

$$\begin{aligned} a_0 & \oplus & a_1x_1 \oplus \cdots \oplus a_nx_n \oplus a_{12}x_1x_2 \oplus a_{13}x_1x_3 \oplus \cdots \\ & \oplus & a_{nn-1}x_nx_{n-1} \oplus \cdots \oplus a_{12\cdots n}x_1x_2\cdots x_n. \end{aligned} \tag{8.2.4}$$

(8.2.3) has only positive literals, so it is called the **positive polarity Reed-Muller expression** *(***PPRM***).*

For a given function f, there is exactly one PPRM. Thus, the PPRM is canonical expansion, and already minimized.

Example 8.2.1 *Let us obtain the PPRM for $f = \bar{x}_1\bar{x}_2\bar{x}_3$. By using the relations $\bar{x}_1 = x_1 \oplus 1$, $\bar{x}_2 = x_2 \oplus 1$, and $\bar{x}_3 = x_3 \oplus 1$, we have*

$$\begin{aligned} f &= (x_1 \oplus 1) \cdot (x_2 \oplus 1) \cdot (x_3 \oplus 1) \\ &= 1 \oplus x_1 \oplus x_2 \oplus x_3 \oplus x_1x_2 \oplus x_2x_3 \oplus x_1x_3 \oplus x_1x_2x_3. \end{aligned}$$

■

Note that this expression has only positive literals. In general, the PPRM of $\bar{x}_1\bar{x}_2\cdots\bar{x}_n$ requires 2^n products

Definition 8.2.2 *For a given function f, suppose that we can use either the positive Davio expansion or the negative Davio expansion for each variable. Then, we have an expression similar to (8.2.3) except that each variable appears always complemented or always un-complemented. This is called the* **fixed polarity Reed-Muller expression** *(***FPRM***).*

Definition 8.2.3 *The* **polarity vector** *for n-variable FPRM is a binary vector $\boldsymbol{a} = (a_1, a_2, \ldots, a_n)$, where $a_i \in \{0, 1\}$. $a_i = 0$ means the positive Davio expansion is used for x_i, and $a_i = 1$ means the negative Davio expansion is used for x_i. FPRMs for an n-variable function have 2^n different polarity vectors, and the polarity vector completely specifies the method of expansion.*

Example 8.2.2 *Consider the two-variable function* $f(x_1, x_2)$. *Suppose that the polarity vector for the FPRM is* $\boldsymbol{a} = (0,1)$, *which specifies that the positive Davio expansion is used for* x_1 *and the negative Davio expansion is used for* x_2. *Thus, when* $\boldsymbol{a} = (0,1)$, *the FPRM is written as* $f = 1 \cdot (1 \cdot f_{01} \oplus \bar{x}_2 f_{02}) \oplus x_1(1 \cdot f_{21} \oplus \bar{x}_2 f_{22})$. ■

Example 8.2.3 *An FPRM for* $f = x_1x_2x_3x_4 \vee \bar{x}_1\bar{x}_2\bar{x}_3\bar{x}_4$ *is derived as follows: Because the two products are mutually disjoint,* f *is also represented as* $f = x_1x_2x_3x_4 \oplus \bar{x}_1\bar{x}_2\bar{x}_3\bar{x}_4$. *Suppose that we use the positive Davio expansions for* x_1 *and* x_2, *and the negative Davio expansions for* x_3 *and* x_4. *By applying the relations* $\bar{x}_1 = x_1 \oplus 1$, $\bar{x}_2 = x_2 \oplus 1$, *and* $x_3 = \bar{x}_3 \oplus 1$, $x_4 = \bar{x}_4 \oplus 1$ *to* f, *we have*

$$\begin{aligned} f &= x_1x_2(\bar{x}_3 \oplus 1)(\bar{x}_4 \oplus 1) \oplus (x_1 \oplus 1)(x_2 \oplus 1)\bar{x}_3\bar{x}_4 \\ &= x_1x_2(1 \oplus \bar{x}_3 \oplus \bar{x}_4 \oplus \bar{x}_3\bar{x}_4) \oplus (1 \oplus x_1 \oplus x_2 \oplus x_1x_2)\bar{x}_3\bar{x}_4 \\ &= x_1x_2 \oplus x_1x_2\bar{x}_3 \oplus x_1x_2\bar{x}_4 \oplus \bar{x}_3\bar{x}_4 \oplus x_1\bar{x}_3\bar{x}_4 \oplus x_2\bar{x}_3\bar{x}_4. \end{aligned}$$

This is a FPRM, with positive polarities for x_1 *and* x_2, *and negative polarities for* x_3 *and* x_4. ■

In general, to represent $x_1x_2 \cdots x_n \vee \bar{x}_1\bar{x}_2 \cdots \bar{x}_n (n = 2r)$, a FPRM requires $2^{(r+1)} - 2$ products.

Definition 8.2.4 *For a given function, for each variable suppose that we can use any one of the three expansions: the positive Davio expansion, the negative Davio expansion, or the Shannon expansion. Then, we have the* **Kronecker expression** *(***KRO***).*

KROs are a generalization of FPRMs. For an n variable function, 3^n different expansion exist for KROs.

Example 8.2.4 *A KRO for* $f = x_1x_2x_3x_4 \vee \bar{x}_1\bar{x}_2\bar{x}_3\bar{x}_4$ *is derived as follows: If we use the Shannon expansions for all the variables, we have* $f = x_1x_2x_3x_4 \oplus \bar{x}_1\bar{x}_2\bar{x}_3\bar{x}_4$, *which is a KRO.* ■

In general, to represent $x_1x_2 \cdots x_n \vee \bar{x}_1\bar{x}_2 \cdots \bar{x}_n$, a KRO requires only two products.

Definition 8.2.5 *Arbitrary products combined with EXOR operators is an* **ESOP** *(***EXOR sum-of-products expression***).*

ESOPs are the most general expressions, and they require the fewest products among the AND-EXOR two-level expressions. No efficient methods are known to obtain exact minimum ESOPs. A heuristic simplification program has been developed[25].

Theorem 8.2.2 *Let $\mathcal{FPRM}$, $\mathcal{KRO}$, and $\mathcal{ESOP}$ denote the set of FPRMs, KROs, and ESOPs, respectively. Then, we have the following relation:*

$$\mathcal{FPRM} \subset \mathcal{KRO} \subset \mathcal{ESOP}.$$

8.3 OPTIMIZATION OF FPRMS

8.3.1 Optimization Strategy

For an n-variable function, there are 2^n different expansions for FPRMs. A minimum FPRM can be found among the 2^n FPRMs.
In this section, we will show an optimization method using extended truth vectors and weight vectors[7]. This method requires memory with size $O(3^n)$ and computation time with $O(3^n)$. Although, it requires more memory than the Gray code method, it is much more efficient. This method simplifies the FPRMs with less than 16 input variables in a reasonable computation time using an ordinary workstation.

8.3.2 EXOR Ternary Decision Diagram

In this section, we introduce EXOR ternary decision trees (ETDTs), which contain all the information necessary to optimize FPRMs[24]. Before showing the formal definition of ETDTs, we start with a simple example.

Example 8.3.1 *Consider a two-variable function* $f(x_1, x_2)$*. Fig.8.3.1 shows the complete ETDT for* $f(x_1, x_2)$*, where*

$$\begin{array}{lll} f_0 = f(0, x_2), & f_1 = f(1, x_2), & f_2 = f_0 \oplus f_1, \\ f_{00} = f(0,0), & f_{01} = f(0,1), & f_{02} = f_{00} \oplus f_{01}, \\ f_{10} = f(1,0), & f_{11} = f(1,1), & f_{12} = f_{10} \oplus f_{11}, \\ f_{20} = f_{00} \oplus f_{10}, & f_{21} = f_{01} \oplus f_{11}, & f_{22} = f_{20} \oplus f_{21}. \end{array}$$

■

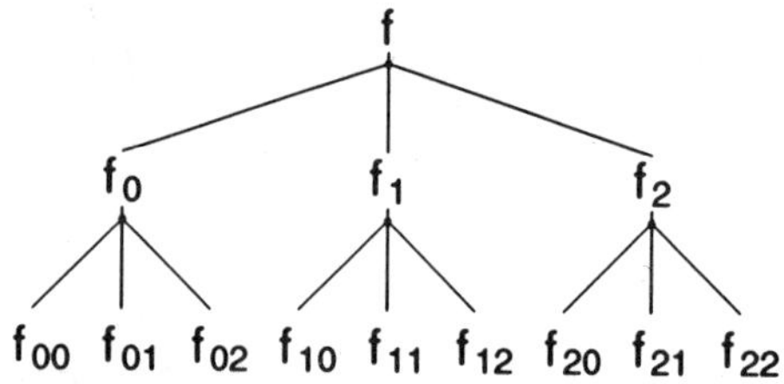

Figure 8.3.1 Complete EXOR ternary decision tree for $f(x_1, x_2)$.

Formally, we can define a complete ETDT for an n-variable function as follows:

Definition 8.3.1 *A complete* **EXOR ternary decision tree** *(***ETDT***) is a tree with vertex set* V *containing two types of nodes: A non-terminal node* v *has as attributes an argument index,* $index(v) \in \{1, 2, \ldots, n\}$*, and three children,* $low(v)$*,* $high(v)$*,* $exor(v) \in V$*. A terminal node* v *has as attribute a value,* $value \in \{0, 1\}$*. The correspondence between a node* v *and the Boolean function* f_v *which is associated with* v *is as follows:*

1. *If* v *is a terminal node:*
   (a) *If* $value(v) = 1$ *then* $f_v = 1$*.*
   (b) *If* $value(v) = 0$ *then* $f_v = 0$*.*
2. *If* v *is a non-terminal node with* $index(v) = i$*, then* f_v *is the function* $f_v(x_1, x_2, \ldots, x_n) = \bar{x}_i f_{low(v)}(x_1, x_2, \ldots, x_{i-1}, x_{i+1}, \ldots, x_n) \oplus x_i f_{high(v)}(x_1, x_2, \ldots, x_{i-1}, x_{i+1}, \ldots, x_n)$*, where* $f_{low(v)}(x_1, x_2, \ldots, x_{i-1}, x_{i+1}, \ldots, x_n) = f(x_1, x_2, \ldots, x_{i-1}, 0, x_{i+1}, \ldots, x_n)$*,* $f_{high(v)}(x_1, x_2, \ldots, x_{i-1}, x_{i+1}, \ldots, x_n) = f(x_1, x_2, \ldots, x_{i-1}, 1, x_{i+1}, \ldots, x_n)$ *and* $f_{exor(v)}(x_1, x_2, \ldots, x_n) = f_{low(v)}(x_1, x_2, \ldots, x_{i-1}, x_{i+1}, \ldots, x_n) \oplus f_{high(v)}(x_1, x_2, \ldots, x_{i-1}, x_{i+1}, \ldots, x_n)$*.*

In a complete ETDT, every path from the root node to a terminal node involves n non-terminal nodes. There are 3^n different terminal nodes.

When the given function is represented by a complete ETDT, the terminal nodes form the extended truth vector.

Definition 8.3.2 *The* **extended truth vector** *of an n-variable function is a binary vector with 3^n components, where each component represents the value of the leaf node in the complete ETDT.*

Example 8.3.2 *Consider the complete ETDT in Fig.8.3.1. The binary vector with four components $[f_{00}, f_{01}, f_{10}, f_{11}]$ is the truth vector of $f(x_1, x_2)$. The binary vector with 9 components $[f_{00}, f_{01}, f_{02}, f_{10}, f_{11}, f_{12}, f_{20}, f_{21}, f_{22}]$ is the extended truth vector of $f(x_1, x_2)$.* ■

The 3^n terminals of the complete ETDT form the extended truth vector: $[f(\alpha_0), f(\alpha_1), \ldots, f(\alpha_{3^n-1})]$, where each element is indexed by a ternary vector $\boldsymbol{\alpha} = (\alpha_1, \alpha_2, \ldots, \alpha_n)$, and $\alpha_i \in \{0, 1, 2\}(i = 1, 2, \ldots, n)$.
Next, we will show a method to compute the extended truth vector. Before showing the formal method, we consider the three-variable case.

Example 8.3.3 *Consider the three-variable function whose truth vector is $[0, 1, 0, 0, 0, 1, 1, 1]$. The extended truth vector is computed as shown in Fig.8.3.2. The column labeled 1 shows the ternary vectors $\boldsymbol{\alpha} = (\alpha_1, \alpha_2, \alpha_3)$. The column labeled 2 shows the truth vector. Note that elements exist only for the rows whose indexes are $\boldsymbol{a} = (a_1, a_2, a_3)$, where $a_i \in \{0, 1\}$. In the column labeled 3, we can calculate the values for the elements whose indexes are $(a_1, a_2, 2)$, where $a_i \in \{0, 1\}$: They are obtained as mod-2 sums of $f(a_1, a_2, 0)$ and $f(a_1, a_2, 1)$. In the column labeled 4, we calculate the values of the elements whose indexes are $(a_1, 2, \alpha_3)$, where $a_1 \in \{0, 1\}$ and $\alpha_3 \in \{0, 1, 2\}$: They are obtained as mod-2 sums of $f(a_1, 0, \alpha_3)$ and $f(a_1, 1, \alpha_3)$. In the column labeled 5, we calculate the values of the elements whose indexes are $(2, \alpha_2, \alpha_3)$, where $\alpha_i \in \{0, 1, 2\}$: They are obtained as mod-2 sums of $f(0, \alpha_2, \alpha_3)$ and $f(1, \alpha_2, \alpha_3)$.* ■

Extended truth vectors for an n-variable function are recursively computed as follows:

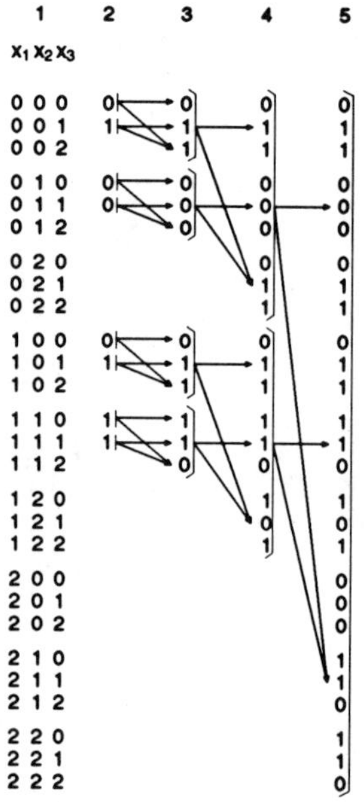

Figure 8.3.2 Computation of the extended truth vector.

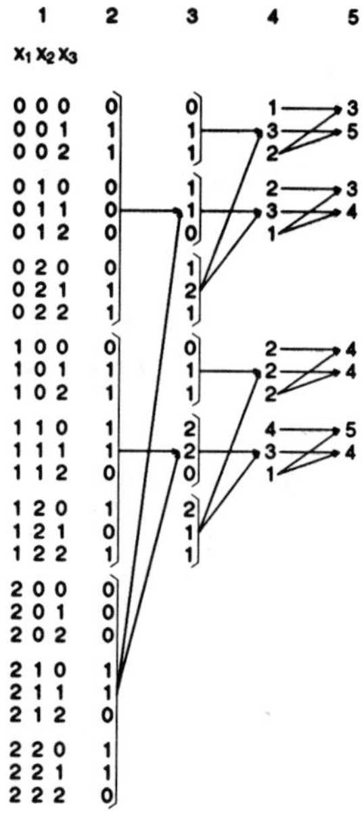

Figure 8.3.3 Computation of the weight vector.

Algorithm 8.3.1 *(Extended truth vector) If $\alpha_i \in \{0, 1\}$ for $i = 1, 2, \ldots,$ and n, then $f(\boldsymbol{\alpha})$ is obtained from the truth vector for f. If $\alpha_i = 2$ for some i, then*

$$f(\alpha_1, \ldots, \overset{i}{2}, \ldots, \alpha_n) =$$
$$f(\alpha_1, \ldots, \overset{i}{0}, \ldots, \alpha_n) \oplus f(\alpha_1, \ldots, \overset{i}{1}, \ldots, \alpha_n).$$

Next, we will show an optimization method for FPRMs using weight vectors.

Definition 8.3.3 *The* **weight vector** *of an n-variable function is a vector with 2^n components, where each component represents the number of the products in the FPRM with polarity vector $\boldsymbol{a}$.*

Example 8.3.4 *Consider the three-variable function in Example 8.3.3. The weight vector is computed from the extended truth vector as shown in Fig.8.3.3.*

The first addition computes the number of products in FPRMs with variable x_1. Note that there are 18 FPRMs in the column labeled 3. The first 9 elements in this column correspond to the positive Davio expansions with respect to x_1, and the second 9 elements correspond to the negative Davio expansions. The second addition computes the number of products in FPRMs with variables x_1 and x_2. This yields 12 FPRMs as shown in the column labeled 4. The first three elements in this column correspond to the positive Davio expansions with respect to x_1 and x_2. The next three elements correspond to the positive Davio expansion with respect to x_1, and the negative Davio expansion with respect to x_2, and so on. The column labeled 5 lists the number of products in all possible FPRMs. Note that there are 8 FPRMs, and these are all the possible FPRMs for the given function. The column labeled 5 corresponds to the weight vector. Each element denotes the number of products in the FPRM with the corresponding polarity.

*The first row in Fig.8.3.3 corresponds to the polarity vector **a**=(0,0,0), which corresponds to the FPRM by using only the positive Davio expansions. In this case, the FPRM has three products: $f = x_3 \oplus x_1x_2 \oplus x_2x_3$. Note that this is a PPRM. The last element in the column labeled 5 of Fig.8.3.3 corresponds to the polarity vector **a**=(1,1,1)((1,1,1) is located in the same row), which shows that the function is expanded by using only the negative Davio expansions. In this case, the FPRM has four products: $f = 1 \oplus \bar{x}_1 \oplus \bar{x}_1\bar{x}_2 \oplus \bar{x}_2\bar{x}_3$. Fig.8.3.3 shows that the when the polarities are (0,0,0) or (0,1,0), FPRMs require three products. When the polarities are (0,1,1), (1,0,0), (1,0,1) and (1,1,1), FPRMs require four products. And when the polarities are (0,0,1) and (1,1,0), FPRMs require five products.* ■

For an n-variable function, the weight vector is recursively computed as follows:

Algorithm 8.3.2 *(The weight vector) Let $\tau(f : \boldsymbol{a})$ be the number of products in the FPRM for f with polarity vector $\boldsymbol{a} = (a_1, a_2, \ldots, a_k)$. For n-variable function $f = \bar{x}_n f_0 \oplus x_n f_1$.*
When $n = 0$, $\tau(0) = 0$ and $\tau(1) = 1$.
When $n = 1$,

$$\tau(f:0) = \tau(f_0) + \tau(f_2), \quad (8.3.1)$$
$$\tau(f:1) = \tau(f_1) + \tau(f_2), \quad (8.3.2)$$

where f_0 and f_1 are binary constants, and $f_2 = f_0 \oplus f_1$.
When $n \geq 2$,

$$\tau(f:\boldsymbol{a}\$0) = \tau(f_0:\boldsymbol{a}) + \tau(f_2:\boldsymbol{a}), \quad (8.3.3)$$

$$\tau(f : \boldsymbol{a}\$1) \quad = \quad \tau(f_1 : \boldsymbol{a}) + \tau(f_2 : \boldsymbol{a}), \qquad (8.3.4)$$

where $f_2 = f_0 \oplus f_1$, and $\boldsymbol{a}\$a_{k+1} = (a_1, a_2, \ldots, a_k, a_{k+1})$. Note that \$ denotes concatenation. (8.3.1) and (8.3.3) show that the positive Davio expansions are used for x_n. (8.3.2) and (8.3.4) show that the negative Davio expansions are used for x_n.

Algorithm 8.3.3 *(Minimum FPRM)*

1) *Compute the extended truth vector.*
2) *Compute the weight vector.*
3) *Find the polarity vector corresponding to a minimum value of the weight vector.*
4) *Obtain the FPRM corresponding to the polarity vector.*

8.4 DATA STRUCTURE AND IMPLEMENTATION

The method in the previous section requires an extended truth vector of size 3^n. Thus, when n is large, the extended truth vector is too large to store. In this section, we present a method to represent extended truth vectors and weight vectors by using EXOR ternary decision diagrams (ETDDs) and multi-terminal binary decision diagrams (MTBDDs), respectively.

8.4.1 Extended Truth Vector

An extended truth vector can be represented by a complete EXOR ternary decision tree (ETDT). A complete ETDT for an n variable function has $(3^{n+1}-1)/2$ nodes. For a function with many inputs, it is impossible to build a complete ETDT. Here, we construct an EXOR ternary decision diagram (ETDD), which contains the same information as a complete ETDT, but requires fewer nodes. An ETDD is generated in a similar way to complete ETDT, but by using the following rules:

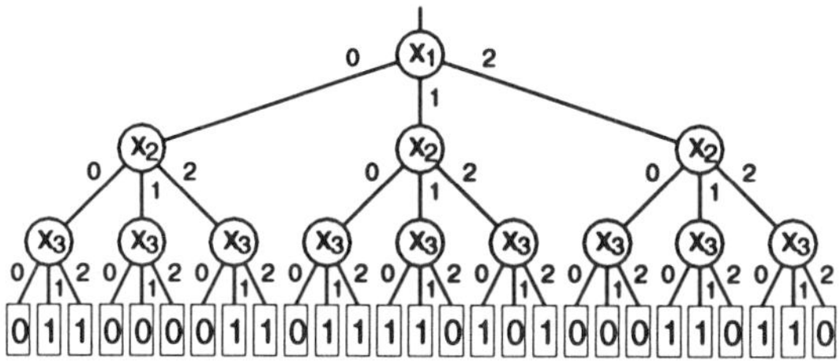

Figure 8.4.1 Complete EXOR ternary decision tree for the function in Example 8.3.3.

1) If a sub-function does not depend on x, then do not expand with respect to x.
2) If sub-functions represent the same function, then only one sub-graph is realized.

The number of nodes in an ETDD is $O(3^n/n)$ [24].

Example 8.4.1 *Fig.8.4.1 show the complete ETDT for the function introduced in Example 8.3.3. The extended truth vectors can be also represented by an ETDD shown in Fig.8.4.2.*

8.4.2 Weight Vector

A weight vector has 2^n integer elements, where each element is the number of products in the FPRM with the corresponding polarity vector. To represent a discrete function having integer values, we use multi-terminal binary decision diagrams (MTBDD)[5]. A complete multi-terminal binary decision tree (complete MTBDT) requires $2^{n+1} - 1$ nodes to represent the weight vector.

The terminals of the MTBDT are integers showing the number of products. A straightforward construction of the MTBDD often requires excessive amount of memory. To reduce memory requirement, the following techniques are used: To minimize the FPRM, only one polarity corresponding to a minimum FPRM is necessary. Let t_0 be the number of products in a minimum FPRM. For the

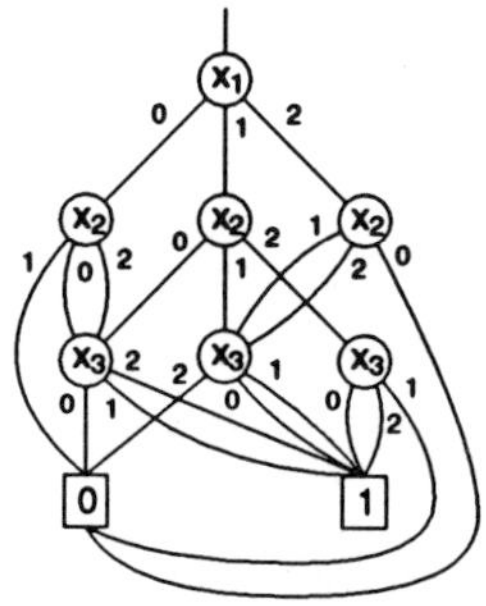

Figure 8.4.2 EXOR ternary decision diagram for the function in Example 8.3.3.

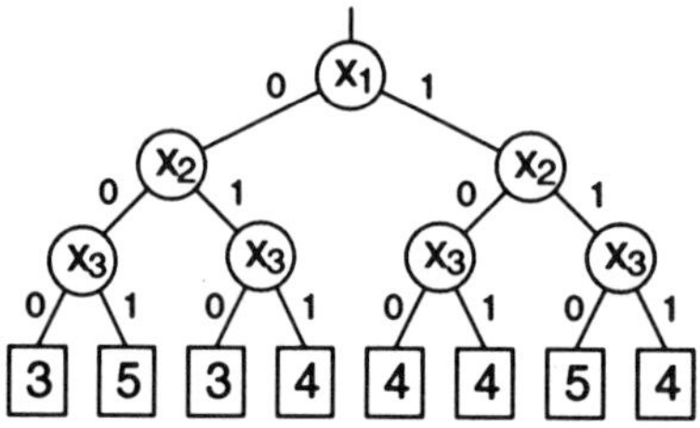

Figure 8.4.3 Complete multi-terminal binary decision tree representing the weight vector in Example 8.3.3.

entries of the weight vector corresponding to polarities other than the minimum, the values need not be the number of the products in FPRMs. These may be any number that are greater than t_0. To reduce the size of the MTBDD, we assign a constant T if the number of the products is greater than T (a threshold value). This technique drastically reduces the memory requirement.

Example 8.4.2 *Fig.8.4.3 shows the complete multi-terminal decision trees representing the weight vector in Example 8.3.3. By using the simplification rules, we have the MTBDD shown in Fig.8.4.4. If we know that the minimized FPRM requires at most three products, then the terminals for 4 and 5 can be merged, and have the MTBDD shown in Fig.8.4.5. This corresponds to setting T=4. The path from the root node to the constant 3 shows the polarity* *vector with the smallest weight. In this case,* $x_1=x_3=0$*. Thus, polarity vectors* $(x_1, x_2, x_3)=(0,0,0)$ *or* $(0,1,0)$ *produce minimum FPRMs.*

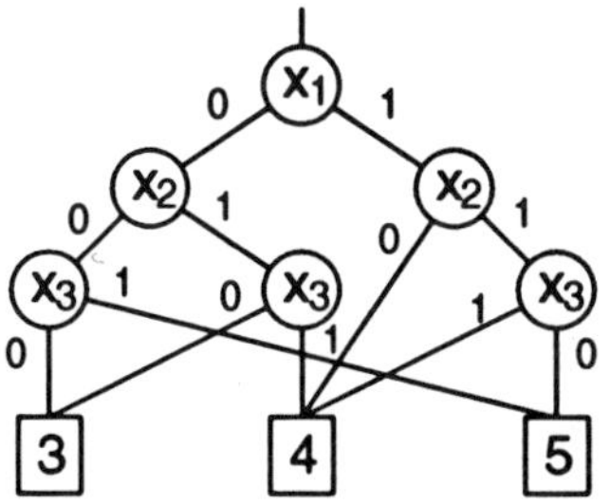

Figure 8.4.4 Multi-terminal binary decision diagram representing the weight vector in Example 8.3.3.

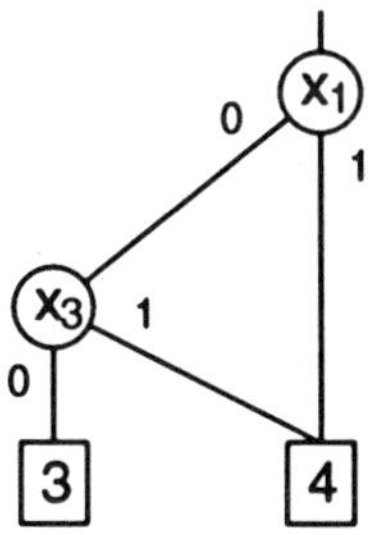

Figure 8.4.5 Multi-terminal binary decision diagram representing the weight vector in Example 8.3.3, when T=4.

As shown in the above example, the MTBDDs with appropriate values of T require many fewer nodes than complete multi-terminal binary decision trees. Thus, the minimum weight in the MTBDD and its corresponding polarity vector can be found by tree traversal rather quickly. Once the best polarity vector is found, the minimum FPRM is derived from the ETDD.

8.4.3 Optimization of Multiple-Output FPRMs

In this section, we show a method to optimize FPRMs for multiple-output functions. Various techniques are used to reduce the memory requirement.

Definition 8.4.1 *An* **extended truth vector** *for an n-input m-output function consists of 3^n elements, where each element is a binary vector of m bits, and each corresponds to the terminal node of the n-variable complete ETDT.*

To generate the weight vector of m-output function, we use the modified extended truth vector.

Definition 8.4.2 *A* **modified extended truth vector** *of an m-output function is obtained from the extended truth vector by replacing zero vectors with 0's, and non-zero vectors with 1's.*

The straightforward generation of modified extended truth vectors often requires excessive memory. To reduce memory, we use the following algorithm:

Algorithm 8.4.1 *(Minimization of multi-output FPRM)*

1. *Construct an ETDD with m root nodes representing m outputs. (i.e., represent a multiple-output function by using a "shared" ETDD.)*
2. *Let $\boldsymbol{EXTRU}(f_i)$ be the extended truth vector for function $f_i (i = 1, 2, \cdots, m)$. Represent $\boldsymbol{EXTRU}(f_i)$ by ETDDs $(i = 1, 2, ..., m)$.*
   Obtain the modified extended truth vector from $\bigvee_{i=1}^{m} \boldsymbol{EXTRU}(f_i)$, where $\vee$ denotes the bitwise OR operation.
3. *Generate the weight vector from the modified extended truth vector. Note that the weight vector is represented by an MTBDD. A straightforward way to construct the weight vector often requires excessive memory. So, if the results of the addition are greater than a threshold value (T), then the results of the addition are set to T.*
4. *Obtain the polarity vector for an expansion with the smallest weight.*

5. Generate the FPRM for the polarity vector.

It is clear that the above algorithm produces a minimum FPRM for multi-output function.

8.5 OPTIMIZATION OF KRONECKER EXPRESSIONS

An n variable function has 3^n different polarities for KROs. By using a similar method to FPRMs, we can optimize KROs. In the case of KROs, we use the extended weight vector, which shows the number of products for 3^n different polarities of expansions. To represent extended weight vectors, we use multi-terminal ternary decision diagrams (MTTDD). Thus, optimization of KROs will require more memory than that of FPRMs.

8.6 EXPERIMENTAL RESULTS

We developed a FORTRAN program for the algorithm in section 8.3, by using linear arrays. When n, the number of inputs, is less than 15, the minimum FPRMs can be easily obtained. The FPRMs and KROs in [24] were optimized by using this method.

For the functions with $n > 15$, we developed a C program the algorithm shown in Section 8.4. Table 8.6.1 shows the experimental results, where the columns denote the followings:
Data: Name of input data.
in: Number of input variables.
out: Number of outputs.
# of products
  FPRM: Number of products in minimum FPRM.
  KRO: Number of products in minimum KRO.
  ESOP: Number of products in near minimum ESOP.
  SOP: Number of products in near minimum SOP.
# of nodes
  BDD: Number of nodes in the BDD for the truth vector.
  ETDD: Number of nodes in the ETDD for the extended truth vector.

MTBDD: Number of nodes in the MTBDD for the weight vector when the threshold is T.

MTTDD: Number of nodes in the MTTDD for the extended weight vector when the threshold is T.

FPRM: maximum number of nodes to obtain a minimum FPRM.

KRO: maximum number of nodes to obtain a minimum KRO.

time

FPRM: CPU seconds to obtain a minimum FPRM.

KRO: CPU seconds to obtain a minimum KRO.

threshold

FPRM: value of T used to obtain a minimum FPRM.

KRO: value of T used to obtain a minimum KRO.

The computation time depends on the value of the threshold as well as the function. The value of threshold must be greater than the number of products in minimum expressions. By using this program, we minimized FPRMs and KROs for the given functions. An HP715 with 256 Mega-bytes of main memory was used.

The minimization algorithm for FPRMs uses two data structures:

1) ETDD to represent an extended truth vector, and

2) MTBDD to represent a weight vector.

ETDDs are larger than BDDs, and ordinary MTBDDs are usually larger than ETDDs.

In this experiment, each value of T was set to $\tau(f)+1$, where $\tau(f)$ denotes the number of products in the minimum expression for f. Thus, the resulting MTBDDs and MTTDDs were quite simple. However, during generation of these DDs, the size of DDs were fairly large.

We failed to generate MTBDD for x6dn due to the memory overflow. For KROs, we used MTTDDs to represent extended weight vectors. In many cases, generation of MTTDDs required more memory than MTBDDs.

Table 8.6.2 compares the maximum number of nodes and computation time for the cases when the thresholds are infinite, $10\tau(f)$, and $\tau(f)+1$, where $\tau(f)$ denotes the number of the products in a minimum FPRM for f. Table 8.6.2 demonstrate that the smaller the threshold, the smaller amount of memory and

Table 8.6.1 Experimental results.

Data	in	out	# of products FPRM	KRO	ESOP	SOP
amd	14	24	156	104	58	66
b9	16	5	105	105	81	119
bc0	26	11	1117	467	168	177
cordic	23	2	6438	1980	776	914
cps	24	109	291	249	137	172
duke2	22	29	255	209	81	86
ex7	16	5	105	105	81	119
gary	15	11	349	242	102	107
in2	19	10	355	262	108	134
in7	26	10	72	72	35	54
intb	15	7	1815	1605	327	629
m181	15	9	67	65	29	41
misg	56	23	104	104	66	69
mish	94	43	53	53	53	82
misj	35	14	17	17	16	35
misex3	14	14	3536	1421	553	690
rckl	32	7	32	32	32	32
ryy6	16	1	64	48	40	112
t1	21	23	232	209	90	103
t481	16	1	13	13	13	481
tial	14	8	3683	2438	487	579
ts10	22	16	432	128	128	128
vg2	25	8	5290		184	110
x6dn	39	5			95	81

computation time are necessary. To obtain the threshold (T), we used heuristic algorithms to find near minimum FPRMs and KROs[13].

Let T be the number of product terms in the solution obtained by the heuristic minimization program. Algorithm 8.4.1 tries to find the solution of less than T products. If the solution obtained by the heuristic minimization program is the exact minimum, then Algorithm 8.4.1 will stop and report that there is no better solution. In this case, the resulting MTBDD will be a terminal node.

8.7 CONCLUSION AND COMMENTS

In this chapter, we presented minimization methods for FPRMs and KROs. In the past, only a near optimum FPRM could be obtained when n is large [2, 15, 16, 21, 26], but now we can obtain exact minimum solutions.

Table 8.6.1 Experimental results (continued).

Data	in	out	# of nodes						time		T	
			BDD	ETDD	MT BDD	MT TDD	FPRM	KRO	FP RM	KRO	FPRM	KRO
amd	14	24	494	1097	15	16	3510	10931	2.4	7.7	157	105
b9	16	5	199	672	18	18	13234	61711	7.8	119	106	106
bc0	26	11	635	16415	20	28	75003	750021	43	7637	1118	468
cordic	23	2	83	144	37	34	2874	19347	2.1	12	6439	1981
cps	24	109	2424	6259	22	26	12007	87390	4.2	76	292	250
duke2	22	29	1008	4028	20	24	20810	335923	30	5921	256	210
ex7	16	5	199	672	18	18	13234	61711	8.8	112	106	106
gary	15	11	540	1694	17	17	14427	200289	6	185	350	243
in2	19	10	2407	3864	20	21	32707	1400250	31	1047	356	263
in7	26	10	171	517	25	28	5428	42886	3.8	66	73	73
intb	15	7	676	4656	26	23	60233	2721243	46	2873	1816	1606
m181	15	9	105	211	17	17	4946	21377	7	15	68	66
misg	56	23	133	453	50	58	12932	150097	12	10828	105	105
mish	94	43	170	263	94	98	315851	414683	920	5239	54	54
misj	35	14	72	114	37	37	1693	2345	2.3	3.5	18	18
misex3	14	14	1317	6517	16	16	39289	617825	22	581	3537	1422
rckl	32	7	217	340	33	34	581	827	1.7	1.8	33	33
ryy6	16	1	25	48	18	22	353	528	1.7	1.7	65	49
t1	21	23	167	252	17	23	5057	82096	3.2	29	233	210
t481	16	1	34	45	18	18	243	376	1.7	1.8	14	14
tial	14	8	701	5306	26	24	49623	1024954	30	962	3684	2439
ts10	22	16	4408	5116	39	23	8431	7223	2.8	4.5	433	129
vg2	25	8	233	802	53		1290997		3968		5291	
x6dn	39	5	253	6116								

Table 8.6.2 Effect of threshold.

T	∞		$10\tau(f)$		$\tau(f)+1$	
	maxnode	time	maxnode	time	maxnode	time
cps	177573	42	69963	24	12007	5.1
cordic	6963	2.9	13585	3	2874	2.2

We successfully minimized FPRMs and KROs with up to 94 inputs by using ETDDs and multi-terminal binary decision diagrams (MTBDDs). The major difficulty using ETDD, MTBDDs, and MTTDDs is that the necessary memory increases quickly with the number of inputs. The results in Table 8.6.1 show the practical upper bounds.

By using exact minimization algorithms, we found that heuristic minimization algorithms can find solutions that are very close to a exact minimum. For example, bc0, cps, and in2 simplified by [13] and [8] are confirmed to be exact minimum. The presented algorithm can be used to prove the minimality of the solutions obtained by a heuristic minimization program.

Acknowledgments

This research was partly supported by the Ministry of Education, Science and Culture of Japan. The author thank Prof. J. T. Butler for his comments.

REFERENCES

[1] Ph. W. Besslich, "Efficient computer method for EXOR logic design," *IEE Proc.*, vol. 130, Part E, pp. 203-206, 1983.

[2] B. Becker, R. Drechsler and N. Göckel"A genetic algorithm for minimization of fixed polarity Reed-Muller expressions," *Proc. International Conference on Artificial Neural Networks and Genetic Algorithm*, April 1995.

[3] G. Bioul, M. Davio and J. P. Deschamps: "Minimization of ring- sum expansions of Boolean functions," *Philips Res. Rpts.*, vol. 28, pp. 17-36, 1973.

[4] R. K. Brayton, G. D. Hachtel, C. T. McMullen, and A. L. Sangiovanni-Vincentelli, *Logic Minimization Algorithms for VLSI Synthesis*, Boston, MA. Kluwer, 1984.

[5] E. M. Clarke, M. Fujita, P. C. McGeer, K. L. McMillan and J. C. Yang, "Multi-terminal binary decision diagrams: An efficient data structure for matrix representation" *Int. Workshop on Logic Synthesis*, May 1993, pp. 6A 1-15. (Also Chapter 4 of this book).

[6] E. M. Clarke, K. L. McMillan, X. Zhao, M. Fujita, P. C. McGeer, and J. C. Yang, "Spectral transforms for large Boolean functions with application to technology mapping," *Int. Workshop on Logic Synthesis*, May 1993, pp. P6b, 1-15.

[7] M. Davio, J.-P. Deschamps, and A. Thayse, *Discrete and Switching Functions*, McGraw-Hill International, 1978.

[8] R. Drechsler, M. Theobald, and B. Becker, "Fast FDD based minimization of generalized Reed-Muller forms," *Proc. European Design Automation Conf.*, 1994.

[9] H. Fujiwara, *Logic Testing and Design for Testability*, The MIT Press, 1985.

[10] D. Green, *Modern Logic Design*, Addison-Wesley Publishing company, 1986.

[11] D. H. Green,"Reed-Muller canonical forms with mixed polarity and their manipulations," *Proc. IEE*, vol. 137. pt.E, No.1, 1990. pp. 110-113.

[12] P. K. Lui and J. Muzio, "Boolean matrix transforms for the parity spectrum and the minimization of modulo-2 canonical expansions," *IEE Proc.* vol. 138, No. 6, pp.411-417.

[13] M. Matsuura and T. Sasao, "Heuristic optimization programs for fixed polarity Reed-Muller expressions and Kronecker expressions," (in preparation).

[14] A. Mukhopadhyay and G. Schmitz, "Minimization of Exclusive OR and logical Equivalence of switching circuits," *IEEE Trans. Comput.*, C-19, pp.132-140, 1970.

[15] U. Kebschull and W. Rosenstiel, "Efficient graph based computation and manipulation of functional decision diagrams," *Proc. EDAC'93*, pp. 43-47, 1993.

[16] M. A. Perkowski, L. Csanky, A. Sarabi, and I. Schaefer, "Fast minimization of mixed-polarity AND-XOR canonical networks," *Proc. ICCD-92*, pp. 33-36, Oct. 1992.

[17] D. K. Pradhan, "Universal test sets for multiple fault detection in AND-EXOR arrays," *IEEE Trans.on Comput.*, Vol. C-27, No. 2, pp. 181-187.

[18] S. M. Reddy, "Easily testable realization for logic functions," *IEEE Trans. on Comput.*, C-21, pp. 1083-1088, 1972.

[19] K. K. Saluja and E. H. Ong, "Minimization of Reed-Muller canonic expansion," *IEEE Trans. on Comput.*, C-28, pp. 535-537,1979.

[20] K. K. Saluja and S. M. Reddy, "Fault detecting test sets for Reed-Muller canonic networks," *IEEE Trans. on Comput.*, Vol. C-24, No. 10, pp. 995-998, 1975.

[21] A. Sarabi and M. A. Perkowski, "Fast exact and quasi-minimal minimization of highly testable fixed polarity AND/XOR canonical networks," *Proc. Design Automation Conference 1992*, June 1992, pp. 20-35.

[22] T. Sasao and P. Besslich, "On the complexity of MOD-2 sum PLA's," *IEEE Trans. on Comput.*, vol. 32, No. 2, pp. 262-266, Feb. 1990.

[23] T. Sasao , "Transformation of multiple-valued input two-valued output functions and its application to simplification of exclusive-or sum-of-products expressions," *Proc. ISMVL-91*, pp. 270-279, May. 1991.

[24] T. Sasao, "AND-EXOR expressions and their optimization," in (Sasao e.d.) *Logic Synthesis and Optimization*, Kluwer Academic Publishers, 1993.

[25] T. Sasao, "EXMIN2: A simplification algorithm for exclusive-OR-Sum-of-products expressions for multiple-valued input two-valued output functions," *IEEE Trans. on Computer-Aided Design of Integrated Circuits and Systems*, Vol. 12, No. 5, May 1993, pp. 621-632.

[26] C. C. Tsai and M. Marek-Sadowska, "Efficient minimization algorithms for fixed polarity and/xor canonical networks," *Great Lake Symp. VLSI*, pp. 76-79, 1993.

[27] S. Yang,"Logic synthesis and optimization benchmark user guide, version 3.0", *MCNC*, Jan. 1991.

9

MULTIPLE DOMAIN LOGIC SYNTHESIS

Jörg BULLMANN[1]
Udo KEBSCHULL[2]

[1] *Wilhelm-Schickard-Institut für Informtik*
Eberhard-Karls-Universität Tübingen
Tübingen, Germany

[2] *Forschungszentrum Informatik*
Karlsruhe, Germany

Abstract— Field programmable gate arrays (FPGA) and other complex programmable devices (CPLD) require new logic minimization techniques since the cost functions used for conventional target implementations are no longer valid. Until now, existing tools are only adapted to the new requirements. However, the underlying approaches for logic minimization and technology mapping remained the same. We present logic minimization techniques that extend classical approaches. Basic function properties as e.g. linearity, monotony, and symmetry of its variables are detected. They are used in decomposition and partial collapsing steps to group variables with common properties within a multi-level Boolean network. The Boolean functions are stored as decision diagrams. Three different Boolean normal forms are used: disjunctive normal form, Reed-Muller expansion, and equivalence polynomial. Therefore, three types of decision diagrams are needed: binary decision diagrams (BDD), functional decision diagrams (FDD), and equivalence decision diagrams (EDD). A multiple domain minimization approach based on decomposition, domain selection, variable ordering, and variable polarity optimization is introduced.

9.1 INTRODUCTION

Recently, several new minimization approaches based on spectral techniques (based on AND-XOR and OR-EQUIV) have been proposed[1, 2, 3, 4, 5, 6, 7, 8, 9, 10]. It has been proven that for certain function classes these techniques lead to significantly better results[11, 12, 13]. For other classes of Boolean functions, behavioral minimization techniques (based on AND-OR) are more suitable.

Therefore, we combined spectral and behavioral techniques within a single logic synthesis tool. The basic requirement for such a universal tool is a common data structure for the representation and manipulation of Boolean functions in all Boolean normal forms. In the case of the disjunctive normal form and the Reed -Muller expansion, these representations are well-known: binary decision diagrams (BDD)[14, 15, 16] and functional decision diagrams (FDD)[1, 17, 18, 19].

For all other canonical forms, like conjunctive, exclusive or, equivalence canonical form, and the equivalence polynomial, neither such representations have been defined, nor have these canonical forms explicitly been used for the minimization of Boolean networks.

We define a graphical data structure for all Boolean canonical forms based on binary decision diagrams, and we show some basic transformations between these representations. We use simple decomposition and collapsing techniques to separate complex functions into smaller parts with special properties like symmetry, monotony, and linearity of variables. Using these special properties, we decide, which kind of representation and which minimization approach are the most promising.

In Section 9.2, basics about Boolean canonical forms and the used graph-based data structures are stated. A new logic minimization approach is introduced in Section 9.3. In Section 9.4, we show some preliminary results. The chapter ends with Section 9.5, containing conclusions and an outlook on further work.

9.2 BASICS

9.2.1 Normal Forms

In the sequel, we consider single output, multiple input Boolean functions. The support *supp*(f) is the set of variables, the function is defined on.

As an extension to Green's[20] definition of minterms m_i and π-terms π_i, we define maxterms M_i and Π-terms Π_i. See Table 9.2.1 for an example of m_i, M_i, π_i, and Π_i for the three variables x_0, x_1, and x_2 in a lexicographical order.

Table 9.2.1 Min-, Max-, π-, and Π-terms.

i	m_i	M_i	π_i	Π_i
0	$\overline{x_2} \wedge \overline{x_1} \wedge \overline{x_0}$	$x_2 \vee x_1 \vee x_0$	1	0
1	$\overline{x_2} \wedge \overline{x_1} \wedge x_0$	$x_2 \vee x_1 \vee \overline{x_0}$	x_0	x_0
2	$\overline{x_2} \wedge x_1 \wedge \overline{x_0}$	$x_2 \vee \overline{x_1} \vee x_0$	x_1	x_1
3	$\overline{x_2} \wedge x_1 \wedge x_0$	$x_2 \vee \overline{x_1} \vee \overline{x_0}$	$x_1 \wedge x_0$	$x_1 \vee x_0$
4	$x_2 \wedge \overline{x_1} \wedge \overline{x_0}$	$\overline{x_2} \vee x_1 \vee x_0$	x_2	x_2
5	$x_2 \wedge \overline{x_1} \wedge x_0$	$\overline{x_2} \vee x_1 \vee \overline{x_0}$	$x_2 \wedge x_0$	$x_2 \vee x_0$
6	$x_2 \wedge x_1 \wedge \overline{x_0}$	$\overline{x_2} \vee \overline{x_1} \vee x_0$	$x_2 \wedge x_1$	$x_2 \vee x_1$
7	$x_2 \wedge x_1 \wedge x_0$	$\overline{x_2} \vee \overline{x_1} \vee \overline{x_0}$	$x_2 \wedge x_1 \wedge x_0$	$x_2 \vee x_1 \vee x_0$

Definition 9.2.1 *For each natural number* $i \in \{0, \ldots, 2^n - 1\}$, *a* Π*-term* Π_i *and a maxterm* M_i *can be defined on the variable set* $\{x_0, \ldots, x_{n-1}\}$. *Having* $i = \sum_{k=0}^{n-1} a_k \cdot 2^k$, *we write*

$$\begin{aligned} \Pi_0 &= 0, \\ \Pi_i &= \bigvee_{a_k=1} x_k, \\ M_i &= \bigvee_{a_k=0} x_k \vee \bigvee_{a_k=1} \overline{x_k}. \end{aligned}$$

Using min-, max-, π-, and Π-terms m_i, M_i, π_i, and Π_i, we define the following normal forms:

Definition 9.2.2 *The Boolean normal forms are expressed as follows:*

$$\begin{aligned} f(x_0, \ldots, x_{n-1}) &= \bigvee_{i=0}^{2^n-1} (a_i \wedge m_i) (\textit{disjunctive normal form}) \\ &= \bigwedge_{i=0}^{2^n-1} (b_i \vee M_i) (\textit{conjunctive normal form}) \\ &= \bigoplus_{i=0}^{2^n-1} (c_i \wedge m_i) (\textit{exclusive or normal form}) \\ &= \equiv_{i=0}^{2^n-1} (d_i \vee M_i) (\textit{equivalence normal form}) \\ &= \bigoplus_{i=0}^{2^n-1} (e_i \wedge \pi_i) (\textit{Reed-Muller expansion}) \end{aligned}$$

$$= \equiv_{i=0}^{2^n-1} (f_i \vee \Pi_i) (\textit{equivalence polynomial})$$

where $a_i, b_i, c_i, d_i, e_i, f_i \in \{0, 1\}$. *The coefficients* a_i *and* c_i *(*b_i*,* d_i*,* e_i*,* f_i*) are called the minterm coefficients (maxterm coefficients,* π*-term, and* Π*-term coefficients, respectively.)*

Theorem 9.2.1 *[21] Let* $f(x_0, ..., x_{n-1})$ *be an n-input and single output Boolean function. Then the following holds:*

$$\forall i \in \{0, .., 2^n - 1\} : a_i = b_i = c_i = d_i.$$

Therefore, the transformation between these normal forms is trivial. The two remaining normal forms, the Reed-Muller expansion, and the equivalence polynomial differ from the above four and from each other. Due to Definitions 9.2.1 and 9.2.2 both forms contain positive literals only. Thus, they are called positive polarity Reed-Muller expansion (PPRM), and positive polarity equivalence polynomial (PPEP), respectively.

Dropping this restriction, we get more general forms, the fixed polarity Reed-Muller expansion (FPRM), and equivalence polynomial (FPEP). To compute these fixed polarity forms, a variable x_i can be substituted by $\overline{x_i} \oplus 1$, or $\overline{x_i} \equiv 0$, respectively. Note, that to change a variable x_i's polarity in a fixed polarity form, *all* of the variable's occurrences have to be replaced.

Definition 9.2.3 (Polarity) *A polarity of a* π*-term (*Π*-term) t is the natural number*

$$p = \sum_{i=0}^{n-1} a_i \cdot 3^i, \quad a_i = \begin{cases} 0, & \textit{if } x_i \textit{ positive,} \\ 1, & \textit{if } x_i \textit{ negative,} \\ 2, & \textit{else.} \end{cases}$$

Definition 9.2.4 *A* π*-term (*Π*-term) with polarity value p is written as* $\pi_i^{<p>}$ *(*$\Pi_i^{<p>}$*).*

Using Definitions 9.2.4 and 9.2.3, we can state the fixed polarity forms of Reed-Muller expansion and equivalence polynomial more formal. Here, the polarity number p consists of $a_i \in \{0, 1\}$ only.

Definition 9.2.5 (FPRM) *The fixed polarity Reed-Muller expansion can be written as*

$$f(x_{n-1},\ldots,x_0) = \bigoplus_{i=0}^{2^n-1} (g_i \wedge \pi_i^{<p>}).$$

Definition 9.2.6 (FPEP) *The fixed polarity equivalence polynomial can be written as*

$$f(x_{n-1},\ldots,x_0) = \mathop{\equiv}_{i=0}^{2^n-1} (h_i \vee \Pi_i^{<p>}).$$

For a function with n variables, 2^n different polarities p (substitutions of variables by their complements) exist. Thus, there are the same number of different representations, which all are normal forms according to a given polarity value p.

More generalized representations, which are no longer normal forms, are *mixed polarity forms* of the Reed-Muller expansion and equivalence polynomial. In the sequel, we use fixed polarity representations only, since we need the normal form property.

9.2.2 Function Expansions

We now describe some expansions of Boolean functions. They are all based on the cofactor, which is the partial evaluation of a Boolean function.

Definition 9.2.7 (Cofactor) *Let f be a Boolean function. We write*

$$f|_{x_i=c} = f(x_{n-1},\ldots,x_{i+1},c,x_{i-1},\ldots x_0)$$

if x_i is fixed to $c \in \{0,1\}$.

The following definition of the Boolean difference is based on Boole [22].

Definition 9.2.8 (Boolean Difference) *For every Boolean function f and $x_i \in supp(f)$ we call*

$$\frac{\partial f(x)}{\partial x_i} = f|_{x_i=1} \oplus f|_{x_i=0}$$

the Boolean difference of f.

Using the cofactor and the Boolean difference, we denote the following expansions of Boolean functions:

Theorem 9.2.2 (Expansions of Boolean functions) *Every Boolean function f can be transformed into:*

$$
\begin{aligned}
f(x_{n-1}, \dots, x_0) &= (x_i \wedge f|_{x_i=1}) \vee (\overline{x_i} \wedge f|_{x_i=0}) && (9.2.1)\\
&= (x_i \wedge f|_{x_i=1}) \oplus (\overline{x_i} \wedge f|_{x_i=0}) && (9.2.2)\\
&= (x_i \wedge \frac{\partial f(x)}{\partial x_i}) \oplus f|_{x_i=0} && (9.2.3)\\
&= (\overline{x_i} \wedge \frac{\partial f(x)}{\partial x_i}) \oplus f|_{x_i=1} && (9.2.4)\\
&= (x_i \vee (f|_{x_i=0} \equiv f|_{x_i=1})) \equiv f|_{x_i=1} && (9.2.5)\\
&= (\overline{x_i} \vee (f|_{x_i=0} \equiv f|_{x_i=1})) \equiv f|_{x_i=0}. && (9.2.6)
\end{aligned}
$$

Proof: (9.2.1) is the Shannon expansion, (9.2.2) is a trivial generalization of the Shannon expansion, (9.2.3) is called the positive Davio expansion, and (9.2.4) is the negative Davio expansion.

We still need to prove (9.2.5) and (9.2.6):
(9.2.5): From the negative Davio expansion we get:

$$
\begin{aligned}
&\overline{x_i} \wedge (f|_{x_i=0} \oplus f|_{x_i=1}) \oplus f|_{x_i=1}\\
&= \overline{\overline{\overline{x_i}} \vee \overline{(f|_{x_i=0} \equiv f|_{x_i=1} \equiv 0)}} \equiv f|_{x_i=1} \equiv 0\\
&= (x_i \vee (f|_{x_i=0} \equiv f|_{x_i=1} \equiv 0 \equiv 0)) \equiv 0 \equiv f|_{x_i=1} \equiv 0\\
&= (x_i \vee (f|_{x_i=0} \equiv f|_{x_i=1})) \equiv f|_{x_i=1}.
\end{aligned}
$$

(9.2.6): From the positive Davio expansion we get:

$$
\begin{aligned}
&x_i \wedge (f|_{x_i=0} \oplus f|_{x_i=1}) \oplus f|_{x_i=0}\\
&= \overline{\overline{x_i} \vee \overline{(f|_{x_i=0} \equiv f|_{x_i=1} \equiv 0)}} \equiv f|_{x_i=0} \equiv 0\\
&= (\overline{x_i} \vee (f|_{x_i=0} \equiv f|_{x_i=1} \equiv 0 \equiv 0)) \equiv 0 \equiv f|_{x_i=0} \equiv 0\\
&= (\overline{x_i} \vee (f|_{x_i=0} \equiv f|_{x_i=1})) \equiv f|_{x_i=0}.
\end{aligned}
$$

□

9.2.3 Function Properties

In this section, we list some basic properties of Boolean functions concerning sets of variables from the support of f.

The following definition describes if a Boolean function does not depend on a variable:

Definition 9.2.9 *A Boolean function f is called independent of x_i iff*

$$f|_{x_i=1} = f|_{x_i=0}.$$

We check the independence of f by using the difference operator:

Theorem 9.2.3 *A Boolean function f is independent of x_i iff*

$$\frac{\partial f(x)}{\partial x_i} = 0.$$

Proof: Follows directly from the definition of the Boolean difference. □

The linearity of a Boolean function f according to a variable x_i is defined now. Note, that using the following definition variables, the function is independent of, are not called linear. We thus define a more strict linearity than usual.

Definition 9.2.10 *A Boolean function f is called linear in x_i, iff*

$$f|_{x_i=1} = \overline{f|_{x_i=0}}$$

holds.

The following theorem describes a simple method to test the criteria above:

Theorem 9.2.4 *[23]. A Boolean function f is linear in x_i, iff*

$$\frac{\partial f(x)}{\partial x_i} = 1.$$

Another class of Boolean functions, which can trivially be minimized, are the monotone functions.

Definition 9.2.11 *A Boolean function f is called monotone increasing in x_i iff*

$$f|_{x_i=0} \Rightarrow f|_{x_i=1}.$$

It is called monotone decreasing in x_i iff

$$f|_{x_i=1} \Rightarrow f|_{x_i=0}.$$

Some functions, like the parity function, only depend on the number of variables which are set to 1. These Boolean functions are called symmetric. Another characterization of symmetric functions is given by the following definition:

Definition 9.2.12 *A Boolean function f is called symmetric in a set of variables $\{x_{i_0}, \ldots, x_{i_k}\} \subseteq \{x_0, \ldots, x_{n-1}\}$ iff any permutation of variables of that set does not change the function values.*

A criteria to find sets of symmetric variables is given by the following theorem:

Theorem 9.2.5 *A Boolean function f is symmetric with respect to variables x_i and x_j iff*

$$f|_{x_i=0,x_j=1} = f|_{x_i=1,x_j=0}.$$

A generalization of this criteria to a set of variables may trivially be done through transitivity of the symmetry property.

9.2.4 Decision Diagrams

Similar to Bryant's [16] definition of binary decision diagrams, we define decision trees for the disjunctive normal form (DNF), the FPRM, and the FPEP of Definitions 9.2.2, 9.2.5, and 9.2.6. In the case of the Reed-Muller expansion this leads to an FDD [17]. Building a decision tree from Π-term coefficients of the EP, we get an equivalence decision diagram (EDD). Tree based representations of the remaining four normal form coefficients result in trees, isomorphic to the corresponding BDD. This is due to the fact that the coefficients of these normal forms are identical (see Theorem 9.2.1).

All three decision diagrams are based on the same construction rules: the coefficients are interpreted as paths and binary trees are built out of them. Then, the following reduction rules are applied until no further improvement is possible:

- If a tree node has two isomorphic successors, delete it and replace it by a reference to the successor.
- If there are two isomorphic (sub-)trees in a tree, delete one of them and replace it by a reference to the other.

For a detailed description of FDDs, EDDs, BDDs, and the applied manipulation algorithms, see [21, 17, 1].

In Fig. 9.2.1, reduced ordered BDDs, FDDs, and EDDs of function $f(x) = m_2 \vee m_4 \vee m_5 \vee m_6$ are shown.

To understand how to transform these graphs into each other, some isomorphisms are helpful. So, we define the following simple tree operations:

Definition 9.2.13 *Writing $*DD$ as an abbreviation for a BDD, FDD, or EDD representation of a function, we define the following notation:*

$*DD^-$: *A decision diagram, where all literals have negative polarity.*

$\overline{*DD}$: *The complement of a function*

$*DD^{1\leftrightarrow 0}$: *A representation of a function, where all leafs are complemented.*

Combining these operations, we get the following notations: $\overline{*DD^-}$, $*DD^{1\leftrightarrow 0/-}$ and $\overline{*DD^{1\leftrightarrow 0/-}}$.

It is trivial to prove that the result does not depend on the order the operations are applied in.

The following theorem shows, which isomorphisms exist:

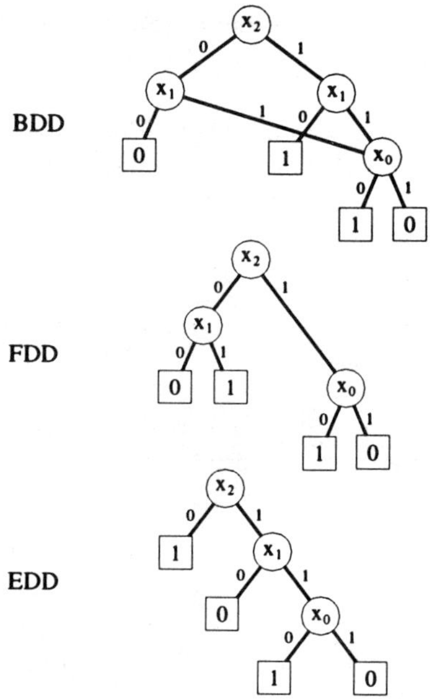

Figure 9.2.1 The different decision diagrams of function $f(x) = m_2 \vee m_4 \vee m_5 \vee m_6$.

Theorem 9.2.6 *[21]. The following combinations of tree operations lead to isomorphic tree representations:*

$$
\begin{array}{lclclcl}
BDD & \Leftrightarrow & \overline{BDD^{1\leftrightarrow 0}}, & \quad & FDD & \Leftrightarrow & \overline{EDD^{1\leftrightarrow 0/-}}, \\
BDD^- & \Leftrightarrow & \overline{BDD^{1\leftrightarrow 0/-}}, & & EDD & \Leftrightarrow & \overline{FDD^{1\leftrightarrow 0/-}}, \\
\overline{BDD} & \Leftrightarrow & BDD^{1\leftrightarrow 0}, & & FDD^- & \Leftrightarrow & \overline{EDD^{1\leftrightarrow 0}}, \\
\overline{BDD^-} & \Leftrightarrow & BDD^{1\leftrightarrow 0/-}, & & EDD^- & \Leftrightarrow & \overline{FDD^{1\leftrightarrow 0}}.
\end{array}
$$

Theorem 9.2.6 may be used to build simple transformation algorithms between different representations.

9.3 THE MULTIPLE DOMAIN MINIMIZATION APPROACH

We now describe our minimization technique. The basic data structures in our synthesis system are BDDs, FDDs, and EDDs. A multilevel circuit is represented as a Boolean net, a directed graph. Every vertex of this net contains a two-level function either as BDD, FDD, or EDD. The connections of the Boolean function's variables to input signals or other intermediate signals are stored in the edges of the net. We implemented all transformation and manipulation algorithms that we need for an efficient calculation of Boolean operations. A direct functional decomposition algorithm is implemented, as well as one for collapsing of Boolean functions.

Our minimization principle is first to isolate the variable based function properties using decomposition and collapsing techniques. We then minimize each function vertex by the best suitable two-level vertex minimization technique. We call this approach *multiple domain minimization*, since BDDs, FDDs, and EDDs are representations of Boolean functions in different function domains.

9.3.1 Property Driven Decomposition and Collapsing

The first step in our minimization system is the elimination of trivial and unnecessary vertices from the multi-level network. A trivial vertex is an inverter or an identity gate. We locate such trivial vertices using an algorithm based on Theorem 9.2.3 and recursively eliminate all predecessor vertices which compute the values of that variable.

Since we want to build internal functions, with properties like symmetry or monotony of their variables, we need a decomposition step.

Definition 9.3.1 *A Boolean function f may be decomposed according to the bound set of variables $B = \{x_0, \ldots, x_{i-1}\}$ and the free set of variables $F = \{x_{i-k}, \ldots, x_{n-1}\}$, with $0 < i < n$, $0 \leq k < i$ if there are j Boolean functions $g_0, \ldots, g_{j-1}$ with $0 < j < i - k$, where f may be represented as*

$$f(x_{n-1}, \ldots, x_0) = h(g_0(x_{i-1}, \ldots, x_0), \ldots, g_{j-1}(x_{i-1}, \ldots, x_0), x_{i-k}, \ldots, x_{n-1}).$$

For $k = 0$ we call it a disjoint decomposition, for $j = 1$ it is called a Ashenhurst decomposition.

For every Boolean function, disjoint decomposition ($k = 0$) is possible. Having selected the bound set B, we get new functions $h, g_0, \ldots, g_{j-1}$. Here, the number j of the g's depends on B. If $j > i = |B|$, function h depends on more variables than f. Wanting to reduce the size of the considered Boolean functions, we try different bound sets until we find one, such that $j < i$ holds. If we do not find an appropriate bound set, no decomposition of the considered Boolean function is performed.

In Fig. 9.3.1, an example of a disjoint decomposition of a BDD is shown. Function f depends on variables $x_0, \ldots, x_5$. A disjoint decomposition is possible, having the bound set $B = \{x_0, x_1, x_2\}$ and free set $F = \{x_3, x_4, x_5\}$. Result of the decomposition are three functions h, g_0, and g_1. Functions g_0 and g_1 depend on variables x_0, x_1, and x_2, whereas function h depends on variables x_3, x_4, x_5, y_0, y_1. After decomposition, f may be replaced by the construct $h|_{y_0=g_0, y_1=g_1}$.

So, for each function vertex in the Boolean net the following steps are performed:

1. *Compute the sets of linear variables L, monotone decreasing and increasing variables M_D and M_A, and all sets of symmetric variables (symmetry groups) $S_1, \ldots, S_s$.*
2. *Select as bound set B the biggest set of the above ones so that decomposition according to Definition 9.3.1 is possible with $k = 0$ and $j < i$.*
3. *If a suitable bound set has been found in the previous step, replace the old function vertex (of function f) by the new function vertices (of functions $h, g_0, \ldots, g_{j-1}$) in the circuit. Otherwise, leave the function vertex unchanged.*

The decomposition step has the following impact on the resulting circuit:

- The number of both function vertices and internal connections will increase.
- The critical path may increase.
- The complexity of a function vertex may decrease.

Opposite effects can be achieved by the counterpart of the decomposition step, the partial collapsing. This means the combination of function vertices based on function properties. Collapsing two Boolean functions f and g in the circuit is the same as constructing the composition function of f and g.

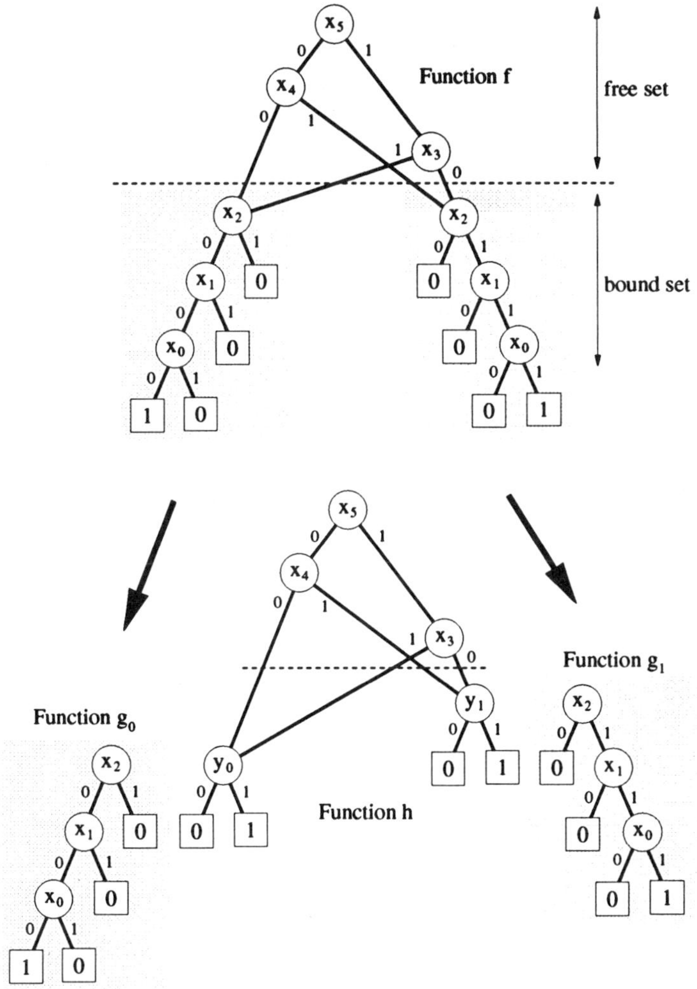

Figure 9.3.1 Example of disjoint decomposition of a Boolean function. The decomposition results in three functions h, g_0, and g_1. $f = h|_{y_0=g_0, y_1=g_1}$ holds.

Definition 9.3.2 *Two Boolean functions f and g can be composed, iff the output of g is used as input variable x_i of f. In this case, we can construct the composition function:*

$$h = f|_{x_i=g} = (f|_{x_i=1} \wedge g) \vee (f_{x_i=0} \wedge \overline{g}).$$

The composition function's support is

$$supp(h) = supp(f) \cup supp(g) \setminus \{g\}.$$

The partial collapsing step has the following impacts on the resulting circuit:

- The number of both function vertices and internal connections will decrease.
- The critical path may decrease.
- The complexity of a function vertex may increase.

In our synthesis system, we apply a mixture of collapsing and decomposition steps until we cannot achieve improvements concerning separability of function properties based on their variables.

9.3.2 Two-Level Vertex Minimization Techniques

Having changed the structure of the Boolean net, two-level minimization steps are applied to each of its function vertices. In our system, it is based on a multiple domain minimization approach. We implemented an exact minimization technique, which applies selection of the most suitable domain, combined with variable ordering and polarity transformation steps. We select the best result, achieved, concerning one of the following minimization criteria:

- The total number of tree nodes in the decision diagram. This minimization goal optimizes the complexity of the internal circuit representation. It is not necessarily the minimum implementation in terms of the following two criteria.
- The total number of literals. In [24], a correlation between the number of CLBs in a FPGA implementation and the number of literals has been proved.
- The total chip area in a multi-level PLA implementation. This simple criteria can easily be computed, and it avoids a complete technology mapping step.

The principle of our two-level vertex minimization algorithm can be stated as follows: for each function vertex in the circuit do

1. *Compute the BDD. Compute the best variable ordering by swapping adjacent variables. Compute the cost function* C_{BDD}.
2. *Compute the FDD. Compute the best variable ordering by swapping adjacent variables. Compute the best polarity value. Compute the cost function* C_{FDD}.
3. *Compute the EDD. Compute the best variable ordering by swapping adjacent variables. Compute the best polarity value. Compute the cost function* C_{EDD}.
4. *Select the optimal domain due to* C_{BDD}, C_{FDD}, *and* C_{EDD}.

Due to the efficiency of our domain transformation algorithms, we are able to compute the full set of domains for the vast majority of all tested benchmarks.

9.4 RESULTS

We implemented all minimization algorithms in C++ and included it into our own multi-level logic synthesis system CLS (Concurrent Logic Synthesis). The underlying data structures are BDDs, FDDs, and EDDs. The benchmarks are performed on a SPARCstation 10/41 with 32 MB of main memory.

Our minimization script works as follows:

1. Read initial network.
2. Redundancy elimination step.
3. Until no further improvements may be achieved:
    - (a) Partial collapse.
    - (b) Compute function properties.
    - (c) Decompose.
4. Two-level minimization for each vertex.
5. Write final network.

We selected combinational multi-level examples from the MCNC 91 benchmark set based on the recommendations for combinational multi-level circuits [25]. Values are measured as the total number of literals. Our starting point is a Boolean network after minimization by SIS 1.1 by applying the SIS minimization standard script.

The improvements of our multiple domain minimization approach are shown in Table 9.4.1. The first column depicts the number of literals of the circuit after reading the initial network. The second column "Decomp./Coll." depicts the impact of the decomposition and partial collapsing step. It includes the computation time in CPU seconds. The column denoted by "Dom. Select." depicts final results after applying the domain selection, and the vertex minimization steps. In the last column, we list the maximum path length from a primary input to a primary output from the initial and the final circuit. It equals the maximum number of function vertices from the input to the output of a circuit. So it is a simple measure for the circuit speed impact of the decomposition and collapsing steps. The values for the number of literals are given as non-overlapping, pair-wise disjoint cubes. This is due to the fact, that we need disjoint cubes for our own technology mapping approach to XILINX FPGAs. Furthermore, this measures the complexity of the underlying data structure.

Table 9.4.2 depicts the results of our minimization approach without performing the SIS minimization step. In this table, we added some bigger examples, where the SIS minimization did not finish within a reasonable time limit, using the standard script. On the other hand, we removed some smaller examples.

Please note the amount of overall computation time for our minimization approach. It shows the efficiency of the underlying data structures, as well as the applicability of our approach for bigger examples. In Table 9.4.3 we listed our minimization results starting from a two-level PLA description. The result is a multi-level implementation which is only generated by our minimization techniques. Table 9.4.4 finally shows the domain distribution of the benchmarks in Table 9.4.1. It illustrates the allocation of internal Boolean functions to the different domains (BDD, FDD, and EDD) and gives an impression of the distribution of function vertices to the different domains. As it can be observed, all three domains are used roughly equally often.

Table 9.4.1 Multi-level minimization results.

Circuit	Size					Pathlength	
	Input	Decomp./Coll.		Dom. Select.		Init.	Fin.
	lit.	lit.	time	lit.	time		
C1355	618	614	0.26	360	0.16	13	14
C499	618	614	0.13	360	0.17	13	14
C5315	5664	2652	20.49	2052	1.23	17	21
C6288	5999	5180	13.17	3636	2.06	59	74
C7552	4071	3046	8.32	2248	1.35	23	30
C880	2058	1054	5.90	715	0.60	13	15
alu2	21031	11233	113.80	7554	19.27	13	20
alu4	55926	23144	368.73	14810	32.45	16	30
apex6	4742	1592	18.03	1293	0.85	6	9
b9	383	268	1.49	227	0.11	6	7
c8	336	280	0.91	215	0.10	7	8
cht	357	282	1.09	246	0.09	4	5
cm138a	60	60	0.01	31	0.02	4	4
cm150a	349	93	1.12	76	0.04	4	6
cm152a	28	28	0.05	28	0.02	3	3
cm162a	111	74	0.20	53	0.02	7	8
cm163a	74	71	0.09	45	0.02	8	9
cm85a	756	88	3.30	72	0.03	5	9
comp	343	236	0.76	172	0.14	8	10
cordic	292	100	0.45	69	0.03	5	9
cu	146	104	0.27	76	0.03	5	7
example2	935	575	2.04	492	0.17	7	8
frg1	4517	1798	19.46	1470	0.58	6	9
frg2	3543	3050	4.61	2508	1.07	11	12
i3	454	278	0.87	212	0.07	4	6
i4	5360	416	11.37	344	0.11	5	13
i6	1191	1066	3.65	631	0.25	4	6
i7	2723	1407	9.34	869	0.33	4	6
i8	24379	2251	77.19	1689	0.58	8	12
i9	5328	1287	18.05	931	0.34	7	11
lal	399	183	0.60	127	0.05	5	6
pcle	177	105	0.53	90	0.04	6	7
pcler8	251	147	0.53	118	0.04	7	9
rot	4006	2469	18.74	1996	1.19	10	15
t481	3298	1540	17.61	1357	0.63	9	15
x1	9328	1636	134.22	1251	1.47	7	10

Table 9.4.2 Direct multi-level minimization results.

Circuit	Size					Pathlength	
	Input	Decomp./Coll.		Dom. Select.		Init.	Fin.
	lit.	lit.	time	lit.	time		
alu2	43380	30577	228.75	21815	21.08	10	14
apex6	3379	1424	10.42	1172	0.54	9	12
b9	365	303	0.21	215	0.08	10	10
c8	441	359	0.71	276	0.10	4	5
cht	261	261	0.54	261	0.09	2	2
cm138a	86	86	0.00	35	0.02	3	3
cm163a	92	76	0.06	53	0.02	5	6
comp	250	255	0.30	180	0.07	7	9
cordic	281	250	0.01	164	0.06	14	11
count	293	293	0.21	158	0.07	18	18
dalu	5528	4842	7.92	3719	1.56	25	31
des	23342	19620	99.63	7207	5.88	6	7
example2	982	610	2.73	504	0.20	7	8
f51m	403	395	0.20	162	0.07	3	2
frg2	4889	3366	11.15	2565	1.20	9	11
i10	6259	5459	0.21	4504	1.89	55	37
i2	2906	2905	0.09	267	0.17	5	5
i3	262	262	0.01	196	0.08	3	3
i4	496	496	0.03	340	0.13	5	5
i6	1352	1278	0.11	963	0.49	4	3
i7	1852	1778	0.14	1237	0.67	4	4
i8	6606	6368	0.35	4388	2.06	9	7
i9	2773	2749	0.18	1429	0.73	8	6
lal	375	375	0.55	293	0.09	6	7
pair	3133	3067	2.18	2215	1.15	19	19
pcle	160	101	0.22	80	0.02	8	10
pcler8	264	141	0.39	112	0.04	9	11
rot	3904	1984	20.56	1598	1.06	11	15
ttt2	936	598	1.65	439	0.17	5	7
unreg	256	256	0.32	160	0.07	3	3
vda	12766	12664	5.86	1432	0.86	3	4
x1	11996	2077	90.33	1495	1.80	2	6
x2	160	149	0.48	93	0.03	3	4
x3	2687	2340	7.08	1684	0.89	10	14
x4	2812	1199	11.72	989	0.51	4	6
z4ml	344	136	0.20	83	0.03	3	4

Table 9.4.3 Two-level minimization results.

Circuit	Size					Pathlength	
	Input	Decomp./Coll.		Dom. Select.		Init.	Fin.
	lit.	lit.	time	lit.	time		
5xp1	363	361	0.27	193	0.08	2	4
9sym	1170	1170	0.08	588	0.04	2	2
add6	3706	3106	6.55	302	0.33	2	7
addm4	1771	1756	1.67	660	0.83	2	3
adr4	506	412	0.81	62	0.08	2	5
apex5	12131	8430	159.00	5225	7.01	2	4
b12	240	205	0.70	151	0.09	2	3
b3	4944	4112	144.67	3283	8.39	2	6
b9	1762	1349	2.57	287	0.17	2	7
chkn	4607	2320	33.10	2291	7.82	2	5
cordic	267279	549	246.88	291	0.14	2	4
ex5	1893	1797	2.33	1064	0.36	2	4
ex7	1762	1349	2.53	287	0.17	2	7
exep	3773	2469	141.98	2303	2.72	2	5
f51m	395	395	0.19	159	0.07	2	2
ibm	4116	3290	20.20	2441	2.79	2	3
in4	5106	4197	166.44	3359	8.88	2	6
in7	758	518	9.80	384	0.42	2	5
intb	9633	9633	9.46	4148	7.57	2	2
life	700	935	0.13	452	0.03	2	3
lin.rom	2903	2902	2.73	1889	1.66	2	6
m181	241	206	0.71	154	0.08	2	3
misg	591	339	1.94	257	0.16	2	4
mish	270	251	0.56	146	0.07	2	5
misj	206	185	0.15	65	0.03	2	5
mlp4	1044	1037	0.77	443	0.42	2	3
mp2d	463	385	2.24	221	0.13	2	4
pdc	7128	4948	49.39	4872	4.85	2	5
pope.rom	1481	1478	1.76	931	0.57	2	4
radd	472	400	0.46	62	0.04	2	5
rd53	168	168	0.04	45	0.01	2	2
rd73	992	992	0.08	189	0.03	2	2
rd84	2266	2266	0.23	352	0.06	2	2
ryy6	1128	43	1.37	28	0.02	2	4
sao2	716	665	0.90	493	0.20	2	3
sqr6	266	265	0.25	158	0.07	2	4
t1	977	680	4.53	496	0.42	2	6
t481	10736	10736	1.48	42	0.07	2	2
wim	85	85	0.07	54	0.03	2	2
x7dn	10097	7980	131.63	6005	6.24	2	4
xor5	80	80	0.01	5	0.00	2	2
z4	340	136	0.25	83	0.03	2	4

Table 9.4.4 The number of BDD, FDD, and EDD function vertices for the benchmarks of Table 9.4.1.

Circuit	Distribution of Domains			
	BDD	FDD	EDD	all
C1355	18	104	44	166
C499	18	104	44	166
C5315	400	125	173	698
C6288	512	436	660	1608
C7552	293	267	135	695
C880	50	26	37	113
alu2	59	16	18	93
alu4	141	28	51	220
apex6	252	18	112	382
b9	26	5	18	49
c8	24	9	18	51
cht	80	0	36	116
cm138a	0	0	11	11
cm150a	20	1	8	29
cm152a	3	0	0	3
cm162a	6	3	13	22
cm163a	1	6	6	13
cm85a	14	6	6	26
comp	8	24	8	40
cordic	9	6	5	20
cu	12	0	6	18
example2	112	8	28	148
frg1	19	1	14	34
frg2	209	18	111	338
i3	20	0	66	86
i4	110	0	48	158
i6	78	0	105	183
i7	142	0	132	274
i8	457	5	219	681
i9	251	1	128	380
lal	17	6	20	43
pcle	24	4	7	35
pcler8	25	6	14	45
pm1	11	1	6	18
t481	298	1	72	371
x1	85	6	48	139

9.5 CONCLUSION

In the past it has been proven, that both operational and spectral minimization techniques have advantages and disadvantages. While other minimization tools consider only either operational or spectral techniques, our approach is a step towards the combination of both within a single logic minimization system.

Using efficient graph based representations of Boolean functions in all used normal forms, we presented decomposition and collapsing techniques, which take special function properties into account. These techniques allow the treatment of multi-level networks. The described vertex minimization techniques are based on an optimal domain selection and application of special domain specific minimization techniques.

Acknowledgment

The work on this chapter was partially supported by DFG project RO 1030/3-1.

REFERENCES

[1] U. Kebschull, E. Schubert, and W. Rosenstiel, "Multilevel Logic Synthesis Based on Functional Decision Diagrams", in *Proc. EDAC*, 1992, pp. 43-47.

[2] G. Lee and J.-Y. Chang, "Synthesis of Multi-level Reed Muller Circuits using Matrix Transformations", in *IFIP WG 10.5 Workshop on Applications of the Reed-Muller Expansion in Circuit Design*, 1993, pp. 61-68.

[3] M.A. Perkowski, "A Fundamental Theorem for EXOR Circuits", in *IFIP WG 10.5 Workshop on Applications of the Reed-Muller Expansion in Circuit Design*, 1993, pp. 52-60.

[4] M.A. Perkowski, A. Sarabi, and F.R. Beyl, "XOR Canonical Forms of Switching Functions", in *IFIP WG 10.5 Workshop on Applications of the Reed-Muller Expansion in Circuit Design*, 1993, pp. 27-32.

[5] I. Schäfer, M.A. Perkowski, and H. Wu, "Multilevel Logic Synthesis for Cellular FPGAs Based on Orthogonal Expansions", in *IFIP WG 10.5 Workshop on Applications of the Reed-Muller Expansion in Circuit Design*, 1993, pp. 42-51.

[6] D. Brand and T. Sasao, "Minimization of AND-EXOR Expressions Using Rewrite Rules", *IEEE Trans. on Comp.*, vol. 42, no. 5, May 1993, pp. 568-576.

[7] T. Sasao, "EXMIN2: A Simplification Algorithm for Exclusive-OR-Sum-of-Products Expressions for Multiple-Valued-Input Two-Valued-Output Functions", *IEEE Trans. on CAD*, vol. 12, no. 5, May 1993, pp. 621-632.

[8] T. Sasao, "An Exact Minimization of AND-EXOR Expressions Using BDDs", in *IFIP WG 10.5 Workshop on Applications of the Reed-Muller Expansion in Circuit Design*, 1993, pp. 91-98.

[9] J.M. Saul, "Logic Synthesis for Arithmetic Circuits Using Reed-Muller Representation", in *Proc. EDAC*, 1992, pp. 109-113.

[10] J.M. Saul, "Towards a Mixed Exclusive-/Inclusive-OR Factored Form", in *IFIP WG 10.5 Workshop on Applications of the Reed-Muller Expansion in Circuit Design*, 1993, pp. 2-5.

[11] T. Sasao and Ph.W. Besslich, "On the Complexity of Mod-2 Sum PLA's", *IEEE Trans. on Comp.*, vol. 39, no. 2, Feb. 1990, pp. 262-266.

[12] U. Rollwage, "The Complexity of Mod-2 Sum PLA's for Symmetric Functions", in *IFIP WG 10.5 Workshop on Applications of the Reed-Muller Expansion in Circuit Design*, 1993, pp. 6-12.

[13] C. Damm, "How much ExOR Improves on OR?", in *IFIP WG 10.5 Workshop on Applications of the Reed-Muller Expansion in Circuit Design*, 1993, pp. 13-19.

[14] S.B. Akers, "Binary Decision Diagrams", *IEEE Trans. on Comp.*, vol. 27, no. 6, June 1978, pp. 509-516.

[15] C. Y. Lee, "Representation of Switching Circuits by Binary-Decision Programs", in *Bell System Tech. Journal*, July 1959.

[16] R.E. Bryant, "Graph-Based Algorithms for Boolean Function Manipulation", *IEEE Trans. on Comp.*, vol. 35, no. 8, Aug. 1986, pp. 677-691.

[17] U. Kebschull and W. Rosenstiel, "Efficient Graph-Based Computation and Manipulation of Functional Decision Diagrams", in *Proc. EDAC 93*, 1993.

[18] E. Schubert, U. Kebschull, and W. Rosenstiel, "Some Optimizations for Functional Decision Diagrams", in *IFIP WG 10.5 Workshop on Applications of the Reed-Muller Expansion in Circuit Design*, 1993, pp. 176-180.

[19] B. Becker, R. Drechsler, and M. Theobald, "On the Implementation of a Package for Efficient Representation and Manipulation of Functional Decision Diagrams", in *IFIP WG 10.5 Workshop on Applications of the Reed-Muller Expansion in Circuit Design*, 1993, pp. 162-169.

[20] D.H. Green, *Modern Logic Design*, Addison Wesley, 1986.

[21] U. Kebschull, "Graphische Darstellung und Transformation von Normalformen Boolescher Funktionen", Tech. Rep. WSI-93-6, Universität Tübingen, 1993.

[22] G. Boole, *An Investigation of the Laws of Thought on which are founded the Mathematical Theories of Logic and Probabilities*, London, 1854.

[23] D. Bochmann and Ch. Posthoff, *Binäre dynamische Systeme*, Oldenburg Verlag München Wien, 1981.

[24] M. Schlag, P.K. Chan, and J. Kong, "Empirical Evaluation of Multilevel Logic Minimization Tools for a Lookup-Table Based Field-Programmable Gate Array", *IEEE Trans. on CAD*, vol. 12, no. 5, May 1993, pp. 713-722.

[25] S. Yang, "Logic Synthesis and Optimization Benchmarks User Guide", Tech. Rep., Microelectronics Center of North Carolina, Jan. 1991.

10

SATISFIABILITY PROBLEMS FOR OFDDS

Ralph WERCHNER[1]
Thilo HARICH[1]
Rolf DRECHSLER[2]
Bernd BECKER[2]

[1] *Computer Science Department*
J.W. Goethe-University
60054 Frankfurt am Main, Germany

[2] *Institute of Computer Science*
Albert-Ludwigs-University
79110 Freiburg im Breisgau, Germany

Abstract— We investigate the complexity of problems on Ordered Functional Decision Diagrams (OFDDs) related to satisfiability, i.e. SAT-ONE, SAT-ALL, and SAT-COUNT. We prove that SAT-ALL has a running time linear in the product of the number of satisfying assignments and the size of the given OFDD. Counting the satisfying assignments in an OFDD is proved to be #*P*-complete, and thus not possible in polynomial time unless P=NP.

10.1 INTRODUCTION

In the context of logic synthesis tools in CAD systems a new type of decision diagram, called *Ordered Functional Decision Diagram* (OFDD), has recently been introduced [8, 7, 1]. While OBDDs result from repeated applications of Shannon expansions, OFDDs are based on the Davio expansion, i.e. an AND/EXOR based decomposition. In the meantime, OFDDs are used in various applications, such as Boolean matching, technology mapping, detection of symmetries, and logic synthesis [11, 13, 14, 6, 4]. But up to now a theoretical background for OFDDs, as it exists for OBDDs with the huge number of results on the properties of branching programs (see e.g. [15, 10]), has not been provided. On the other hand, this background is necessary to direct further research on this topic. A first step in this direction has been given in [2]. There the relation between OBDDs and OFDDs has been investigated from the theoretical point of view. It has been shown that there is an exponential gap between OBDDs and OFDDs in both directions, i.e. there are Boolean functions that can only be represented by OBDDs of exponential size but each OFDD is of polynomial size and vice versa. Additionally, it has been proved that the AND-synthesis of two OFDDs can result in an output of size exponential in the input size. Thus, deciding $f \leq g$, for f and g given as OFDDs, cannot be done in polynomial time by the AND-synthesis of f and $\bar{g}$, as it is the case for OBDDs. On the other hand, no hardness result is known for this problem.

Further operations turn out to be important in practical applications. One important class is the set of *satisfiability problems*, i.e. problems that are related to the satisfiability of a Boolean function f. Let $f^{-1}(1) = \{\mathbf{x} | f(\mathbf{x}) = 1\}$ be the satisfiability set of f and denote its cardinality by $\sharp(f)$. There are three major problems related to the satisfiability set [3, 16]:

SAT-ONE: Decide whether $f^{-1}(1) \neq \emptyset$ and, if so, find one $\mathbf{x} \in f^{-1}(1)$.

SAT-ALL: Enumerate all $\mathbf{x} \in f^{-1}(1)$.

SAT-COUNT: Compute $\sharp(f)$.

These problems arise in many applications in logic synthesis and verification. The complexity of these problems depends on the representation f is given in. For a representation REP, *e.g.* OBDD, OFDD or 3CNF (conjunction of clauses of 3 literals), we append the abbreviation REP to the abbreviation of the problem. Thus the problem of counting the satisfying assignments for a given OFDD is called SAT-COUNT-OFDD, and asking for a satisfying assignment for a 3CNF is SAT-ONE-3CNF, commonly called 3SAT.

In [2] it has been shown that deciding whether a given OFDD represents the constant 0 function is trivial if the OFDD is reduced. The two remaining satisfiability problems for OFDDs were left open.

We show how to compute the lexicographically first satisfying assignment of a reduced OFDD in time linear in the number of variables. For SAT-ALL-OFDD we give an algorithm with running time $O(|G| \cdot \sharp(f))$ and space requirement $O(|G|)$, where $|G|$ denotes the size of the given OFDD. For SAT-COUNT-OFDD we show that the problem cannot be solved in polynomial time unless $NP = P$.

Table 10.1.1 Time complexities of typical operations.

Operation	OBDD	OFDD
$f \oplus g$	$O(\|G_f\| \cdot \|G_g\|)$	$O(\|G_f\| \cdot \|G_g\|)$
$f \wedge g$	$O(\|G_f\| \cdot \|G_g\|)$	$2^{O(\|G_f\|+\|G_g\|)}$
$f = g$?	$O(\min(\|G_f\|, \|G_g\|))$	$O(\min(\|G_f\|, \|G_g\|))$
$f \leq g$?	$O(\|G_f\| \cdot \|G_g\|)$	?
$f\|_{x_i=c}$	$O(\|G_f\|)$	$O(\|G_f\|^2)$
SAT-ONE	O(n)	O(n)
SAT-COUNT	$O(\|G_f\|)$	$\#P$-complete
SAT-ALL	$O(n \cdot \sharp(f))$	$O((\|G_f\| + n) \cdot \sharp(f))$

Table 10.1.1 summarizes the time complexities of some important operations on Boolean functions f and g on n variables represented by OBDDs or OFDDs G_f and G_g.

10.2 FUNDAMENTAL CONCEPTS AND DEFINITIONS

In the following, *Ordered Functional Decision Diagrams* (with mixed polarity variables) are defined. The core of the data structures is a *Decision Diagram* (DD), which is a labeled directed acyclic graph with some additional properties.

Definition 10.2.1 A *DD* over $X_n = \{x_1, x_2, \ldots, x_n\}$ is a rooted directed acyclic graph $G = (V, E)$ with vertex set V containing two types of vertices, *non-terminal* and *terminal* vertices. A non-terminal vertex v is labeled with a variable from X_n, called the *decision variable* for v, and has exactly two successors denoted by $low(v), high(v) \in V$. A terminal vertex v is labeled with a 0 or 1 and has no successors.

The size of a DD G, denoted by $|G|$, is given by its number of non-terminal nodes. If DDs are to be used as a data structure in design automation, it turns out that further restrictions on their structure will be necessary. One such restriction is defined below:

Definition 10.2.2 A DD is *ordered* if the variables are encountered at most once and in the same order on every path in the DD from the root to a terminal vertex. (In the following, the letter O will be used to denote ordered DDs.)

It is possible to define certain reductions on the DDs in order to reduce their size. Two reduction types will be used in the following:

Type I: Delete a node v' whose label and successors are identical to the label and successors of another node v and redirect the edges pointing to v' to point to v.

Type D: Delete a node v whose successor $high(v)$ is the terminal 0 and connect the incoming edges of v to $low(v)$.

Definition 10.2.3 A DD is *reduced* if no reductions can be applied to the DD. Two DDs, G_1 and G_2, are called *equivalent* iff G_2 results from G_1 by repeated applications of reductions and inverse reductions. A DD G_2 is called the *reduction* of a DD G_1, if G_1 and G_2 are equivalent and G_2 itself is reduced.

A careful analysis of the proofs in [3, 12] shows that the following lemma is valid for DDs:

Lemma 10.2.1 The reduction of an ordered DD G is uniquely determined and can be computed in time linear in the size of G.

Now we define how DDs can be interpreted as Boolean functions. To do this, the following notions are helpful. Let $f : \mathbf{B}^n \to \mathbf{B}$ be a Boolean function over the variable set X_n. Then f_i^0 denotes the *cofactor* of f with respect to $x_i = 0$, defined by $f_i^0(\mathbf{x}) := f(x_1, .., x_{i-1}, 0, x_{i+1}, .., x_n)$ for $\mathbf{x} = (x_1, x_2, \ldots, x_n) \in \mathbf{B}^n$. Similarly, f_i^1 denotes the cofactor for $x_i = 1$. Finally, f_i^2 is defined as $f_i^2 := f_i^0 \oplus f_i^1$. (Note that the three functions f_i^0, f_i^1, f_i^2 can naturally be interpreted as Boolean functions from $\mathbf{B}^{n-1}$ to $\mathbf{B}$ defined over the variables $x_1, \ldots, x_{i-1}, x_{i+1}, \ldots, x_n$.) Using the above definitions, the following decompositions can be proven for an arbitrary Boolean function f:

$$f = f_i^0 \oplus x_i f_i^2 \qquad \textit{positive Davio (pD)} \qquad (10.2.1)$$

$$f = f_i^1 \oplus \overline{x}_i f_i^2 \qquad \textit{negative Davio (nD)} \tag{10.2.2}$$

Now, *Ordered Functional Decision Diagrams* (OFDDs) can formally be defined as follows:

Definition 10.2.4 An *OFDD over* X_n is given by an ordered DD G over X_n together with a decomposition type list (DTL) $\mathbf{d} = (d_1, \ldots, d_n)$, $d_i \in \{\text{pD}, \text{nD}\}$. The function $f_G^d : \mathbf{B}^n \to \mathbf{B}$ represented by G with DTL $\mathbf{d}$ is defined as:

1. If G consists of a single node labeled with 0 (1), then G is an OFDD for $f = 0$ ($f = 1$).
2. If G has a root v with label x_i, then G is an OFDD for

$$f = \begin{cases} f^d_{G_{low(v)}} \oplus x_i f^d_{G_{high(v)}} & : \ d_i \text{ is pD} \\ f^d_{G_{low(v)}} \oplus \overline{x}_i f^d_{G_{high(v)}} & : \ d_i \text{ is nD} \end{cases},$$

   where $f^d_{G_{low(v)}}$ ($f^d_{G_{high(v)}}$) is the function represented by the DD rooted at $low(v)$ ($high(v)$).

Utilizing reductions, it is possible to define canonical representations of functions if the decomposition types of the variables and the ordering remain fixed.

Lemma 10.2.2 Fixing the variable ordering and the DTL there is a unique reduced OFDD for every Boolean function.

Investigating the complexity of the three satisfiability problems for OFDDs, it is sufficient to concentrate on the special case that for every variable the decomposition type is positive (pD). Algorithms for satisfiability problems for positive decomposition types can easily be modified to cope with arbitrary decomposition types. Let G be a DD and let f be the Boolean function represented by G with only positive decomposition types. Let $\mathbf{d} = (d_1, \ldots, d_n)$ be an arbitrary DTL and let $\tilde{f}$ be the Boolean function represented by G with DTL $\mathbf{d}$. Define the bijection $\varphi(x_1, \ldots, x_n) = (y_1, \ldots, y_n)$ by setting $y_i = x_i$ if $d_i = \text{pD}$ and $y_i = \overline{x}_i$ if $d_i = \text{nD}$. Then

$$\tilde{f}^{-1}(1) = \{\varphi(x_1, \ldots, x_n) | (x_1, \ldots, x_n) \in f^{-1}(1)\}.$$

Thus, f is satisfiable iff $\tilde{f}$ is satisfiable and SAT-COUNT yields the same result on f and $\tilde{f}$. The enumeration of $\tilde{f}^{-1}(1)$ can be done by enumerating $f^{-1}(1)$ and switching certain bits in each assignment. Therefore, the OFDDs considered in the following have DTL $\mathbf{d} = (pD, \ldots, pD)$ and for simplicity f_G is used instead of f_G^d to denote the function represented by an OFDD G with positive decomposition types.

10.3 COMPUTING SATISFYING ASSIGNMENTS

From Lemma 10.2.1 it follows that an OFDD represents the constant 0 function iff it can be reduced to a single node, the terminal 0 node. Thus, the question $f^{-1}(1) \neq \emptyset$ is trivial if the given OFDD is reduced. Otherwise it can be decided in time linear in the size of the given OFDD. The lexicographically first satisfying assignment can be found as follows: Starting from the root of the reduced OFDD, we continue following the *low*-pointer, unless it points to the terminal 0 node, in which case we follow the *high*-pointer. This path ends at the terminal 1 node. Setting all Boolean variables to 1 where we followed the *high*-pointer and the others to 0 results in the lexicographically first satisfying assignment according to the variable order of the OFDD.

Now we give an algorithm for SAT-ALL-OFDD. As before, let G be an OFDD with positive decomposition types and $f = f_G$. Assume w.l.o.g. that the order of the variables is given by $(x_1, x_2, \ldots, x_n)$. The algorithm enumerates the satisfying assignments of f lexicographically with x_n being the most significant bit. To avoid running over all 2^n assignments, we use a simple criterion to decide whether an assignment to $(x_i, \ldots, x_n)$ can be extended to a complete satisfying assignment.

The algorithm consists of the recursive procedure *Sat* which is initially invoked as *Sat(n)*. There is a global variable $l(v)$ for each node v of G and a variable a_i for $1 \leq i \leq n$. These variables take values in $\{0, 1, *\}$. C is a global integer variable. Let V_i be the set of nodes marked with the variable x_i. Whenever *Sat(i)* is invoked for some $1 \leq i \leq n$, the algorithm will make sure that the following holds: For $j > i$, we have $a_j \in \{0, 1\}$. For any node $v \in V_j$ with $j > i$ the label $l(v)$ is the value of the Boolean function represented by v for the assignment $(a_j, \ldots, a_n)$. Furthermore, $l(v) = *$ and $a_j = *$ for $v \in V_j$, $j \leq i$. Finally, $C = |\{(u, v) \in E | l(u) = *, l(v) = 1\}|$.

```
procedure Sat(i)

for a_i ∈ {0,1} do
    for v ∈ V_i do
        l(v) := l(low(v)) ⊕ a_i l(high(v))
        C := C − l(low(v)) − l(high(v)) + ingrad(v) * l(v)
    if i = 1 and l(root(G)) = 1 then print (a_1, ..., a_n)
    if i > 1 and C > 0 then Sat(i-1)
    for v ∈ V_i do
        C := C + l(low(v)) + l(high(v)) − ingrad(v) * l(v)
a_i = *
for v ∈ V_i do
    l(v) = *
```

Figure 10.3.1 Procedure for SAT-ALL-OFDD.

It can be shown that by the time of a call to *Sat(i)* the partial assignment $(a_{i+1}, \ldots, a_n)$ can be extended to a complete satisfying assignment iff there is an edge (u, v) in G with $u \in V_1 \cup \ldots \cup V_i$, $v \in V_{i+1} \cup \ldots \cup V_n$, and $l(v) = 1$. These edges are counted by C.

The algorithm consists of a single call to the procedure *Sat(n)* (see Figure 10.3.1). All a_i and $l(v)$ for non-terminal nodes v are initialized to $*$. For the two terminal nodes of G, the value of l is set to their label. C is initialized with the number $ingrad(v)$ of ingoing edges of the terminal node v with label 1.

Since the algorithm performs a constant amount of work for each satisfying assignment and each node of G, we get:

Theorem 10.3.1 The algorithm for SAT-ALL-OFDD for a function f represented by an OFDD G has running time $O((|G|+n)\cdot\sharp(f))$ and space requirement $O(|G|+n)$.

10.4 COUNTING SATISFYING ASSIGNMENTS

We investigate the complexity of SAT-COUNT-OFDD in comparison to SAT-COUNT-OBDD. Let us review the algorithm to compute the number of satisfying assignments for a complete OBDD: Let f be the Boolean function represented by a node v of a complete OBDD and let f_{low} and f_{high} be the functions represented by the *low*-successor and the *high*-successor of v. Since $f^{-1}(1)$ is the disjoint union of $f_{low}^{-1}(1)$ and $f_{high}^{-1}(1)$, we have $\sharp(f) = \sharp(f_{low}) + \sharp(f_{high})$. This enables us to solve SAT-COUNT-OBDD in linear time by a single bottom-up parse of the given graph computing a single number for each node.

For OFDDs the situation is more complicated. The satisfiability set of the cofactor f^1 is the symmetric difference of $f_{low}^{-1}(1)$ and $f_{high}^{-1}(1)$. Thus, knowledge of $\sharp(f_{low})$ and $\sharp(f_{high})$ is no longer sufficient to compute $\sharp(f)$. In a naive approach to the SAT-COUNT-OFDD problem we cannot bound the amount of information to be computed for each node.

Of course, the above argument does not rule out the existence of a different and more efficient approach to the SAT-COUNT-OFDD problem. However, we will show that SAT-COUNT-OFDD cannot be solved in polynomial time unless $P = NP$. In fact, we show an even stronger result: SAT-COUNT-OFDD is complete for $\#P$, a complexity class briefly introduced in the next paragraph (see Chapter 2 in [9]). Therefore, the existence of a polynomial time algorithm for SAT-COUNT-OFDD would imply $P = \#P$, which would in turn imply $P = NP$. Since this is generally not believed to be true, we provide strong evidence against the existence of a polynomial time SAT-COUNT-OFDD algorithm.

A typical problem in NP is to decide if there exists an object in some (usually exponentially large) search space with a specific property, *e.g.* if there is an assignment to Boolean variables with the property that a given 3CNF is satisfied by this assignment. A typical problem in the class $\#P$ (pronounced *number P* or *sharp P*) is to compute the number of objects in the search space with this specific property, *e.g.* to compute the number of assignments satisfying the given 3CNF, or to compute the number of Hamiltonian cycles in a given graph.

SAT-COUNT-OFDD is naturally contained in $\#P$. To show the completeness for $\#P$, we have to prove that each problem in $\#P$ can be reduced to SAT-COUNT-OFDD. That means a polynomial time procedure for SAT-COUNT-OFDD could be used to design a polynomial time algorithm for each problem in $\#P$. Since it is well-known that SAT-COUNT-3CNF is complete for $\#P$, we just have to reduce SAT-COUNT-3CNF to SAT-COUNT-OFDD. This is done in the remainder of this section.

Let $\mathbf{x} = (x_1, \ldots, x_m)$ denote a vector of Boolean variables and let $f(\mathbf{x}) = \bigwedge_{1 \leq i \leq n} c_i(\mathbf{x})$ be the given 3CNF, *i.e.* each c_i is a clause of 3 literals over the variables $\{x_1, \ldots, x_m\}$.

To reduce SAT-COUNT-3CNF to SAT-COUNT-OFDD, we could try to construct an OFDD G of size polynomial in n and m so that $\sharp(f) = \sharp(f_G)$. Then, a polynomial time algorithm solving SAT-COUNT-OFDD could easily be used to solve SAT-COUNT-3CNF. Unfortunately, such a construction of an OFDD out of a 3CNF cannot exist unless $P = NP$, because it would imply SAT-ONE-3CNF $\in P$. However, what we can achieve, is a construction of a polynomial size OFDD G with $\sharp(f_G) = 2^{3n+m+3} + 2^{n+1}\sharp(f)$. Given $\sharp(f_G)$, the value of $\sharp(f)$ can easily be reconstructed. Thus, we have a polynomial time reduction from SAT-COUNT-3CNF to SAT-COUNT-OFDD.

The transformation of the 3CNF for f into a corresponding OFDD G can be divided into two major steps. The first step transforms the AND-operation on the clauses into the EXOR-operation on corresponding functions $g_1, \ldots, g_{2n+1}$. In the second step, the inherent ability of an OFDD to compute the EXOR-operation is used to combine the OFDDs representing the g_i's into a single OFDD G with $\sharp(f_G)$ determining $\sharp\left(\bigoplus_{1 \leq i \leq 2n+1} g_i\right)$.

For the first step, let $\mathbf{y} = (y_1, \ldots, y_n)$ be a vector of new Boolean variables. For $1 \leq j \leq n$, define the symmetric Boolean function $E_j(\mathbf{y})$ to be 1 iff exactly j of the variables y_i are set to 1. Defining the functions $g_1, \ldots, g_{2n+1}$ as

$$
\begin{aligned}
\mathrm{ODD}(\mathbf{y}) &= \bigvee_{\substack{0 \leq j \leq n \\ j \text{ odd}}} E_j(\mathbf{y}) \\
\mathrm{EVEN}(\mathbf{y}) &= \bigvee_{\substack{0 \leq j \leq n \\ j \text{ even}}} E_j(\mathbf{y}) \\
g_i(\mathbf{x}, \mathbf{y}) &= \mathrm{ODD}(\mathbf{y}) \wedge y_i \wedge c_i(\mathbf{x}) \qquad (1 \leq i \leq n) \\
g_{n+i}(\mathbf{x}, \mathbf{y}) &= \mathrm{EVEN}(\mathbf{y}) \wedge y_i \wedge c_i(\mathbf{x}) \qquad (1 \leq i \leq n) \\
g_{2n+1}(\mathbf{x}, \mathbf{y}) &= \mathrm{EVEN}(\mathbf{y})
\end{aligned}
$$

we have

Theorem 10.4.1

$$\sharp\left(\bigoplus_{1\le i\le 2n+1} g_i(\mathbf{x},\mathbf{y})\right) = 2^{m+n-1} + 2^{n-1}\sharp(f(\mathbf{x})).$$

The proof of Theorem 10.4.1 is rather technical and postponed to Section 10.6.

Now we show how to construct OFDDs for the functions $g_1,\ldots,g_{2n+1}$ and how to combine these OFDDs into a single OFDD G with $\sharp(f_G)$ determining $\sharp\left(\bigoplus_{1\le i\le 2n+1} g_i(\mathbf{x},\mathbf{y})\right)$.

Figure 10.4.1 shows OFDDs for the functions $\mathrm{ODD}(\mathbf{y})\wedge y_i$ and $\mathrm{EVEN}(\mathbf{y})\wedge y_i$ as is easily seen by bottom-up computing the subfunctions for all nodes. The left (right) outgoing edge of each node v points to $low(v)$ ($high(v)$). Every clause $c_i(\mathbf{x})$ can be represented by an OFDD of constant size since it essentially depends on no more than three variables. Given two OFDDs H and H' over disjoint sets of variables, an OFDD for $f_H \wedge f_{H'}$ is obtained by identifying the source node of H' with the 1-sink of H. This way, OFDDs $G_1,\ldots,G_{2n+1}$ for the functions $g_1,\ldots,g_{2n+1}$ can be constructed, each of size $O(n)$.

The OFDDs $G_1,\ldots,G_{2n+1}$ are combined building the two OFDDs G' and G'' as shown in Figure 10.4.2. For $(s_1,\ldots,s_{2n+2})\neq(1,1,\ldots,1)$ we have

$$f_{G'}(s_1,\ldots,s_{2n+2},\mathbf{x},\mathbf{y}) = \neg f_{G''}(s_1,\ldots,s_{2n+2},\mathbf{x},\mathbf{y})$$

and

$$\begin{aligned} f_{G'}(1,1,\ldots,1,\mathbf{x},\mathbf{y}) &= f_{G''}(1,1,\ldots,1,\mathbf{x},\mathbf{y}) \\ &= \bigoplus_{1\le i\le 2n+1} g_i(\mathbf{x},\mathbf{y}). \end{aligned}$$

Thus

$$\begin{aligned} \sharp(f_{G'})+\sharp(f_{G''}) &= (2^{2n+2}-1)\,2^{n+m} + 2\,\sharp\left(\bigoplus_{1\le i\le 2n+1} g_i(\mathbf{x},\mathbf{y})\right) \\ &= 2^{3n+m+2} + 2^n\sharp(f(\mathbf{x}))\,. \end{aligned}$$

For the OFDD G constructed out of G' and G'' as shown in Figure 10.4.3 holds

$$\begin{aligned} f_G(t_1,t_2,t_3,s_1,\ldots,s_{2n+2},\mathbf{x},\mathbf{y}) &= (\overline{t_1}\wedge\overline{t_3}\wedge f_{G'})\vee(t_1\wedge t_2\wedge f_{G''}) \\ \Longrightarrow\ \sharp(f_G) &= 2\,(\sharp(f_{G'})+\sharp(f_{G''})) \\ &= 2^{3n+m+3} + 2^{n+1}\sharp(f(\mathbf{x}))\,. \end{aligned}$$

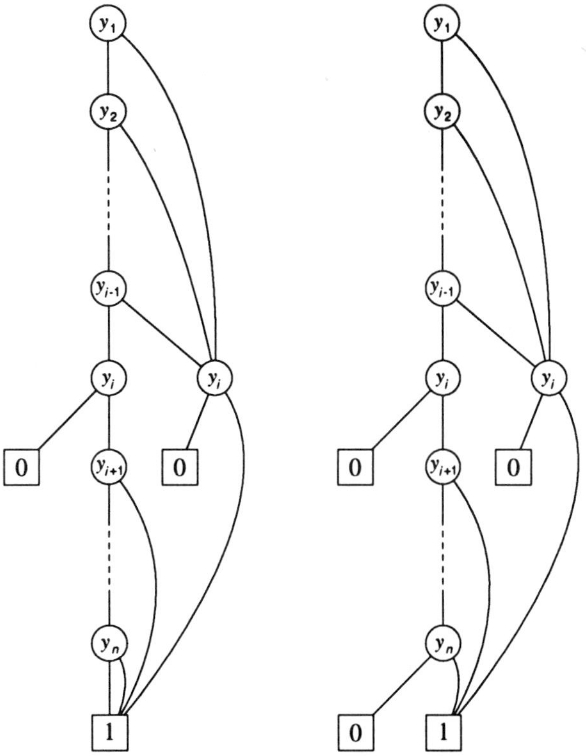

Figure 10.4.1 OFDDs representing ODD(Y) $\wedge\, y_i$ and EVEN(Y) $\wedge\, y_i$.

We note that the size of G is $O(n^2)$ and that, given the clauses $c_1, \ldots, c_n$, G can be constructed in polynomial time. The answer to the original SAT-COUNT-3CNF problem is obtained by applying the function $x \mapsto \left(x - 2^{3n+m+3}\right)/2^{n+1}$ to the answer of a SAT-COUNT-OFDD algorithm on input G. Thus we have:

Theorem 10.4.2 SAT-COUNT-OFDD is $\#P$-complete.

For the more general problem SAT-COUNT-FFDD of computing the number of satisfying assignments for a free FDD the $\#P$-hardness result is implicitly contained in the NP-hardness proof given in [2].

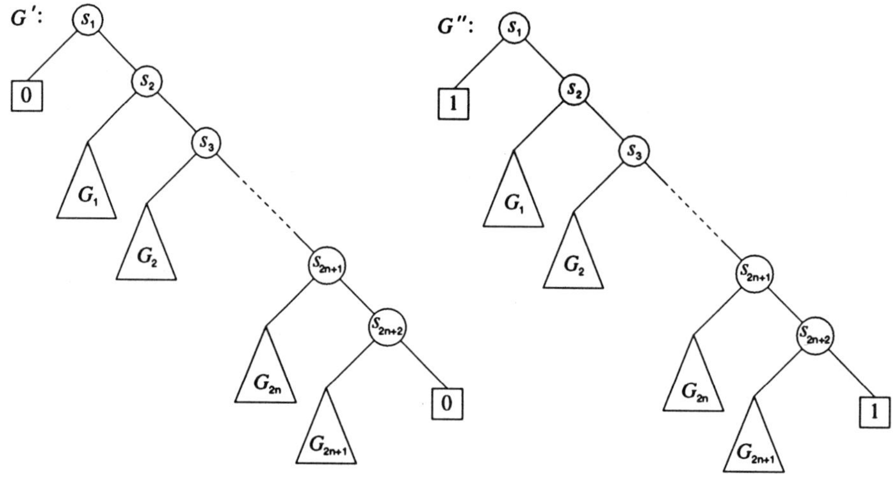

Figure 10.4.2 Combined OFDDs.

10.5 CONCLUSIONS

We investigated the complexity of satisfiability problems on *Ordered Functional Decision Diagrams* (OFDDs), i.e. the complexity of SAT-ONE, SAT-ALL and SAT-COUNT.

We noted that SAT-ONE has linear running time while SAT-ALL can be solved in time $O(|G| \cdot \sharp(f))$. Counting the satisfying assignments in an OFDD has been shown to be $\#P$-complete.

The results obtained in this chapter together with the results from [2] (almost completely) determine the complexity of the set of desireable functions proposed in [16] for OFDDs. As mentioned in the introduction it still remains open whether the inclusion relation can be solved in polynomial time for OFDDs.

The results directly transfer to more general data structures, where both (positive and negative) Davio decompositions are used (see e.g. [5]).

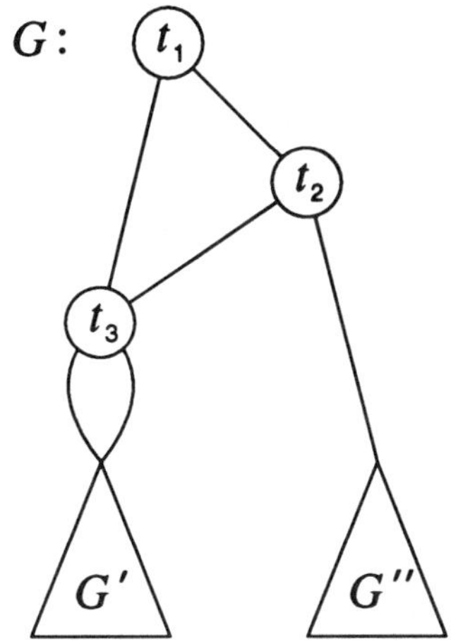

Figure 10.4.3 Final OFDD.

10.6 PROOF OF THEOREM 2

For $1 \leq i \leq n$, let

$$g_{i,j}(\mathbf{x},\mathbf{y}) = E_j(\mathbf{y}) \wedge y_i \wedge c_i(\mathbf{x}).$$

Then we have

Lemma 10.6.1

$$\sum_{1\leq j\leq n} (-1)^{j+1}\sharp\left(\bigoplus_{1\leq i\leq n} g_{i,j}(\mathbf{x},\mathbf{y})\right) = 2^{n-1}\sharp(f(\mathbf{x}))$$

Proof: For $0 \leq l \leq n$, let S_l be the set of assignments to $\mathbf{x}$ satisfying exactly l clauses $c_i(\mathbf{x})$ and let $s_l = \#S_l$. Obviously $\sharp(f) = s_n$. The number $a_{l,j}$ of assignments to $\mathbf{y}$ with $\bigoplus_{1\leq i\leq n} g_{i,j}(\bar{\mathbf{x}},\mathbf{y}) = 1$ is the same for all $\bar{\mathbf{x}} \in S_l$. $a_{l,j}$ does only depend on l and j. Assuming w.l.o.g. that $\bar{\mathbf{x}}$ satisfies exactly the first l clauses, we have

$$a_{l,j} \quad = \quad \#\left\{\mathbf{y}\middle| \bigoplus_{1\leq i\leq n} g_{i,j}(\bar{\mathbf{x}},\mathbf{y})\right\}$$

$$
\begin{aligned}
&= \#\left\{\mathbf{y}\middle| E_j(\mathbf{y}) \wedge \#\{i|1 \le i \le l \wedge y_i\} \text{ odd}\right\} \\
&= \sum_{\substack{0\le k\le l \\ k \text{ odd}}} \#\left\{\mathbf{y}\middle| E_j(\mathbf{y}) \wedge \#\{i|1 \le i \le l \wedge y_i\} = k\right\} \\
&= \sum_{\substack{0\le k\le l \\ k \text{ odd}}} \binom{l}{k}\binom{n-l}{j-k}.
\end{aligned}
$$

Note that for $n \ge 1$ we have

$$
\sum_{\substack{0\le i\le n \\ i \text{ even}}} \binom{n}{i} = \sum_{\substack{0\le i\le n \\ i \text{ odd}}} \binom{n}{i}.
$$

We then prove the lemma as follows:

$$
\begin{aligned}
&\sum_{1\le j\le n} (-1)^{j+1} \,\sharp\left(\bigoplus_{1\le i\le n} g_{i,j}(\mathbf{x},\mathbf{y})\right) \\
&= \sum_{1\le j\le n} (-1)^{j+1} \sum_{0\le l\le n} \sum_{\bar{\mathbf{x}}\in S_l} \#\left\{\mathbf{y}\middle| \bigoplus_{1\le i\le n} g_{i,j}(\bar{\mathbf{x}},\mathbf{y})\right\} \\
&= \sum_{1\le j\le n} (-1)^{j+1} \sum_{0\le l\le n} s_l a_{l,j} \\
&= \sum_{1\le j\le n} (-1)^{j+1} \sum_{0\le l\le n} s_l \sum_{\substack{0\le k\le l \\ k \text{ odd}}} \binom{l}{k}\binom{n-l}{j-k} \\
&= \sum_{0\le l\le n} s_l \sum_{\substack{0\le k\le l \\ k \text{ odd}}} \binom{l}{k} \sum_{1\le j\le n} (-1)^{j+1}\binom{n-l}{j-k} \\
&= \sum_{0\le l\le n} s_l \sum_{\substack{0\le k\le l \\ k \text{ odd}}} \binom{l}{k} \delta_{n,l} \\
&= s_n\, 2^{n-1}.
\end{aligned}
$$

($\delta_{n,l} := 1$, iff $n = l$, and $\delta_{n,l} := 0$, iff $n \neq l$.) □

Since no assignment to $\mathbf{x}, \mathbf{y}$ can simultaneously satisfy $g_{i,j}$ and $g_{i',j'}$ for $j \neq j'$, the alternating sum in Lemma 10.6.1 can easily be translated into a disjoint disjunction of Boolean functions.

We have

$$\begin{aligned}
\bigoplus_{1\le i\le 2n+1} g_i(\mathbf{x},\mathbf{y}) &= \left(\mathrm{ODD}(\mathbf{y}) \wedge \bigoplus_{1\le i\le n} (y_i \wedge c_i(\mathbf{x}))\right) \\
&\quad \oplus \left(\mathrm{EVEN}(\mathbf{y}) \wedge \left(1 \oplus \bigoplus_{1\le i\le n} (y_i \wedge c_i(\mathbf{x}))\right)\right) \\
&= \Big(\bigvee_{\substack{0\le j\le n\\ j \text{ odd}}} E_j(\mathbf{y}) \wedge \bigoplus_{1\le i\le n} (y_i \wedge c_i(\mathbf{x}))\Big) \\
&\quad \vee \Big(\bigvee_{\substack{0\le j\le n\\ j \text{ even}}} E_j(\mathbf{y}) \wedge \neg \bigoplus_{1\le i\le n} (y_i \wedge c_i(\mathbf{x}))\Big).
\end{aligned}$$

For the number of satisfying assignments follows

$$\begin{aligned}
&\sharp\left(\bigoplus_{1\le i\le 2n+1} g_i(\mathbf{x},\mathbf{y})\right) \\
&= \sum_{\substack{0\le j\le n\\ j \text{ odd}}} \sharp\left(\bigoplus_{1\le i\le n} g_{i,j}(\mathbf{x},\mathbf{y})\right) + 2^{m+n-1} - \sum_{\substack{0\le j\le n\\ j \text{ even}}} \sharp\left(\bigoplus_{1\le i\le n} g_{i,j}(\mathbf{x},\mathbf{y})\right) \\
&= 2^{m+n-1} + 2^{n-1}\sharp(f(\mathbf{x}))\,.
\end{aligned}$$

This completes the proof of Theorem 10.4.1. □

REFERENCES

[1] B. Becker, R. Drechsler, and M. Theobald, "On the implementation of a package for efficient representation and manipulation of functional decision diagrams," *IFIP WG 10.5 Workshop on Applications of the Reed-Muller Expansion in Circuit Design*, pp. 162–169, Sept. 1993.

[2] B. Becker, R. Drechsler, and R. Werchner, "On the relation between BDDs and FDDs," *LATIN, LNCS 911*, pp. 72–83, Apr. 1995.

[3] R.E. Bryant, "Graph - based algorithms for Boolean function manipulation," *IEEE Trans. Comput.*, vol. C-35, pp. 677–691, Aug. 1986.

[4] R. Drechsler and B. Becker, "Sym*pathy*: Fast exact minimization of Fixed Polarity Reed-Muller expressions for symmetric functions," *Proc. European Design & Test Conf.*, pp. 91–97, Mar. 1995.

[5] R. Drechsler, A. Sarabi, M. Theobald, B. Becker, and M.A. Perkowski, "Efficient representation and manipulation of switching functions based on Ordered Kronecker Functional Decision Diagrams," *Proc. Design Automation Conf.*, pp. 415–419, June 1994.

[6] R. Drechsler, M. Theobald, and B. Becker, "Fast OFDD based minimization of Fixed Polarity Reed-Muller expressions," to be published in *IEEE Trans. Comput.*, 1996.

[7] U. Kebschull and W. Rosenstiel, "Efficient graph-based computation and manipulation of functional decision diagrams," *Proc. European Conf. on Design Automation*, pp. 278–282, Mar. 1993.

[8] U. Kebschull, E. Schubert, and W. Rosenstiel, "Multilevel logic synthesis based on functional decision diagrams," *Proc. European Conf. on Design Automation*, pp. 43–47, Mar. 1992.

[9] J. Van Leeuven, editor, *Handbook of Theoretical Computer Science, Algorithms and Complexity.* The MIT Press, 1990.

[10] C. Meinel, "Modified Branching Programs and their Computational Power," *LNCS 370*, Apr. 1989.

[11] E. Schubert, U. Kebschull, and W. Rosenstiel, "FDD based technology mapping for FPGA," In *Proc. EUROASIC*, pp. 14–18, Mar. 1992.

[12] D. Sieling and I. Wegener, "Reduction of BDDs in linear time," *Information Processing Letters*, 48(3):139–144, 1993.

[13] C.C. Tsai and M. Marek-Sadowska, "Boolean matching using generalized Reed-Muller forms," *Proc. Design Automation Conf.*, pp. 339–344, June 1994.

[14] C.C. Tsai and M. Marek-Sadowska, "Detecting symmetric variables in Boolean functions using generalized Reed-Muller forms," In *Int'l Symp. Circ. and Systems*, 1994.

[15] I. Wegener, *The Complexity of Boolean Functions.* John Wiley & Sons Ltd., and B.G. Teubner, Stuttgart, 1987.

[16] I. Wegener, "Efficient data structures for Boolean functions," Discrete Mathematics, 136:347–372, Dec. 1994.

11

COMPLEXITY THEORETICAL ASPECTS OF OFDDS

Beate BOLLIG
Martin LÖBBING
Martin SAUERHOFF
Ingo WEGENER

Department of Computer Science
University of Dortmund
Dortmund, Germany

Abstract— Experimental results have shown that OFDDs (ordered functional decision diagrams) are a representation of Boolean functions which are sometimes superior to OBDDs (ordered binary decision diagrams). Most of the complexity theoretical problems have been solved for OBDDs. Here some results for OFDDs are proved. It is NP-complete to decide whether a function represented by some OFDD can be represented by an OFDD of size s using another variable ordering. Given an OFDD representation for an incompletely specified function, it is NP-hard to compute an optimal OFDD cover for this function respecting the same variable ordering. The replacement of variables by constants may cause an exponential blow-up of the OFDD size. Finally, it is investigated how a local change of the variable ordering may change the OFDD size. This leads to simulated annealing algorithms to improve variable orderings.

11.1 INTRODUCTION

OBDDs (ordered binary decision diagrams) introduced by Bryant [7] are the most common representation of Boolean functions in many applications, in particular in hardware verification and model checking. Most of the complexity theoretical problems concerning OBDDs are already solved.

Kebschull, Rosenstiel and Schubert [13] have introduced OFDDs (ordered functional decision diagrams) which have turned out to be sometimes superior to OBDDs. OFDDs are based on the Reed-Muller decomposition of Boolean functions, namely

$$f = f_{|x_1=0} \oplus x_1 \left(f_{|x_1=0} \oplus f_{|x_1=1}\right).$$

An OFDD respecting the variable ordering $x_1, \ldots, x_n$ is based on a recursive application of this decomposition rule. The function $f_{|x_1=0}$ is represented at the low-successor of the source representing f and the function $f_{|x_1=0} \oplus f_{|x_1=1}$ is represented at the high-successor of the source. The decomposition directly leads to a complete binary tree of depth n having exponential size. This size can be reduced by two reduction rules:

- merge nodes which are sources of isomorphic subgraphs;
- eliminate nodes whose high-successor is the 0-sink.

As usual we only consider reduced OFDDs which are a canonical form, if the variable ordering is fixed.

Becker, Drechsler and Werchner [2] have already investigated complexity theoretical properties of OFDDs. Among others the following results are known:

- an OFDD G can be evaluated for an input a in linear time $O(|G|)$;
- the satisfiability test is possible in linear time;
- counting the number of satisfying inputs is #P-complete;
- negation and replacing a variable by the constant 0 is possible in linear time;
- the $\oplus$-synthesis of the OFDDs G_1 and G_2 and replacing a variable by the constant 1 is possible in quadratic time $O(|G_1||G_2|)$ resp. $O(|G|^2)$;
- the $\wedge$-synthesis or the $\vee$-synthesis may lead to an exponential blow-up of the OFDD size, the same holds for the replacement of variables by functions and for unification.

We add some results to this list by solving some open problems. For circuits the satisfiability problem is NP-complete. Hence, it is NP-complete to decide for a circuit whether the represented function can be represented by an OFDD of size 1. This implies that the variable ordering problem is hard for OFDDs. But the situation may change, if the function f is already represented by an OFDD. We consider the problem to decide for some s whether f can be represented by some OFDD respecting an arbitrary variable ordering such that the OFDD size is bounded by s. We prove in Section 11.2 that this problem is NP-complete. This destroys the hope to construct optimal OFDDs, if the function is already represented by some OFDD of moderate size.

In particular, in model checking tools (see e.g. [8] and [16]) one has to work with incompletely specified Boolean functions $[f^{\text{on}}, f^{\text{care}}]$. The functions f^{on} and f^{care} are represented by OFDDs respecting the same variable ordering, f^{care} describes the care-set and f^{on} is some cover of the considered incompletely specified function f, i.e.

$$f(a) = f^{\text{on}}(a) \text{ for all } a \text{ such that } f^{\text{care}}(a) = 1.$$

If the OFDD becomes too large, one may ask whether some cover of f can be represented by an OFDD of size s respecting the given variable ordering. In Section 11.3 we prove that this problem is NP-hard. Even the approximate solution of the optimization version of the problem turns out to be hard.

In Section 11.4 we add a final negative result. Replacing variables by constants is a common operation and the OBDD size can only be reduced by this operation. We present an OFDD of polynomial size. After the replacement of a small number of variables, more precisely $O(\log n)$ variables, by the constant 1 we obtain a function whose OFDD size and even whose FFDD size (free functional decision diagrams) is exponential.

The final Section 11.5 contains positive results. The improvement of variable orderings is done either by local search approaches like sifting [14] or by simulated annealing algorithms. All these algorithms are based on local changes of the variable ordering. For all these local operations we determine how much the OFDD size may decrease and increase.

At the end of this Introduction we describe the structure of the reduced OFDD for a function f. For $S \subseteq \{1, \ldots, n\}$ we denote by f^S the EXOR-sum of all $f_{|x_1=a_1,\ldots,x_n=a_n}$, where $a_i \in \{0,1\}$, if $j \in S$, and $a_j = 0$ otherwise. For $n = 4$, e.g. we get $f^{\{1,4\}} = f_{|x_1=0,x_2=0,x_3=0,x_4=0} \oplus f_{|x_1=1,x_2=0,x_3=0,x_4=0} \oplus f_{|x_1=0,x_2=0,x_3=0,x_4=1} \oplus f_{|x_1=1,x_2=0,x_3=0,x_4=1}$.

Theorem 11.1.1 *The reduced OFDD representing f with the variable ordering $x_1, \ldots, x_n$ contains as many x_i-nodes as there are different functions f^S, $S \subseteq \{1, \ldots, i-1\}$, depending essentially on x_i (i.e. $f^S_{|x_i=0} \oplus f^S_{|x_i=1} \not\equiv 0$).*

11.2 IMPROVING THE VARIABLE ORDERING OF OFDDS IS NP-COMPLETE

The complexity of the problem OPTIMAL OFDD is investigated.

Definition 11.2.1 (OPTIMAL OFDD)

Input: *An OFDD G and a size bound s.*
Output: *Yes, if the function represented by G can be represented by an OFDD G^* (respecting an arbitrary ordering of the variables) with at most s nodes. No, otherwise.*

The NP-completeness of the corresponding OBDD problem has been proved by Bollig and Wegener [6]. Since several problems have a different flavor for OFDDs than for OBDDs, it is not clear that OPTIMAL OFDD is also NP-complete. We prove the NP-completeness of OPTIMAL OFDD. We try to follow the lines of the OBDD proof but there are several situations where we need new arguments for OFDDs. So the reduction is a little bit different. We describe it carefully.

Theorem 11.2.1 *The problem OPTIMAL OFDD is NP-complete.*

(Sketch of the proof) The problem OPTIMAL OFDD is contained in NP. The OFDD G^* can be guessed. The equivalence of OFDDs with different orderings of the variables can be verified in polynomial time. Using the τ-operator introduced in [1] the problem to check the equivalence of OFDDs can be solved with the equivalence test for OBDDs of Fortune, Hopcroft and Schmidt [10].

From the set of known NP-complete problems (see [11]) we have chosen the problem OLA (OPTIMAL LINEAR ARRANGEMENT) defined in the following way.

Input: An undirected graph $H = (V = \{1, \ldots, n\}, E)$ and a bound B.
Output: Yes, if there is a permutation π on $\{1, \ldots, n\}$ such that

$$\text{cost}(\pi) := \sum_{\{u,v\} \in E} |\pi(u) - \pi(v)| \leq B.$$

No, otherwise.

The cost of π measures the length of all edges, if the nodes of H are arranged in linear order with respect to π.

We present a polynomial time reduction from OLA to OPTIMAL OFDD. Let $H = (V, E)$ and B be given. Let $m := |E|$. W.l.o.g. we assume that n and m are sufficiently large and there are no isolated nodes in H.

We have to transform the input (H, B) for OLA into an input (G, s) for OPTIMAL OFDD such that the minimal OFDD size of the function represented by G is at most s if and only if the cost of an optimal linear arrangement for H is at most B. One key idea is the use of phage functions (this notion has been introduced by Tani, Hamaguchi, and Yajima [17]). The $\{u, v\}$-phage on the variables $(z_1, \ldots, z_n)$ is the function

$$p(z_1, \ldots, z_n) = (z_u \oplus z_v) \wedge \bigwedge_{r \notin \{u,v\}} z_r.$$

Using the τ-operator it is easy to see that the OFDD structure of a phage function is very similar to the OBDD structure with respect to the same variable ordering. OFDDs for the $\{u, v\}$-phage have $n + dist(u, v) - 1$ inner nodes (see Fig. 11.2.1), if the distance between z_u and z_v in the variable ordering equals $dist(u, v)$. Hence, the OFDD size of the $\{u, v\}$-phage reflects the cost of the edge $\{u, v\}$, if $|\pi(u) - \pi(v)| = dist(u, v)$.

We have to consider phages for all edges. In order to obtain OFDDs these phages have to be combined to one function. This will be done by selection variables. If the phages work on the same set of variables, the representations of the different phages can be merged. For that reason we use (like [17]) disjoint sets of variables for the different phages, namely $x_{1j}, \ldots, x_{nj}$ for the j-th phage representing the j-th edge of H. This leads to the problem that we can choose a suitable variable ordering for each phage independently from the orderings for other phages. Hence, the OFDD size will not correspond to the cost of a linear arrangement. Therefore, we add another function called bb to the phages. The OFDD size of bb will be larger than the OFDD sizes for all phages but it will be the same for all blockwise orderings. A blockwise ordering is an ordering where the variables of each block B_i, namely $x_{i1}, \ldots, x_{im}$ which represent the i-th node of H, are tested consecutively. We will show that the OFDD size for a blockwise ordering $B_{\pi(1)}, \ldots, B_{\pi(n)}$ corresponds to the cost of the linear arrangement π. Moreover, the function bb will ensure that only blockwise orderings are optimal orderings.

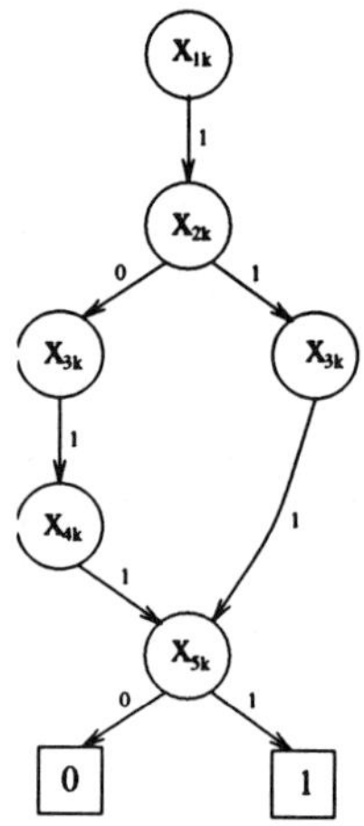

Figure 11.2.1 The phage p_k representing the edge $\{2,4\}$ where $n = 5$. The missing edges are leading to the 0-sink.

Now we make these ideas precise. Let (H, B) be given. We define the input (G, s) for OPTIMAL OFDD such that (H, B) is transformed to (G, s). We do not define the semantics of f, the function represented by G, but only its syntax, namely its OFDD. The function f depends on $mn+m$ variables, namely $x_{ij}, 1 \le i \le n, 1 \le j \le m$, and the selection variables $s_j, 1 \le j \le m$. The selection variables are arranged as in Fig. 11.2.2. They point to descriptions of the phages $p_1, \ldots, p_m$ and of bb.

The phage p_k represents the k-th edge of H. If this edge combines the nodes u and v,

$$p_k(x_{1k}, \ldots, x_{nk}) := (x_{uk} \oplus x_{vk}) \wedge \bigwedge_{r \notin \{u,v\}} x_{rk}.$$

For the definition of bb we introduce the symmetric function b_l working on l variables. As an example the (non-reduced) OFDD for b_{12} is shown in Fig. 11.2.3. The function bb is obtained by replacing the i-th level in an OFDD for b_n by copies of OFDDs for b_m on the variables in B_i. In the OFDD for b_n, $n = 12$, in Fig. 11.2.3 the nodes marked with $*$ can be eliminated by reduction rules. This does not hold for those copies for b_m replacing some nodes, e.g. near the top of the OFDD for b_n. Therefore, it is crucial to work with the non-reduced OFDD in Fig. 11.2.3.

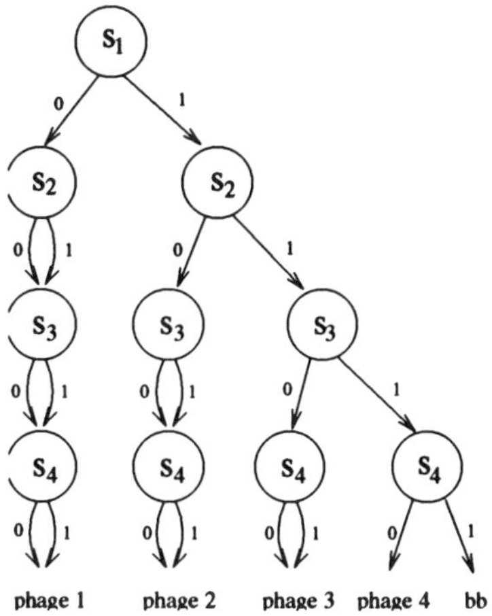

Figure 11.2.2 The structure of f, if $m = 4$.

The transformation computes the OFDD G representing f with respect to the ordering $s_1, \ldots, s_m, B_1, \ldots, B_n$. The ordering of the variables in B_i is the natural one, namely $x_{i1}, \ldots, x_{im}$.

We discuss the construction of G and compute $|G|$, the number of nodes of G without its sinks. The part with the selection variables consists of $m(m+1)/2$ nodes. The phage p_k for the k-th edge $\{u, v\}$, where $u < v$, consists of one x_{ik}-node, if $i \leq u$ or $i \geq v$, and two x_{ik}-nodes, if $u < i < v$. An example is shown in Fig. 11.2.1. Hence, the number of nodes for all phages equals $nm + \text{cost(id)} - m$, where id is the identical permutation $\text{id}(i) := i$. It is easy to compute the OFDD size $S(m, n)$ for *bb* and blockwise orderings in polynomial time.

Since exactly one node on the last level can be saved by the merging rule, the OFDD size of f with respect to the ordering $s_1, \ldots, s_m, B_1, \ldots, B_n$ equals

$$m(m-1)/2 + nm + \text{cost(id)} + S(m, n) - 1.$$

This can be computed in polynomial time.

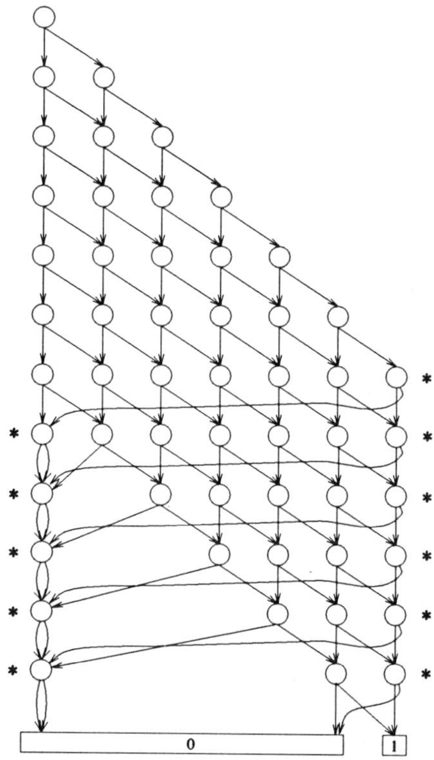

Figure 11.2.3 An OFDD for b_{12}. The left edges are low-edges and the right edges are high-edges.

Finally, we define the size bound s for the transformed input (G, s) which is constructed from (H, B):

$$s := m(m-1)/2 + nm + B + S(m, n) - 1.$$

Now we have to prove that H has a linear arrangement whose cost is bounded by B if and only if f can be represented by an OFDD with at most s nodes.

The only-if-part is easy. For each permutation π we can use a procedure similar to the one discussed above and can represent f with

$$m(m-1)/2 + nm + \text{cost}(\pi) + S(m, n) - 1.$$

nodes. If $\text{cost}(\pi) \leq B$, the size of this OFDD is bounded by s.

The if-part is more involved. A blockwise ordering of the variables $s_1, \ldots, s_m$, $x_{11}, \ldots, x_{nm}$ is an ordering $s_1, \ldots, s_m, B_{\pi(1)}, \ldots, B_{\pi(n)}$ for some permutation π. By our considerations above it is sufficient to prove that some optimal ordering of the variables of f is a blockwise ordering. Here we omit the technical proof. □

11.3 MINIMAL OFDD COVERS

In this section, we are concerned with the problem of finding a minimum cover for an incompletely specified function. This problem arises in typical model checking algorithms, and these constitute an important field of applications for decision diagram techniques.

A standard operation of model checking tools is the computation of reachable states of some finite state machine (FSM). This is needed, for example, for checking the equivalence of two FSMs, which can be done by a breadth first traversal of the reachable states of the product machine. We take a closer look at one step of the algorithm for the computation of reachable states. Given the set of states S_k reachable with k transitions, the set of states reachable with one further transition is obtained by

$$S_{k+1} = S_k \cup \{s' \mid s' = \delta(s, x),\ s \in S,\ x \text{ some input}\},$$

where δ denotes the transition function of the FSM. Normally, S is simply chosen to be the set of states reached for the first time in the last step, i.e. $S = S_k - S_{k-1}$. In state-of-the-art implementations of this algorithm, the states are not enumerated explicitly. The involved sets and relations are encoded as OBDDs and manipulated by standard OBDD operations. It has turned out to be advantageous to choose the state set S more carefully than indicated above. Since it is not forbidden to reconsider states already reached, we can choose any S with $S_k - S_{k-1} \subseteq S \subseteq S_k$. Under all these sets, we want to find one with the smallest representation as an OBDD (this is called "frontier set simplification" and was originally described by Coudert, Berthet and Madre [9]).

We state this problem in just another way. Given two sets S_{on} and S_{care}, find some set S with $S \cap S_{\text{care}} = S_{\text{on}} \cap S_{\text{care}}$ which has the smallest representation. In our special case, we can choose $S_{\text{care}} = \overline{S_{k-1}}$ and $S_{\text{on}} = S_k$. In practice, these sets are represented by their characteristic functions (and these in turn by

OBDDs). The characteristic functions f^{on} of S_{on} and f^{care} of S_{care} describe an incompletely specified function (denoted by $[f^{\text{on}}, f^{\text{care}}]$ in the following). This function computes $f^{\text{on}}(x)$ for inputs x with $f^{\text{care}}(x) = 1$, and is "unspecified" otherwise. We are looking for a (complete) *cover* f^* of this function, that is a completely specified function f^* with $f^* \wedge f^{\text{care}} = f^{\text{on}} \wedge f^{\text{care}}$. Furthermore, we want to choose the cover with the smallest possible representation as an OBDD.

We have seen now why finding minimum covers for incompletely specified functions is a desirable operation in a package for decision diagram manipulation. If we want to replace OBDDs by OFDDs as a decision diagram data structure, we therefore have to think about an efficient realization of this operation. For the sake of clarity, we give an exact definition of our problem in the version for OFDDs.

Definition 11.3.1 (MC-OFDD)

Input: *An OFDD G^{on} for a function f^{on}, an OFDD G^{care} for f^{care}, both respecting the same variable ordering; a natural number s.*
Question: *Is there an OFDD G^* respecting the same variable ordering as G^{on} and G^{care}, which represents a cover f^* of $[f^{\text{on}}, f^{\text{care}}]$ and for which $|G^*| \leq s$?*

In the case of OBDDs, the problem was considered to be "difficult" by several researchers and only heuristical solutions had been found (see [8], [12], [16]). Sauerhoff and Wegener finally proved that it is indeed NP-complete [15]. It has turned out now, that for OFDDs this problem is at least as difficult as for OBDDs.

Theorem 11.3.1 *MC-OFDD is NP-hard.*

(Sketch of the proof) We give only a rough outline of the proof here, for a complete version, see [5]. We construct a reduction from the following problem to MC-OFDD.

Definition 11.3.2 (Graph colorability, GC)

Input: *An undirected graph $G = (V = \{v_1, \ldots, v_n\}, E)$ and a number k.*
Question: *Is there a k-coloring $\varphi: V \to \{1, \ldots, k\}$ of G, i.e. a function with $\varphi(v) \neq \varphi(w)$ for all $\{v, w\} \in E$?*

Graph colorability is a well-known NP-complete problem, for further information see for example [11]. For the proof, we map an input instance (G, k) of

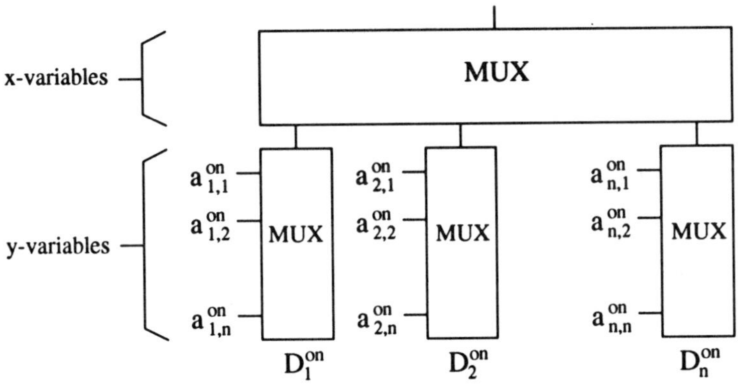

Figure 11.3.1 The OFDD D^{on}.

GC to an input instance for MC-OFDD consisting of two OFDDs, called D^{on}, D^{care}, and a size bound s. This has to be done in polynomial time. We have to show further that this mapping is indeed a reduction, i.e. the thus constructed OFDDs D^{on}, D^{care} have a cover not larger than s nodes iff G has a k-coloring.

We describe now how we can "encode" the problem GC in form of D^{on} and D^{care}. Figure 11.3.1 gives an overview on the structure of D^{on}. D^{care} is identical to D^{on} up to an exchange of the values a_{ij}^{on} by a_{ij}^{care}. These values are chosen as below.

Both OFDDs depend on the variables $x_1, \ldots, x_n$ and $y_1, \ldots, y_n$, which also are to be tested in this order. The graphs hidden in the "black boxes" called "MUX" are multiterminal OFDDs which all compute the function MUX: $\{0,1\}^n \rightarrow \{1, \ldots, n\}$. The only fact about this function we need to know here is that for each value i, $1 \leq i \leq n$, there is an appropriate input vector a_i with $\text{MUX}(a_i) = i$, the value computed for a_i by the OFDD according to the evaluation rules is exactly that at the i-th sink.

For $i = \text{MUX}(x)$, $j = \text{MUX}(y)$, the functions computed by D^{on} and D^{care} are $f^{\text{on}}(x, y) = a_{ij}^{\text{on}}$ and $f^{\text{care}}(x, y) = a_{ij}^{\text{care}}$, respectively. We choose the following

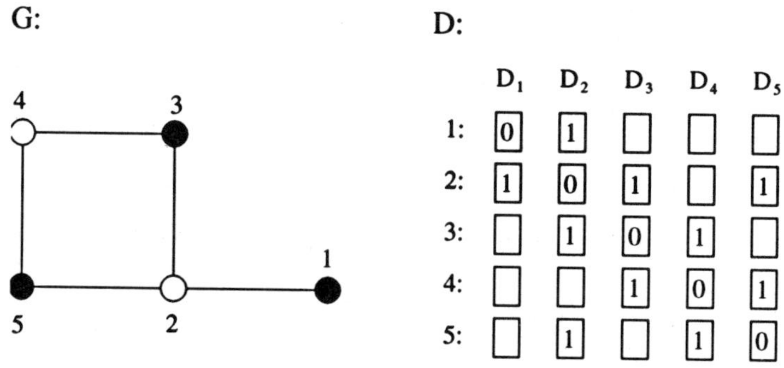

Figure 11.3.2 Inputs for GC and MC-OFDD (Example).

values for the constants at the sinks.

$$a_{ij}^{\mathrm{on}} := \begin{cases} 1, & \text{if } \{v_i, v_j\} \in E; \\ 0, & \text{otherwise.} \end{cases}$$

$$a_{ij}^{\mathrm{care}} := \begin{cases} 1, & \text{if } j = i \vee \{v_i, v_j\} \in E; \\ 0, & \text{otherwise.} \end{cases}$$

The incompletely specified function f represented by D^{on} and D^{care} behaves as follows. $f(x, y)$ equals 1, if $\{v_i, v_j\} \in E$ (where $i = \mathrm{MUX}(x)$, $j = \mathrm{MUX}(y)$), equals 0, if $j = i$, and is unspecified otherwise.

The vertices $v_1, v_2, \ldots, v_n$ of the graph G are mapped to the subgraphs D_1^{on}, $D_2^{\mathrm{on}}, \ldots, D_n^{\mathrm{on}}$ (see Fig. 11.3.1) by our reduction. We will demonstrate now why solutions of the graph coloring problem correspond to solutions of the minimum cover problem for the OFDDs constructed above and vice versa.

Let us consider the example in Fig. 11.3.2. The graph G depicted on the left side is an input for GC. The matrix on the right side is a symbolic representation of the sink values of D^{on} and D^{care}, merged into a new, auxiliary graph called D. In this representation, a "0" or "1" says that $a_{ij}^{\mathrm{on}} = 0$ or $a_{ij}^{\mathrm{on}} = 1$, resp., and $a_{ij}^{\mathrm{care}} = 1$; an empty box says that the value of the function is unspecified for this input combination, i.e. $a_{ij}^{\mathrm{care}} = 0$.

Obviously, the graph G has a 2-coloring. We now show how we can compute this 2-coloring with the aid of the graph D. For the problem MC-OFDD, we are looking for a cover of the incompletely specified function, i.e. we have to fill the empty boxes with values 0 and 1. The goal is to do this in such a way that in the corresponding OFDDs as many nodes as possible can be merged.

As we see in the figure, the subgraphs D_2, D_4 and D_1, D_3, D_5 have a common cover (all corresponding boxes in the same line can be put atop of each other without collisions of zeros and ones). These subgraphs belong to the node sets $\{2, 4\}$ and $\{1, 3, 5\}$ in the graph G. And indeed, exactly these nodes form the two color classes of a legal 2-coloring. The resulting complete OFDD (where the empty boxes are filled with values) has only two subgraphs with the structure of the D_i. In general, k color classes of G correspond to a cover of D^{on}, D^{care} in which all subgraphs in the bottom part can be merged into only k distinct copies.

Unfortunately, the construction needed for the complete proof is a little bit more complicated than indicated in this simplified version here. First, the realization of the sub-OFDDs for the function MUX is more difficult than it seems at the first glance. Second, the correspondence between mergeable subgraphs and color classes is not as direct as suggested. It is possible that subgraphs are partially mergeable, even if they do not correspond to nodes with the same color. A remedy for this is to add a certain amount of "dummy nodes" at the top of the subgraphs D_i^{on} and D_i^{care} which can only be merged if the subgraphs can be merged completely. □

We do not know whether MC-OFDD $\in$ NP. Of course, we could guess some OFDD D^* for a function f as a candidate cover. But the verification of $f \wedge f^{\text{care}} = f^{\text{on}} \wedge f^{\text{care}}$ seems to be difficult, since the $\wedge$-operation may result in an exponential blow-up for OFDDs.

Anyway, we have seen that we cannot hope for an efficient algorithm which solves MC-OFDD exactly. As an alternative, we could try to find an efficient approximation algorithm which is not always able to find a minimal solution, but can be guaranteed to be worse than the minimum only by a small factor. To be more precise, let $R_A(I)$ denote the *performance ratio* for an approximation algorithm A and an input I for the considered problem, i.e. the ratio of the value of the solution computed by the algorithm to the value of the optimal solution. Then we are interested in algorithms for which the *worst case performance ratio*

$$R_A(n) := \sup\{R_A(I) \mid I \text{ input instance of size } n\}$$

is a constant or a slowly growing function. For some NP-complete problems, efficient approximation algorithms of this type are known, e.g. for the BIN PACKING problem, the well-known "First Fit" heuristic has a worst case performance ratio of $R_{\mathrm{FF}}(n) = 1.7$ (for further details, see [11]). Unfortunately, it turns out that MC-OFDD has no such approximation algorithm, if P $\neq$ NP.

Theorem 11.3.2 *If P $\neq$ NP, every polynomial time approximation algorithm for MC-OFDD has a worst case performance ratio of $\Omega\left(s^{1/28}\right)$, where s is the size of the input. In particular, this means that there is no polynomial time approximation algorithm with constant worst case performance ratio.*

(Sketch of proof) The complete proof of Theorem 11.3.1 shows that we can use our polynomial time reduction to convert any "good" approximation algorithm for MC-OFDD into a "good" approximation algorithm for the graph coloring problem. On the other hand, if P $\neq$ NP, no such "good" approximation algorithm for graph coloring exists, as recent results show [3]. □

11.4 AN EXPONENTIAL BLOW-UP BY THE REPLACEMENT OF VARIABLES BY CONSTANTS

The replacement of variables by constants is an important operation. It is used to consider subcircuits or subfunctions, to replace variables by functions or to unify variables, i.e. to represent $(\exists x_i)f$ or $(\forall x_i)f$. For OBDDs, this operation can only reduce the size and the same holds for OFDDs and the replacement of variables by the constant 0. If we like to replace the variable at the source, say x_1, by the constant 1, we have to represent the function $f_{|x_1=1}$ which can be obtained as $\oplus$-synthesis of the low-successor and the high-successor of the source. The size of the OFDD G may grow but is limited by $|G|^2$. If the size of the given OFDD is polynomial, this holds also after the replacement of constantly many variables by the constant 1. But already the replacement of logarithmically many variables may cause an exponential blow-up of the OFDD size. In the following theorem we prove an even stronger statement.

Theorem 11.4.1 *There exists a Boolean function f on $O(n^2)$ variables which can be represented by an OFDD of polynomial size but the subfunction obtained by replacing $O(\log n)$ variables by 1 can be represented only by OFDDs or FFDDs of exponential size.*

(Proof) The function $1\text{-}cl_{n,3}$ is defined on $\binom{n}{2}$ variables deciding which of the possible edges are contained in an undirected graph of n vertices. The function outputs 1 iff the graph contains exactly three edges building a triangle. In [2] it is proved that the OFDD size and even the FFDD size of $1\text{-}cl_{n,3}$ grows exponentially.

Let $\Delta_{i,j,k}$ be the minterm computing 1 iff the graph contains exactly the three edges building the triangle $\{i, j, k\}$. Minterms have linear OFDD size for each variable ordering. Moreover, $1\text{-}cl_{n,3}$ is the $\oplus$-sum of all $\Delta_{i,j,k}$. Now we describe the function f syntactically by its OFDD. On top we have a complete tree of $O(\log n)$ new variables. For each $\Delta_{i,j,k}$ there is one leaf of the tree, where $\Delta_{i,j,k}$ is represented. The other leaves represent 0. The OFDD for f has polynomial size. If we replace the additional variables by ones, we obtain the $\oplus$-sum of all $\Delta_{i,j,k}$ and, therefore, $1\text{-}cl_{n,3}$ whose FFDD size grows exponentially. □

11.5 THE EFFECT OF LOCAL CHANGES OF THE VARIABLE ORDERING

For all used functions we like to have good variable orderings so that the OFDDs representing the functions are small. If the function is given as a circuit, there are a lot of heuristics to produce suitable variable orderings, but almost always these orderings have to be improved. To achieve better orderings one uses local search algorithms like the sifting algorithm [14] or simulated annealing.

We present three types of local changes: swaps, exchanges and jumps (distinguished into jump-down and jump-up). Given an OFDD for the initial variable ordering π (for $i < j$)

$$x_1, \ldots, x_{i-1}, x_i, x_{i+1}, \ldots, x_{j-1}, x_j, x_{j+1}, \ldots, x_n,$$

the operation SWAP$(i, i+1)$ computes an OFDD respecting the variable ordering where two adjacent variables are exchanged, resulting in the ordering

$$x_1, \ldots, x_{i-1}, x_{i+1}, x_i, \ldots, x_{j-1}, x_j, x_{j+1}, \ldots, x_n.$$

A more general operation is EXCHANGE(i, j) which produces an OFDD for the ordering

$$x_1, \ldots, x_{i-1}, x_j, x_{i+1}, \ldots, x_{j-1}, x_i, x_{j+1}, \ldots, x_n.$$

At last the operation JUMP(i, j) lets one variable jump to a different position, while the relative order of all other variables remains unchanged. For $i < j$,

JUMP(i, j) is called jump-down, which results in the ordering

$$x_1, \ldots, x_{i-1}, x_{i+1}, \ldots, x_{j-1}, x_j, x_i, x_{j+1}, \ldots, x_n,$$

while the operation JUMP(j, i) is called jump-up resulting in the ordering

$$x_1, \ldots, x_{i-1}, x_j, x_i, x_{i+1}, \ldots, x_{j-1}, x_{j+1}, \ldots, x_n.$$

The most restricted operation is SWAP, it is a special case of exchange as well as a special case of jump. In most implementations it is realized by local rearrangements of the layers i and $i+1$, all other layers remain the same. The disadvantage of this operation is that on average $\Theta(n^2)$ swaps are necessary to get an arbitrary variable ordering. For the exchange and the jump operation, only $\Theta(n)$ steps are sufficient. The more powerful of these operations is jump: Each exchange can be simulated by two jumps, but to simulate one jump there are $|j - i|$ exchanges necessary.

A jump operation JUMP(i, j) can be realized in two ways. The first one is for example used by the sifting algorithm: a series of at least $|j - i|$ swaps is performed to get finally a jump. The second one uses commonly known synthesis algorithms, preserving the original OFDD. It bases on the fact that the OFDD for $f_{|x_i=c}$ $(c \in \{0, 1\})$ respects the original variable ordering π as well as the new variable ordering π', since it does not contain any x_i-node.

Algorithm 11.5.1

Input: *G, the OFDD for f respecting π.*

Output: *G', the OFDD for f respecting π'.*

1. *Construct the OFDD G_0 for $g_0 = f_{|x_i=0}$.*

   A new OFDD is built identical to G, but edges pointing to x_i-nodes are led to their low-successors.

2. *Construct the OFDD G_1 for $g_1 = f_{|x_i=0} \oplus f_{|x_i=1}$.*

   A new OFDD is built identical to G, but edges pointing to x_i-nodes are led to their high-successors and edges crossing the x_i-layer are led to the 0-sink.

3. *Construct the OFDD G_1' for $g_1' = x_i \cdot g_1$ respecting π'.*

Every edge crossing the new x_i-layer and not leading to the 0-sink must point to an x_i-node whose low-successor is the 0-sink and whose high-successor is the node the edge pointed to.

4. *Construct the OFDD G' for $g' = g_0 \oplus g_1'$.*

Altogether, the needed time is of the same magnitude as a $\oplus$-synthesis. To assess the different operations, we determine the maximal increase and decrease of the OFDD after one local change. We only prove lower and upper bounds, but these bounds are almost tight in the worst case.

Theorem 11.5.1 *Let G be an OFDD and let G' be the result after one of the local changes. The following holds:*

swap:	$\|G\|/2 \leq \|G'\| \leq 2 \cdot \|G\|.$
exchange:	$\sqrt{\|G\|} \leq \|G'\| \leq \|G\|^2.$
jump-up:	$\sqrt{\|G\|} \leq \|G'\| \leq 2 \cdot \|G\|.$
jump-down:	$\|G\|/2 \leq \|G'\| \leq \|G\|^2.$

This theorem follows from the following results. Let us first investigate the jump-up operation JUMP(j,i). Given the OFDD G respecting the variable ordering $\pi = (x_1, \ldots, x_{i-1}, x_i, x_{i+1}, \ldots, x_{j-1}, x_j, x_{j+1}, \ldots, x_n)$, let s_k be the number of x_k-nodes in G. For the resulting OFDD G' according to the ordering $\pi' = (x_1, \ldots, x_{i-1}, x_j, x_i, x_{i+1}, \ldots, x_{j-1}, x_{j+1}, \ldots, x_n)$ let s_k' be the number of x_k-nodes in G'.

Theorem 11.5.2 *The size of the OFDD after the jump-up operation* JUMP(j,i) *is bounded above by $2(s_1 + \ldots + s_{j-1}) + 1 + s_{j+1} + \ldots + 2$ and also by $s_1 + \ldots + s_{i-1} + 3(s_i + \ldots + s_{j-1}) + s_j + s_{j+1} + \ldots + s_n + 2$. In particular, the size of an OFDD can be at most doubled by a jump-up operation.*

(Proof) For $k < i$ or $j < k$ the x_k-layer is not changed, this follows from Theorem 11.1.1. Hence, $s_k' = s_k$. For $i \leq k < j$ s_k is the number of different functions f^S, $S \subseteq \{1, \ldots, k-1\}$, which essentially depend on x_k. The number of different functions $f^{S'}$, $S' = \{1, \ldots, k-1, j\}$, which essentially depend on x_k is at most twice as much, hence $s_k' \leq 2s_k$.

There are two upper bounds on s'_j, leading to two bounds for the total number of nodes in the OFDD. Firstly, each x_j-node in the new OFDD has to be reached from at least one node above, so $s'_j \leq s_1 + \ldots s_{i-1} + 1$. The second bound for s'_j equals $s_i + \ldots + s_j$; this is proved by an injective mapping from every x_j-node in the new OFDD, which represents a function f^S, $S \subseteq \{1, \ldots, i-1\}$ depending essentially on x_j, to an x_k-node $(i \leq k \leq j)$ in the original OFDD representing the same function. □

Corollary 11.5.1 *The size of the OFDD after the* SWAP$(i, i+1)$ *operation is bounded by* $2(s_1 + \ldots + s_i) + 1 + s_{i+2} + \ldots + s_n + 2$ *and also by* $s_1 + \ldots + s_{i-1} + 3s_i + s_{i+1} + s_{i+2} + \ldots + s_n + 2$.

For the jump-down operation JUMP(i, j) let s'_k be the number of x_k-nodes in the new OFDD respecting to the variable ordering $\pi' = (x_1, \ldots, x_{i-1}, x_{i+1}, \ldots, x_{j-1}, x_j, x_i, x_{j+1}, \ldots, x_n)$.

Theorem 11.5.3 *The size of the OFDD after a jump-down operation* JUMP(i, j) *is bounded above by* $s_1 + \ldots + s_{i-1} + (s_{i+1} + \ldots + s_n + 2)^2$ *and also by* $s_1 + \ldots + s_n + (2^{j-i+1} - 2)s_i + 2$. *In particular, the size of an OFDD can be at most squared by a jump-down operation.*

(Proof) Only the ideas of the proof are presented. Again, for $k < i$ or $k > j$ the layer x_k is not changed, this follows from Theorem 11.1.1. Hence, $s'_k = s_k$. For the x_k-layers with $i \leq k \leq j$ the node is produced by a $\oplus$-synthesis, according to Algorithm 11.5.1. This means that each node corresponds to a pair of nodes from G_0 labelled x_l and from G'_1 labelled x_m with $i < l, i < m$. One of these nodes may be a sink. There are less than $(s_{i+1} + \ldots + s_n + 2)^2$ such pairs, which leads to the first bound. The second bound is obtained, if the jump operation is simulated by a series of swaps: SWAP$(i, i+1)$, SWAP$(i+1, i+2)$, ..., SWAP$(j-1, j)$. □

Exchange operations can be simulated by a jump-down (the size is at most squared) immediately followed by a jump-up (then the size is at most doubled). The direct proof is omitted that the total size is at most squared.

(Proof of Theorem 11.5.1) The upper bound for each operation was proved separately and the lower bound follows from the upper bound of the inverse operation. The inverse operation of a swap or exchange is a swap or an exchange again. The inverse operation of a jump-down is a jump-up and vice versa. □

Jump-up operations look quite promising, because the OFDD can grow only moderately, but can shrink a lot. For simulated annealing algorithms it is important that neighborhoods are symmetric, so that jump-down must be used, too. Experiments for OBDDs [4] showed that one should use all possible local operations.

11.6 CONCLUSION

Several open problems on complexity theoretical aspects of OFDDs have been solved. It is difficult to improve variable orderings and to find better covers for incompletely specified functions. Even such a simple operation like the replacement of logarithmically many variables by the constant 1 may lead to an exponential blow-up in the OFDD size. With respect to local changes of the variable ordering OFDDs behave like OBDDs and simulated annealing algorithms may often compute good variable orderings. The NP-hardness results are no arguments against the use of OFDDs, which have turned out to have many applications. The results only imply that certain problems cannot be expected to be exactly solvable by efficient algorithms and therefore have to be solved heuristically.

Acknowledgment

This work was supported by DFG grant 1066/7-3.

REFERENCES

[1] B.Becker and R.Drechsler, "On the computational power of functional decision diagrams." Technical report, Universität Frankfurt, 1993.

[2] B.Becker, R.Drechsler, and R.Werchner, "On the relation between BDDs and FDDs," *Latin American Theoretical Informatics, LNCS 911*, pp. 72–83, 1995.

[3] M.Bellare, O.Goldreich, and M.Sudan. "Free bits, PCP and non-approximability – towards tight results," *IEEE Symp. on Foundations of Computer Science*, pp. 422–431, 1995.

[4] B.Bollig, M.Löbbing, and I.Wegener, "Simulated annealing to improve variable orderings for OBDDs," *International Workshop on Logic Synthesis*, 1995.

[5] B.Bollig, M.Löbbing, M.Sauerhoff and I.Wegener, "Complexity theoretical aspects of OFDDs," *Proc. of the Workshop on Applications of the Reed-Muller Expansion in Circuit Design, IFIP WG 10.5*, pp. 198–205, 1995

[6] B.Bollig and I.Wegener, "Improving the variable ordering of OBDDs is NP-complete," submitted to IEEE Trans. Computers, 1994.

[7] R.E.Bryant, "Graph-based algorithms for Boolean function manipulation," *IEEE Trans. Computers*, Vol. C-35, No. 8, pp. 677–691, 1986.

[8] S.-C.Chang, D.I. Cheng, and M.Marek-Sadowska, "Minimizing ROBDD size of incompletely specified multiple output functions," *Proc. of the European Design Automation Conference*, pp. 620–624, 1994.

[9] O.Coudert, C.Berthet, and J.C.Madre, "Verification of sequential machines using Boolean functional vectors," *IMEC/IFIP International Workshop on Applied Formal Methods for Correct VLSI Design*, pp. 111–128, 1989.

[10] S.Fortune, J.Hopcroft, and E.M. Schmidt, "The complexity of equivalence and containment for free single variable program schemes," *Proc. ICALP, LNCS 62*, pp. 227–240, 1978.

[11] M.R.Garey and D.S.Johnson, "Computers and Intractability – A Guide to the Theory of NP-Completeness," Freeman and Company, New York, 1979.

[12] A.J.Hu, G.York, and D.L.Dill, "New techniques for efficient verification with implicitly conjoined BDDs," *Proc. of the 31st ACM/IEEE Design Automation Conference*, pp. 276–282, 1994.

[13] U.Kebschull, E.Schubert, and W.Rosenstiel, "Multilevel logic synthesis based on functional decision diagrams," *Proc. of the European Design Automation Conference*, pp. 43–47, 1992.

[14] R.L.Rudell, "Dynamic variable ordering for ordered binary decision diagrams," *Proc. of the ACM/IEEE Int. Conf. on Computer Aided Design*, pp. 42–47, 1993.

[15] M.Sauerhoff and I.Wegener, "On the complexity of minimizing the OBDD size for incompletely specified functions," submitted to *IEEE Trans. on CAD of Int. Circuits and Systems*, 1994.

[16] T.R.Shiple, R.Hojati, A.L.Sangiovanni-Vincentelli, and R.K.Brayton, "Heuristic minimization of BDDs using don't cares," *Proc. of the 31st ACM/IEEE Design Automation Conference*, pp. 225–231, 1994.

[17] S.Tani, K.Hamaguchi, and S.Yajima, "The complexity of the optimal variable ordering problems of shared binary decision diagrams," *Proc. of the 4th Int. Symp. on Algorithms and Computation, LNCS 762*, pp. 389–398, 1993.

12

TERNARY DECISION DIAGRAMS AND THEIR APPLICATIONS

Tsutomu SASAO

Department of Computer Science and Electronics
Kyushu Institute of Technology
Iizuka, Japan

Abstract— This chapter presents various ternary decision diagrams (TDDs) to represent logic functions. A path from the root node to the terminal node representing a constant 1 is called a 1-path. The 1-paths for a Reduced Ordered AND TDD (ROATDD) and prime TDD (PTDD) represent the sets of the implicants and the prime implicants (PIs) of the functions, respectively. The 1-paths for a Reduced Ordered TDD (ROTDD) represent a sum-of-products expression (SOP). For any function, there is a unique ROATDD and a unique PTDD. For any SOP, there is a unique ROTDD. An arbitrary function of n variables can be represented by a TDD with $O(3^n/n)$ nodes. A symmetric function can be represented by a TDD with $O(n^3)$ nodes. A program to generate all the PIs by using PTDDs is developed and shown to be much faster than conventional methods.

12.1 INTRODUCTION

Various methods have been developed for representing and manipulating logic functions. A truth table is the most straightforward method. A sum-of-products expression (SOP) is an another method, and has a direct relation to an AND-OR two-level logic network. A Binary Decision Diagram (BDD) is suitable for representing a complex function with many variables [1, 7]. These methods have their own applications.

This chapter presents Ternary Decision Diagrams (TDDs) as new methods to represent logic functions. TDDs are similar to BDDs, except that each nonterminal node has three children. We propose Quasi Reduced Ordered AND TDDs (QROATDDs), Reduced Ordered AND TDDs (ROATDDs) and Prime TDDs (PTDDs). These TDDs are unique for a given function. We also propose Reduced Ordered TDDs (ROTDDs), which are unique for given SOPs. A path from the root node to the constant 1 node is called a 1-path. The 1-paths for a QROATDD, a PTDD, and an ROTDD represent the complete sum of implicants, the complete sum of prime implicants (PIs), and an SOP, respectively. Especially, the PTDD represents all the PIs of the given function, and useful for PI generation.

Various methods to generate all the PIs are known [2, 18, 19, 25, 24, 16, 20]. However, these methods need to store all the PIs in a memory of a computer, and require memory proportional to

$$(\text{the number of the PIs}) \times (\text{size of memory to represent a PI}).$$

Therefore, when the number of the PIs is very large, it is impossible to use these methods for PI generation.

Recently, Coudert and Madre [8] developed an implicit method to generate all the PIs using BDDs. They successfully generated all the PIs for the functions with millions of PIs. This chapter also considers the similar method, but our method is easier to understand because each PI corresponds to a 1-path of the PTDD. The program to generate all the PIs by using PTDDs is much faster than conventional methods.

We also analyze the numbers of nodes in QROATDDs and PTDDs to represent a function of n variables. The number of nodes to represent a function is $O(3^n/n)$ for an arbitrary function and $O(n^3)$ for an arbitrary symmetric function. There exist a $2n$-variable function whose ROTDD (PTDD) requires only $2n^2+2$ nodes, while whose ROBDD requires 2^{n+1} nodes.

12.2 DEFINITIONS AND BASIC PROPERTIES

In this chapter, f represents a logic function, $\mathcal{F}$ represents an SOP, and n represents the number of the input variables. We assume that an SOP does not contain two or more copies of the the same products. Under this assumption, we consider that two SOPs are the same if one can be obtained from the other by permuting the order of the products in the SOP.

Definition 12.2.1 *A* ***Binary Decision Diagram (BDD)*** *is a rooted directed graph with node set V containing two types of nodes. A* ***nonterminal node*** *v has as attributes an argument index* ***index***$(v) \in \{1,\ldots,n\}$*, and two children* ***low***(v)*,* ***high***$(v) \in V$*. A* **terminal node** *v has as attribute a value* ***value***$(v) \in \{0,1\}$*. For any nonterminal node v, if* ***low***(v) *is also nonterminal, then* ***index***$(v) <$ ***index***$($***low***$(v))$*. Similarly, if* ***high***(v) *is nonterminal, then* ***index***$(v) <$ ***index***$($***high***$(v))$*. A logic function f and a BDD are related as follows:*
For v a terminal node:
If ***value***$(v) = 1$ *then $f_v = 1$. If* ***value***$(v) = 0$ *then $f_v = 0$.*
For v a nonterminal node: If ***index***$(v) = i$*, then f_v is a function such that* $f_v(x_1, x_2, \ldots, x_n) = \bar{x}_i f_{\mathbf{low}(v)}(x_1, x_2, \ldots, x_n) \vee x_i f_{\mathbf{high}(v)}(x_1, x_2, \ldots, x_n)$.
Note that the root node represents the function f itself.

Definition 12.2.2 *A BDD is a* ***Quasi Reduced Ordered Binary Decision Diagram (QROBDD)*** *if it does not contain distinct nodes v_1 and v_2 such that the subgraphs rooted by v_1 and v_2 are isomorphic, and if every path from the root node to the terminal nodes involves all the variables.*

Example 12.2.1 *Fig. 12.2.1 shows the QROBDD for $f = xy \vee yz \vee zx$.* ■

Definition 12.2.3 *A path from the root node to the terminal node representing a constant 1 is called a* ***1-path****.*

Theorem 12.2.1 *For any function f, there is a unique QROBDD. The number of nodes needed to represent an arbitrary n variable function is $O(2^n/n)$. The set of 1-paths for a QROBDD represents the* ***canonical SOP*** *(minterm expansion) for f.*

(Proof) The uniqueness of the representation can be proved in a similar way to ROBDDs [7]. As for the number of nodes, see [13]. Every 1-path of QROBDD contains nodes for all indexes, so it corresponds to a minterm. Thus, the set of 1-path represents a canonical SOP. □

Definition 12.2.4 *A BDD is a* ***Reduced Ordered Binary Decision Diagram (ROBDD)*** *if it contains no node v with* ***low***$(v) =$ ***high***(v)*, and if it does*

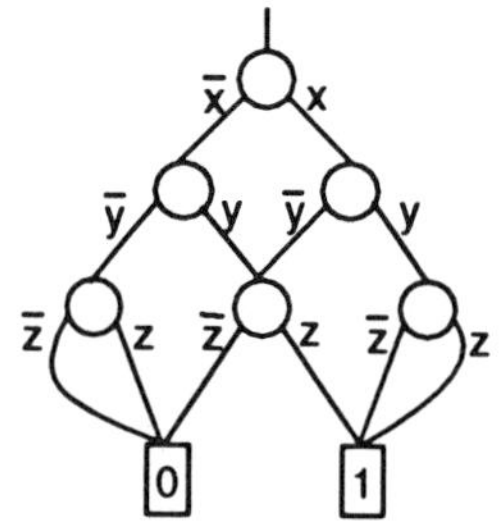

Figure 12.2.1 QROBDD.

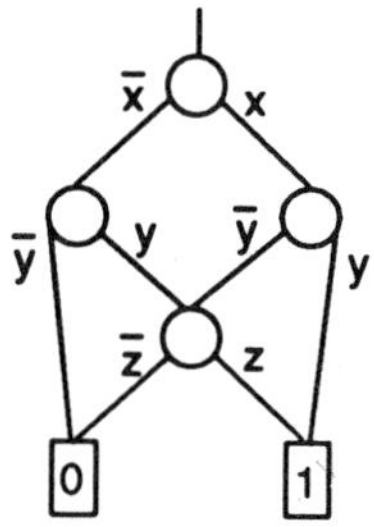

Figure 12.2.2 ROBDD.

not contain distinct nodes v_1 and v_2 such that the subgraphs rooted by v_1 and v_2 are isomorphic.

Example 12.2.2 *Fig. 12.2.2 shows the ROBDD for $f = xy \vee yz \vee zx$.* ■

Theorem 12.2.2 *[1, 7, 13] For any function, there is a unique ROBDD. The number of nodes to represent an arbitrary n variable function is $O(2^n/n)$. The set of 1-paths for an ROBDD represents the* ***disjoint SOP*** *(DSOP) for f.*

Example 12.2.3 *The set of 1-paths in Fig. 12.2.1 denotes the canonical SOP: $\mathcal{F}_1 = \bar{x}yz \vee x\bar{y}z \vee xy\bar{z} \vee xyz$, while that of Fig. 12.2.2 denotes a DSOP: $\mathcal{F}_2 = \bar{x}yz \vee x\bar{y}z \vee xy$.* ■

12.3 AND TDDS

In this section, we define an AND TDD which presents all the implicants of a function. We also enumerate the number of the nodes in AND TDDs.

Definition 12.3.1 *An* ***AND Ternary Decision Diagram*** *(ATDD) is a rooted directed graph with a set of nodes V containing two types of nodes. A nonterminal node v has as attributes an argument index* **index**$(v) \in \{1, \ldots, n\}$ *and three children* **low**(v), **high**(v), **and**$(v) \in V$. *A terminal node v has as attribute a value* **value**$(v) \in \{0, 1\}$. *For any nonterminal nodes v, if* **low**(v) *is also nonterminal, then* **index**$(v) <$ **index**$($**low**$(v))$. *Similarly, if* **high**(v) *is*

nonterminal, then $\mathbf{index}(v) < \mathbf{index}(\mathbf{high}(v))$. *Also, if* $\mathbf{and}(v)$ *is nonterminal, then* $\mathbf{index}(v) < \mathbf{index}(\mathbf{and}(v))$. *The correspondence between an ATDD and a function f is defined as follows:*
For v a terminal node: If $\mathbf{value}(v) = 1$ *then* $f_v = 1$. *If* $\mathbf{value}(v) = 0$ *then* $f_v = 0$.
For v a nonterminal node: if $\mathbf{index}(v) = i$, *then* f_v *and* $f_{and(v)}$ *are functions such that*
$f_v(x_1, x_2, \ldots, x_n) = \bar{x}_i f_{low(v)}(x_1, x_2, \ldots, x_n) \vee x_i f_{high(v)}(x_1, x_2, \ldots, x_n)$, *and*
$f_{and(v)}(x_1, x_2, \ldots, x_n) = f_{low(v)}(x_1, x_2, \ldots, x_n) f_{high(v)}(x_1, x_2, \ldots, x_n)$.
If the relation $f_{and(v)} = f_{low(v)} f_{high(v)}$ *does not hold, then there is no function for the ATDD.*

Definition 12.3.2 *A TDD is a* ***Quasi Reduced Ordered AND Ternary Decision Diagram*** *(QROATDD), if it does not contain distinct nodes* v_1 *and* v_2 *such that the subgraphs rooted by* v_1 *and* v_2 *are isomorphic, and if every path from the root node to the terminal nodes involves all the variables.*

Example 12.3.1 *Fig. 12.3.1 shows the QROATDD for* $f = xy \vee yz \vee zx$. ■

Lemma 12.3.1 *For every node v in a QROATDD, the subgraph rooted by v is itself a QROATDD.*

(Proof) It follows directly from the definition of a QROATDD. □

Theorem 12.3.1 *For any function, there is a unique QROATDD.*

(Proof) Clearly, an arbitrary function can be represented by a QROATDD. As for the proof of uniqueness, we will use mathematical induction. Clearly, the theorem is true for 0 and 1 variable functions. Suppose that the theorem is true for the functions with k or fewer variables. Any function f of $k+1$ variables can be expanded as $f = \bar{x}_1 f_0 \vee x_1 f_1$. Consider the TDD for f with the root node v, where the $\boldsymbol{low}(v)$, $\boldsymbol{high}(v)$, and $\boldsymbol{and}(v)$ represent f_0, f_1, and $f_0 f_1$, respectively. Because these are functions with k or fewer variables, there are unique QROATDDs for them. The QROATDD for f is obtained by reducing isomorphic subgraphs from this TDD, and so it is also unique. □

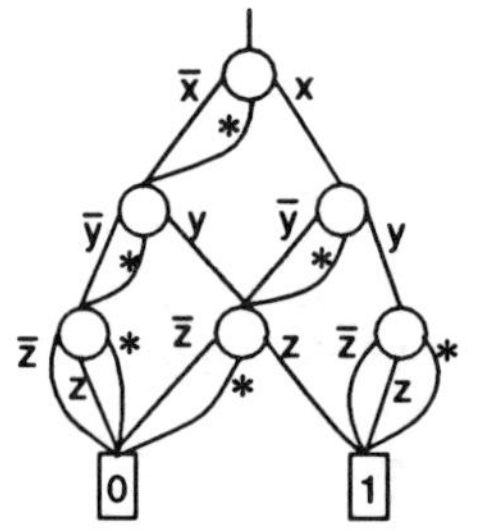

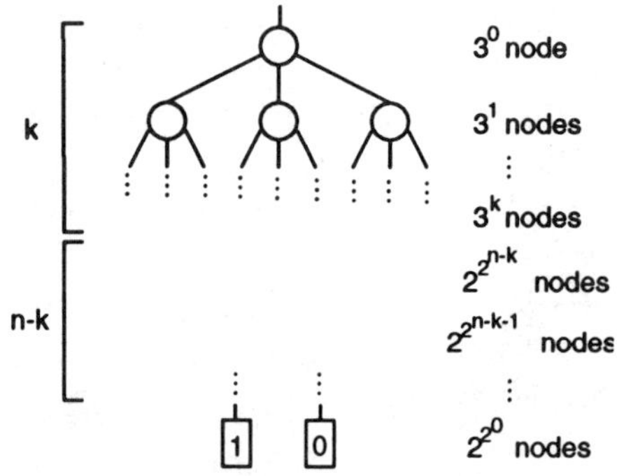

Figure 12.3.1 QROATDD. **Figure 12.3.2** QROATDD.

Lemma 12.3.2 *In a QROATDD, nodes with the same index represent distinct functions.*

(Proof) If two nodes with the same index represent same function, then, by Theorem 12.3.1, the subgraphs rooted by these nodes are isomorphic, and they can be reduced. □

Theorem 12.3.2 *The number of nodes in a QROATDD for an n-variable function is at most*

$$\min_{k=1}^{n}\left(\frac{3^{k+1}-1}{2}+2^{2^{n-k}}+2^{2^{n-k-1}}+2^{2^{n-k-2}}+\cdots+2^{2^1}+2^{2^0}\right).$$

(Proof) A given QROATDD can be divided into the upper and lower blocks as Fig. 12.3.2, where the upper block has the first k variables and the lower block has the other $(n-k)$ variables.

(1) For the upper block: Consider the complete ternary decision tree of k variables. The number of nodes with index i is at most 3^j. Therefore, the number of nodes in the upper block is at most

$$\sum_{j=0}^{k} 3^j = \frac{3^{k+1}-1}{2}.$$

(2) For the lower block: Consider the QROATDDs of $(n-k)$ variables. By Lemma 12.3.2, the nodes with the same index represent distinct functions. There are $2^{2^{n-k}}$ functions, so the number of the nodes with index $(n-k)$ is at most $2^{2^{n-k}}$. Similarly, for the nodes with index $(n-k-1)$, the number of nodes is at most $2^{2^{n-k-1}}$, and so on. From this argument, we can show that the number of nodes in the lower block is at most

$$2^{2^{n-k}} + 2^{2^{n-k-1}} + \cdots + 2^{2^1} + 2^{2^0}$$

Because (1) and (2) hold for any k $(k = 1, 2, \ldots, n)$, we have the theorem. □

Set $k = n - \log_3 n$ in Theorem 12.3.2, and we have the following:

Corollary 12.3.1 *The number of nodes in a QROATDD for an n-variable function is $O(3^n/n)$.*

Theorem 12.3.3 *The set of all the 1-paths of a QROATDD for f denotes the set of all the implicants for f.*

(Proof) The proof will be done by mathematical induction. Clearly, the theorem is true for 0 and 1 variable functions. Suppose that the theorem is true for the functions with k or fewer variables. Consider a QROATDD for a $k+1$ variable function f. It can be expanded as $f = \bar{x}_1 \cdot f_0 \vee x_1 \cdot f_1$. Let v be the root node, then ***low***(v), ***high***(v), and ***and***(v) represent f_0, f_1, and $f_0 \cdot f_1$, respectively. By the hypothesis of the induction, in a subgraph rooted ***low***(v), ***high***(v), or ***and***(v), each 1-path represents an implicant of the sub-function. Thus, each 1-path of the QROATDD for f represents one of the followings:

$$\begin{aligned} &\bar{x}_1 && \cdot(\text{an implicant of } f_0), \\ &x_1 && \cdot(\text{an implicant of } f_1), \text{or} \\ &1 && \cdot(\text{an implicant of } f_0 \cdot f_1). \end{aligned}$$

On the other hand, any implicant of f can be represented by one of the above three. Also, above three represents only the implicants for f. Therefore, we have the theorem. □

Example 12.3.2 *In Fig. 12.3.1, the SOP represented by the 1-paths is*

$$\begin{aligned} \mathcal{F}_3 &= xy(\bar{z} \vee z \vee 1) \vee x(\bar{y} \vee 1)z \vee (\bar{x} \vee 1)yz \\ &= xy\bar{z} \vee xyz \vee xy \vee x\bar{y}z \vee xz \\ &\quad \vee \bar{x}yz \vee yz. \end{aligned}$$

Note that $\mathcal{F}_3$ contains all the implicants for f. ■

Definition 12.3.3 *An ATDD is a* ***Reduced Ordered AND Ternary Decision Diagram*** *(ROATDD) if it does not contain a node v with* **low**(v) = **high**(v) = **and**(v), *and if does not contain distinct nodes v_1 and v_2 such that the subgraphs rooted by v_1 and v_2 are isomorphic.*

Example 12.3.3 *Fig. 12.4.1 shows the ROATDD for $f = xy \vee yz \vee zx$.* ■

Theorem 12.3.4 *For any function, there is a unique ROATDD. The set of 1-path for an ROATDD represents a sum of implicants for f.*

(Proof) We can prove this in a similar way to the case of QROATDDs. □

Example 12.3.4 *In Fig. 12.4.1, the SOP represented by the 1-paths is*

$$\begin{aligned} \mathcal{F}_4 &= xy \vee x(1 \vee \bar{y})z \vee (\bar{x} \vee 1)yz \\ &= xy \vee xz \vee x\bar{y}z \vee \bar{x}yz \vee yz. \end{aligned}$$

Note that $\mathcal{F}_4$ contains some of the implicants for f. ■

12.4 REDUCED ORDERED TDD AND SOP

In this section, we define a reduced ordered TDD that represents a given sum-of-products expression. We also enumerate the number of the nodes in the reduced ordered TDDs.

Definition 12.4.1 *A* ***Reduced Ordered Ternary Decision Diagram*** *(ROTDD) for an SOP $\mathcal{F}$ is a TDD that is recursively defined as follows: For an SOP $\mathcal{F}$, consider the expansion $\mathcal{F} = \bar{x} \cdot \mathcal{F}(|x=0) \vee x \cdot \mathcal{F}(|x=1) \vee 1 \cdot \mathcal{F}(|x=*)$, where $\mathcal{F}(|x=0)$, $\mathcal{F}(|x=1)$, and $\mathcal{F}(|x=*)$ denote the SOPs without the literals of x. Let v denote the node for $\mathcal{F}$, then* **low**(v), **high**(v), *and* **and**(v) *denote $\mathcal{F}(|x=0)$, $\mathcal{F}(|x=1)$, and $\mathcal{F}(|x=*)$, respectively. And, the TDD does not contain distinct nodes v_1 and v_2 such that the subgraphs rooted by v_1 and v_2 are isomorphic, and it does not have a node v with* **low**(v) = **high**(v) = **and**(v).

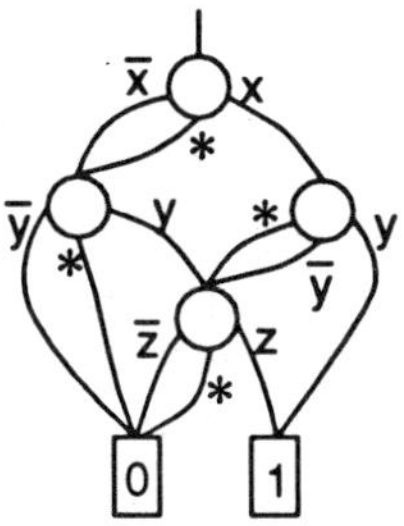

Figure 12.4.1 ROATDD.

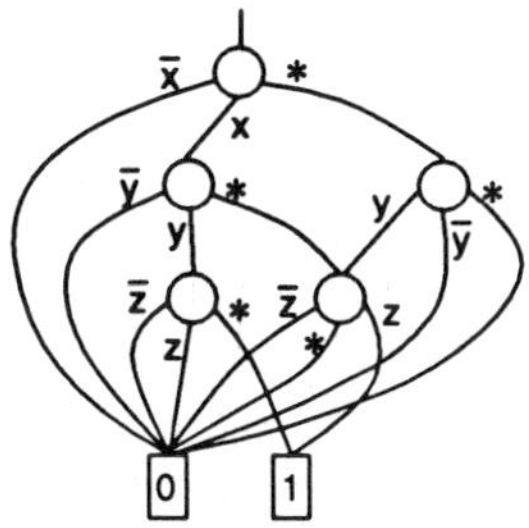

Figure 12.4.2 ROTDD.

Example 12.4.1 *Fig. 12.4.2 shows the ROTDD for the SOP $\mathcal{F} = xy \vee yz \vee zx$. Note that $\mathcal{F}$ can be represented as*

$$\begin{aligned}\mathcal{F} &= \bar{x}[0] \vee x[y \vee z] \vee 1[yz] \\ &= \bar{x}[0] \vee x[\bar{y}(0) \vee y(1) \vee 1(z)] \vee 1[\bar{y}(0) \vee y(z) \vee 1(0)].\end{aligned}$$

■

Theorem 12.4.1 *For any SOP, there is a unique ROTDD.*

(Proof) Clearly, an arbitrary SOP can be represented by an ROTDD. To prove uniqueness, we use mathematical induction. Clearly, the theorem is true for SOPs with 0 and 1 variable. Suppose that the theorem is true for SOPs with k or fewer variables. Any SOP $\mathcal{F}$ of $k+1$ variables can be expanded as $\mathcal{F} = \bar{x}_1 \cdot \mathcal{F}(|x = 0) \vee x_1 \cdot \mathcal{F}(|x = 1) \vee 1 \cdot \mathcal{F}(|x = *)$, where $\mathcal{F}(|x = 0)$, $\mathcal{F}(|x = 1)$, and $\mathcal{F}(|x = *)$ are SOPs with k or fewer variables. By the inductive hypothesis, there are unique ROTDDs for these three SOPs. Consider a TDD for $\mathcal{F}$ with the root node v, where ***low***(v), ***high***(v), and ***and***(v) denote the ROTDD for $\mathcal{F}(|x = 0)$, $\mathcal{F}(|x = 1)$, and $\mathcal{F}(|x = *)$, respectively. The ROTDD for $\mathcal{F}$ is obtained from the TDD for f by reducing the isomorphic subgraphs, and it is also unique. □

Theorem 12.4.2 *The set of 1-paths for an ROTDD represents an SOP.*

Theorem 12.4.3 *Let p be the number of the products in an SOP $\mathcal{F}$. Then, the number of the nodes in the ROTDD for $\mathcal{F}$ is at most $np + 2$, where $p \geq 1$.*

(Proof) We will use mathematical induction for the proof. Clearly, the theorem is true for 0 variable functions, where we use two terminal nodes for constants 0 and 1. Suppose that the theorem is true for SOPs with k or fewer variables. Any SOP $\mathcal{F}$ with $k+1$ variables can be expanded as $\mathcal{F} = \bar{x}_1\mathcal{F}_0 \vee x_1\mathcal{F}_1 \vee 1\cdot\mathcal{F}_2$, where $\mathcal{F}_0$, $\mathcal{F}_1$, and $\mathcal{F}_2$ are SOPs with k or fewer variables. Let the inductive numbers of products for $\mathcal{F}_0$, $\mathcal{F}_1$, and $\mathcal{F}_2$ be p_0, p_1, and p_2, respectively. By the hypothesis, $\mathcal{F}_0$, $\mathcal{F}_1$, and $\mathcal{F}_2$ can be represented by ROTDDs with at most kp_0, kp_1, and kp_2 nonterminal nodes, respectively. The ROTDD for $\mathcal{F}$ is a reduced graph with these three ROTDDs as subgraphs. Thus, the number of nodes is at most $1+kp_0+kp_1+kp_2+2 = k(p_0+p_1+p_2)+3 = kp+3 \leq (k+1)p+2$. □

Let p be the number of products in an irredundant SOP $\mathcal{F}$. Because any n-variable function is represented by an irredundant SOP with at most 2^{n-1} products, we have $p \leq 2^{n-1}$. Thus, for an irredundant SOP, an upper bound on the number of nodes of an ROTDD is $n2^{n-1}+2$. On the other hand, for an ROBDD, number of nodes is $O(2^n/n)$. However, if p is small, then the number of nodes for ROTDD is often smaller than ROBDD. For example, for the SOP $\mathcal{F} = x_1x_{p+1} \vee x_2x_{p+2} \vee \cdots \vee x_px_{2p}$, the ROBDD requires 2^{p+1} nodes, while the ROTDD requires only $2p^2+2$ nodes. Since $\mathcal{F}$ is the sum of all the prime implicants, the size of the PTDD is also $2p^2+2$.

12.5 PRIME TDD AND GENERATION OF PRIME IMPLICANTS

In this section, we define a prime TDD which presents all the prime implicants of the function. We also enumerate the nodes in the prime TDDs.

Definition 12.5.1 *The set of the prime implicants for f is denoted by PI(f).*

Definition 12.5.2 *A **Prime Ternary Decision Diagram** (PTDD) for PI(f) is an ROTDD for $\mathcal{F} = PI(f)$.*

Example 12.5.1 *Fig. 12.5.1 shows the PTDD for $f = xy \vee yz \vee zx$.* ■

The SOP consisting all the PIs is called the complete sum-of-PIs. It is also called a Blake canonical form [6].

Lemma 12.5.1 *For a function f, the set of all the PIs for f is unique.*

Lemma 12.5.2 *Let $f = \bar{x}f_0 \vee xf_1$, then*

$$\begin{aligned} PI(f) &= \bar{x}\{PI(f_0) - PI(f_0f_1)\} \cup x\{PI(f_1) - PI(f_0f_1)\} \\ &\quad \cup PI(f_0f_1) \end{aligned}$$

(Proof) (1) For a PI of f, exactly one of the following holds:

A) The PI includes $\bar{x}$ as a literal,
B) The PI includes x as a literal, or
C) The PI include neither $\bar{x}$ nor x as a literal.

These PIs are obtained, respectively, as follows:

D) A product of $\bar{x}$ and a PI for f_0.
E) A product of x and a PI for f_1.
F) A PI for $g = f_0 \cdot f_1$.

From this argument, it is clear that the left-hand side of the equation contains all the PIs for f. Note that if p is a PI of g, then, neither $\bar{x} \cdot p$ nor $x \cdot p$ are PIs of g.
(2) Next, let $\bar{x} \cdot q$ be an arbitrary element of $\bar{x}\{\text{PI}(f_0) - \text{PI}(f_0 \cdot f_1)\}$, where q does not contain a literal of x. Suppose that $\bar{x} \cdot q$ is not a PI of f. Then, there exists another PI r that properly contains $\bar{x} \cdot q$. If r does not contain $\bar{x}$ as a literal, then r must be a PI of g. However, this contradicts to the condition of q. Also, if r contains $\bar{x}$ as a literal, then q is not a PI of f_0, which also contradicts to the assumption. Therefore, each element in $\bar{x}\{\text{PI}(f_0) - \text{PI}(f_0 \cdot f_1)\}$ is a PI for f. In a similar way, we can prove that each element in $x \cdot \{PI(f_1) - PI(f_0 \cdot f_1)\}$ is a PI for f. Next, let s be an arbitrary element of $PI(f_0 \cdot f_1)$. Suppose that s is not a PI of f. Then there exists another PI that properly contains s. However, this contradicts the assumption that s is an element of $PI(f_0 \cdot f_1)$. Therefore, the elements in the left side hand of the equation are all PIs of f.
From (1) and (2), we have the lemma. □

Theorem 12.5.1 *For any function, there is a unique PTDD.*

(Proof) By Lemma 12.5.1, for any function f, there is a unique PI(f). By Theorem 12.4.1, for the given complete sum-of-PIs, there is a unique ROTDD.
□

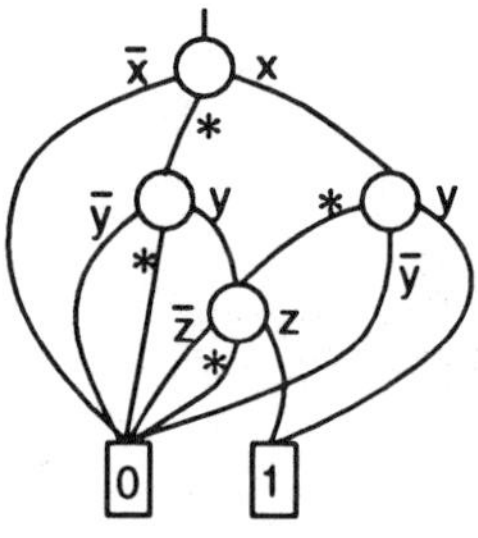

Figure 12.5.1 PTDD.

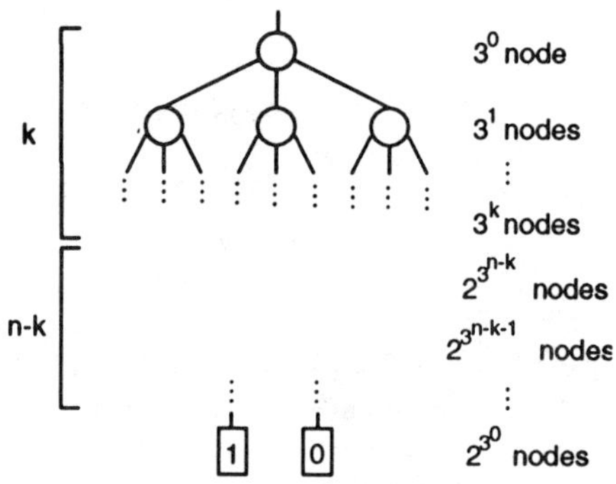

Figure 12.5.2 PTDD.

Lemma 12.5.3 *In a PTDD, nodes with the same index represent distinct SOPs.*

(Proof) If two nodes with the same index represent the same SOP, then, by Theorem 12.4.1, the subgraphs rooted these nodes are isomorphic and they can be reduced. □

Theorem 12.5.2 *The number of nodes in a PTDD is at most*

$$\min_{k=1}^{n}\left(\frac{3^{k+1}-1}{2}+2^{3^{n-k}}\right).$$

(Proof) A given PTDD can be divided into upper and lower blocks as shown in Fig. 12.5.2, where the upper block has first k inputs and the lower block has the remaining $(n-k)$ variables.
(1) For the upper block: Consider the complete ternary decision tree of k variables. The number of nodes with index i is at most 3^j. Therefore, the number of nodes in the upper block is at most $\sum_{j=0}^{k} 3^j = \frac{3^{k+1}-1}{2}$.
(2) For the lower block: Consider the PTDDs of $(n-k)$ variables. By Lemma 12.5.3, the nodes with the same index represent distinct SOPs. There are at most $2^{3^{n-k}}$ distinct SOPs for an $(n-k)$-variable function. However, among them, $2^{3^{n-k-1}}$ do not contain the literal of x_{n-k}. Therefore, the number of the nodes with index $(n-k)$ is at most $2^{3^{n-k}} - 2^{3^{n-k-1}}$. Similarly, for the nodes with

index $(n-k-1)$, the number of nodes is at most $2^{3^{n-k-1}} - 2^{3^{n-k-2}}$, and so on. From this argument, we can see that the number of nodes in the lower block is at most $(2^{3^{n-k}} - 2^{3^{n-k-1}}) + (2^{3^{n-k-1}} - 2^{3^{n-k-2}}) + \cdots + (2^{3^1} - 2^{3^0}) + 2^{3^0} = 2^{3^{n-k}}$. Because (1) and (2) hold for any k $(k = 1, 2, \ldots, n)$, we have the theorem. □

Set $k = n - \log_3 n$ in Theorem 12.5.2, and we have the following:

Corollary 12.5.1 *The number of nodes in a PTDD for an n variable function is $O(3^n/n)$.*

12.6 BDDS AND TDDS FOR SYMMETRIC FUNCTIONS

In this section, we enumerate the number of the nodes in the ROBDDs and TDDs for symmetric functions.

Definition 12.6.1 *An n-variable* ***fundamental symmetric function*** *$S(n, k)$ takes the value 1 iff the number of 1's in the inputs is exactly $k (k = 0, 1, 2, \ldots, n)$.*

Lemma 12.6.1 *All the symmetric functions of n variables can be represented by an ROBDD with $2^{n+2} - 2n - 2$ nodes.*

(Proof) The proof is done by mathematical induction. For $n = 0$, all the symmetric functions (constants 0 and 1) are realized by a BDD with 2 nodes. For $n = 1$, all the symmetric functions (constants 0 and 1, $\bar{x}$, x) are realized by a BDD with 4 nodes as shown in Fig. 12.6.1. Suppose that all the symmetric functions of $(k-1)$ variables can be represented by an ROBDD with $2^{k+1} - 2(k-1) - 2$ nodes. Except for the constant functions, there are $2^{k+1} - 2$ different symmetric functions of k variables and each of them can be represented as $S(k) = \bar{x}_k S_a(k-1) \vee x_k S_b(k-1)$, where $S_a(k-1)$ and $S_b(k-1)$ are symmetric functions of $(k-1)$ variables. Therefore, $S(k)$ can be represented by the ROBDD in Fig. 12.6.2. Note that the total number of nodes is $(2^{k+1} - 2) + 2^{k+1} - 2(k-1) - 2 = 2^{k+2} - 2k - 2$. □

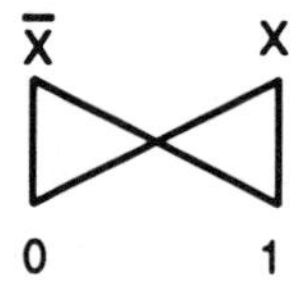

Figure 12.6.1 ROBDD for one-variable function.

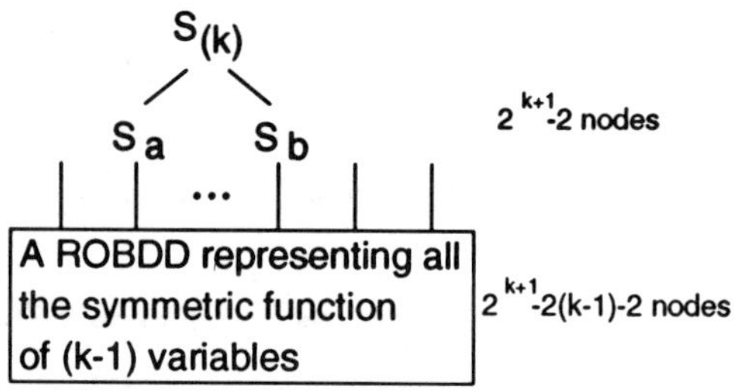

Figure 12.6.2 ROBDD representing all the k-variable symmetric functions.

Theorem 12.6.1 *An arbitrary n-variable symmetric function can be represented by an ROBDD with at most*

$$BN(n, k) = \frac{(k+1)(k+2)}{2} + 2^{n-k+1} - 2(n-k) - 2$$

nodes, where $1 \leq k \leq n$.

(Proof) Consider the BDD shown in Fig. 12.6.3, where the upper block has the first k variables, and the lower block has the remaining $(n-k)$ variables. The nodes with index k in the upper block correspond to the fundamental symmetric functions $S(k, j)$, $(j = 0, 1, \ldots, k)$. Fig. 12.6.3 shows that the total number of nodes is $\text{BN}(n, k)$. □

Set $k = n$ in Theorem 12.6.1, and we have the following:

Corollary 12.6.1 *The number of nodes in an ROBDD for an n-variable symmetric function is $O(n^2)$.*

Lemma 12.6.2 *All the symmetric functions of n variables can be represented by an ROATDD with $2^{n+2} - 2n - 2$ nodes.*

(Proof) Similar to the proof of Lemma 12.6.1. □

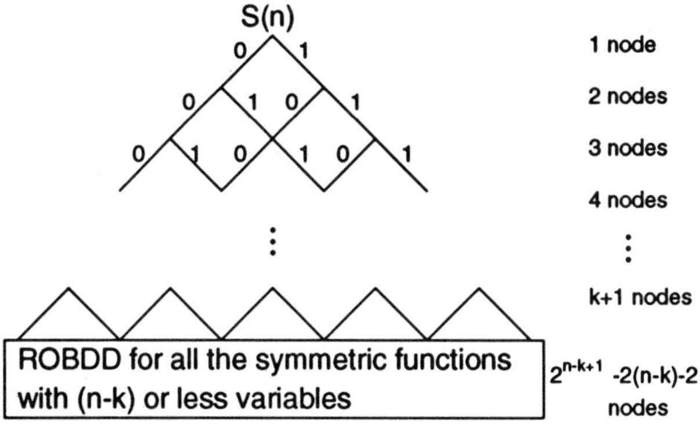

Figure 12.6.3 ROBDD for symmetric function of n-variables.

Theorem 12.6.2 *An arbitrary n-variable symmetric function can be represented by an ROATDD with at most*

$$TN(n,k) = \frac{k(k+1)(k+5)}{6} + (k+1) + 2^{n-k+2} - 2(n-k) - 2$$

nodes, where $1 \leq k \leq n$.

(Proof) Consider the complete ternary decision tree of k variables in Fig. 12.6.4. The different number of functions generated in the lowest level of the tree is derived as follows:

1. Because f is completely symmetric, the permutation of the subscripts of f will not change the function: e.g., f_{021120} is equal to f_{001122}. So, the number of k-variable functions generated by the complete ternary decision tree is equal to "the number of ways to select k objects from 3 distinct objects with repetition".

2. "The number of ways to select k objects from p distinct objects with repetition" is $C(p+k-1,k)$ [14].

3. So, the number of the different symmetric functions is $(k+1)(k+2)/2$. The total number of nodes in the ROATDD is

$$\sum_{i=0}^{n} \frac{(i+1)(i+2)}{2} = \frac{k(k+1)(k+5)}{6} + (k+1).$$

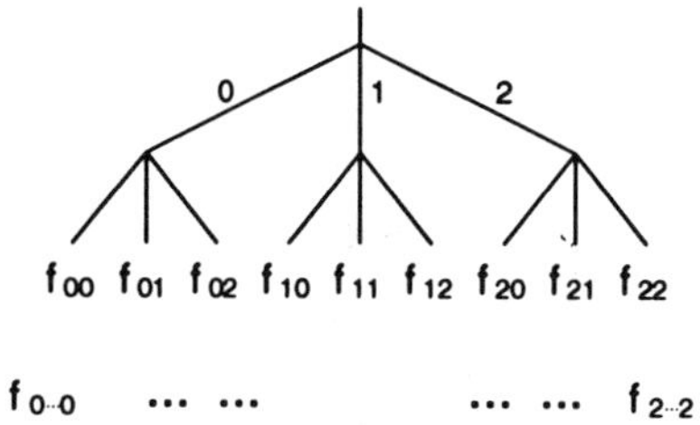

Figure 12.6.4 Complete ternary decision tree for k-variable function.

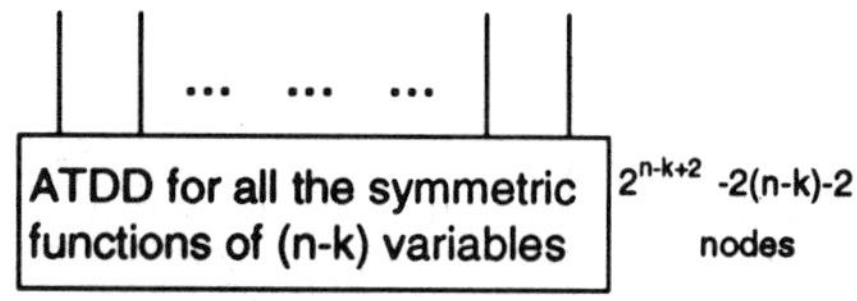

Figure 12.6.5 ATDD for all the symmetric functions of $(n-k)$ variables.

Consider the ATDD that generates all the symmetric functions of $(n-k)$ variables. By combining the ROATDD derived from Fig. 12.6.4, and the ATDD in Fig. 12.6.5, we can derive an ATDD which represents an arbitrary symmetric function of n variables.

Therefore, we have the theorem. □

Set $k = n$ in Theorem 12.6.2, and we have the following:

Corollary 12.6.2 *An arbitrary symmetric function of n variables can be represented by an ROATDD with $O(n^3)$ nodes.*

Theorem 12.6.3 *An arbitrary n-variable symmetric function can be represented by a PTDD with at most $O(n^3)$ nodes.*

(Proof) The number of nodes in a PTDD for a symmetric function does not exceed that of the complete ternary decision tree of the symmetric functions described in the proof of Theorem 12.6.2. So, we need at most $(n+1)(n+5)/6+(n+1)$ nodes. □

There is a symmetric function with $O(3^n/n)$ PIs[9]. However, all the PIs for the function can be represented by a PTDD with only $O(n^3)$ nodes.

12.7 EXPERIMENTAL RESULTS

12.7.1 Generation of Prime Implicant Using PTDDs

We coded the PI generation algorithm in C. The outline of the algorithm is as follows:

1) Convert a PLA format into an ROBDD.
2) Convert an ROBDD into an ROATDD.
3) Convert an ROATDD into a PTDD.
4) Enumerate all the 1-paths.

The most important step is 3), where we used Lemma 12.5.2.

A. Number Nodes for Computation: We randomly generated functions of n variables with 2^{n-1} minterms: For each n, 50 functions when $n \leq 14$, and one function for $n = 15$ and 16. Fig. 12.7.1 shows the average number of nodes for ROBDDs, ROATDDs and PTDDs. It also shows WORK, the maximal number of nodes during computation for PTDDs. The number of nodes for ROBDDs is almost proportional to $2^n/n$, while the numbers for the nodes in ROATDDs are less than $3^n/n$.

We also generated PIs for the complete specified functions in the ESPRESSO book [4]. Table 12.7.1 shows the number of PIs (the 5th column), the number of nodes for ROATDDs (center of the 6th column), WORK (the maximal number of nodes during computation), and cputime (the 7th column, milliseconds by an HP720 workstation with 64 Mega Bytes of main memory). In many cases, PIs can be generated in a short time. Blank entries for the column with a label "Primes" denote that we aborted the computation because of the memory overflow. In addition to the ESPRESSO functions, we successfully enumerated the PIs for large functions such as mainpla (# of PIs=87692), misj (# of PIs=139103), and ts10 (# of PIs=524280). Conventional methods could not generate all the PIs for these functions [21].

B. Influence of the Ordering of the Variables: For a randomly generated function of 8 variables, we permuted the input variables in all possible (8!=40320) ways. For each permutation, we obtained the number of nodes for the ROBDD, the PTDD and WORK (maximal number of nodes during computation). Table 12.7.2 shows the minimum, the average, and the maximum number of nodes. About 10 percent of the nodes can be reduced by optimizing the ordering of the input variables.

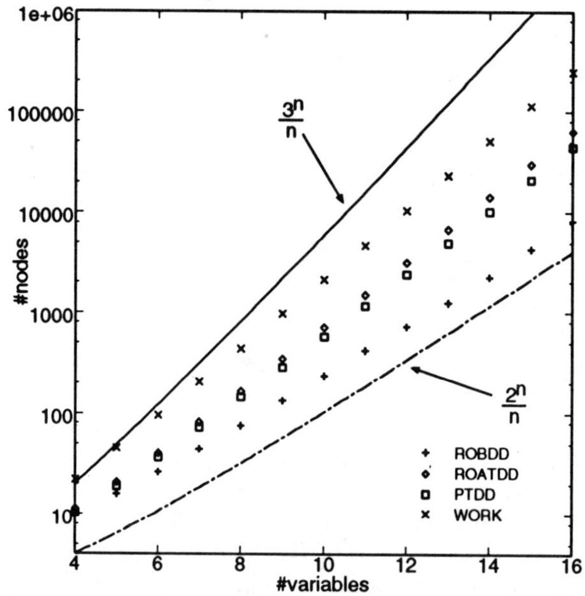

Figure 12.7.1 Number of nodes.

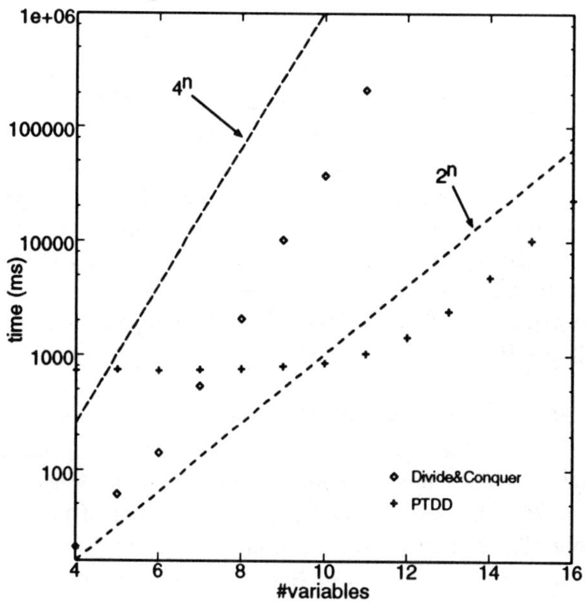

Figure 12.7.2 Computation time.

Table 12.7.1 Experimental result.

Name	In/Out	Terms	SOP	Primes	BDD/PTDD /WORK	Time	Cou.	ROTDD /WORK	Time	s_t/s_c
add6	12/7	1092	355	8568	598/2399/12381	1.4	0.2	695/4616	0.2	1.89
adr4	8/5	255	75	397	123/261/1007	0.6	0.2	178/752	0	2.66
alu1	12/8	19	19	780	1003/872/5809	0.9		129/326	0	6.56
bc0	26/11	419	177	6596	983/8278/82253	7.0	9.2	1646/4854	0.2	5.92
chkn	29/7	153	141	671	2103/1424/16260	1.9	1.0	869/5104	0.2	3.75
co14	14/1	14	14	14	29/29/56	0.5	0.2	29/211	0	1.87
dc1	4/7	15	9	22	22/38/74	0.5		26/67	0	3.85
dc2	8/7	58	39	173	207/275/845	0.6		122/432	0	3.50
dist	8/5	255	121	401	145/466/1819	0.6	0.5	303/1378	0	2.80
f51m	8/8	255	76	561	511/872/2884	0.7	0.3	165/851	0	2.43
gary	15/11	214	107	706	510/1567/10602	1.2	1.2	704/1968	0	5.58
in1	16/17	110	104	928	617/3035/11061	1.1	2.9	635/2421	0	4.88
in2	19/10	137	135	666	3839/1388/24040	3.6	1.1	785/3626	0.1	4.40
in3	35/29	75	75	1114	494/1748/12976	1.7	4.1	1319/3311	0.1	8.95
in4	32/20	234	212	3076	3548/5441/56061	5.5	4.2	1364/7125	0.4	3.53
in5	24/14	62	62	1067	1267/2575/17456	1.8	1.8	627/2156	0	6.74
in6	33/23	54	54	6174	18793/16510/116092	20.1	2.7	1002/2313	0	9.99
in7	26/10	84	55	2112	1182/3325/22149	2.1	1.1	408/1806	0	4.72
jbp	36/57	166	126				204	1713/4948	0.2	6.34
misg	56/23	75	69				0.4	941/5334	0.2	5.09
mish	94/43	91	82				2.7	2241/8558	0.4	6.60
mlp4	8/8	225	126	606	337/769/3319	0.7	0.7	321/1442	0	2.85
opa	17/69	342	79	477	313/998/2937	0.7	6.7	904/1927	0	7.63
radd	8/5	120	75	397	108/186/705	0.5	0.2	144/709	0	2.15
rckl	32/7	96	32	302	120/1346/9920	1.1	0.9	530/1585	0	9.46
rd53	5/3	31	31	51	21/30/75	0.5	0.2	31/204	0	1.27
rd73	7/3	147	127	211	36/61/183	0.5	0.2	61/934	0	0.56
risc	8/31	74	27	43	73/111/325	0.5		139/342	0	5.15
root	8/5	255	57	152	101/243/967	0.6	0.3	172/655	0	3.38
sqn	7/3	84	84	75	87/133/544	0.5		101/372	0	1.40
sqr6	6/12	63	51	205	127/371/965	0.5	0.3	139/507	0	3.18
ti	47/72	241	215				56	4087/13591	1.2	7.39
tial	14/8	640	579	7145	2877/47521/153753	12.2		1701/8906	0.4	2.65
vg2	25/8	110	110	1188	1100/1340/13642	1.3	0.4	215/1835	0.1	1.30
x1dn	27/6	112	110	1220	2417/1080/14510	1.6	0.7	286/2413	0.1	1.65
x2dn	82/56	112	104				9.7	2528/9822	0.5	6.48
x6dn	39/5	121	81	916	362/1627/8375	1.0	0.9	1170/3182	0.1	7.22
x7dn	66/15	622	538				29	1720/17879	2.7	1.07
x9dn	27/7	120	120	1272	2121/1297/15859	1.7	0.7	273/2901	0.1	1.45
z4	7/4	127	59	167	64/144/579	0.5	0.2	98/508	0	1.94
5xp1	7/10	75	67	390	255/769/2310	0.6	0.3	133/601	0	2.22
9sym	9/1	87	87	1680	35/62/167	0.5	0.2	198/874	0	2.45

Terms : number of products terms in the original data
SOP : number of products in an SOP
Time : CPU Sec, HP9000/720, 64MB main memory
Cou. : Coudert-Madre CPU Sec, SUN SPARC2, 48MB main memory

Table 12.7.2 Distribution of nodes for ROBDD, PTDD, WORK.

	min	average	max
ROBDD	69	74.78	79
PTDD	130	140.92	154
WORK	384	437.76	487

C. Comparison with Divide and Conquer Method: The methods developed by Morreale [16] and Reusch [20] are also called a divide and conquer method [4]. Before the appearances of Coudert-Madre's algorithm [8], the divide and conquer method was considered to be the most efficient way for PI generation.

We compared the performance of the presented algorithm with an old PI generation program based on divide and conquer method. The old program was developed for minimization of SOPs with multiple-valued inputs by using FORTRAN [22]. Fig. 12.7.2 compares the computation time for PI generation for two methods. Because the data structure and computers are different, only the rate of the increase for the CPU time is important. The divide and conquer method requires computation time nearly proportional to 4^n, while the PTDD method requires CPU time nearly proportional to 2^n. Note that the CPU time for PTDD method only contains the time for conversion from ROBDDs to PTDDs. To generate all the PIs, we need additional time to trace all the 1-paths. On the other hand, the CPU time for divide and conquer method contains the time for PI generation.

In this experiment, we are mainly interested in the statistical properties of the TDDs rather than developing a fast tool. So, we did not use any special technique to speed up the computation. However, the program is much faster than conventional methods. It is possible to speed up the program by using more sophisticated programming techniques [3, 11, 15].

D. Comparison with Coudert-Madre's Method: Coudert-Madre's meta-product uses two variables to represent a variable of the original function. On the other hand, TDD uses one variable with three-value to represent same thing. To eliminate the redundancy of the meta-product, Coudert-Madre introduced implicit prime set (IPS). It is a very efficient way to represent a set of prime implicants. In Table 1, the 7th column shows the CPU time of their method. In

Table 12.7.3 Representation of SOP by ROTDD.

Name	In/Out	Terms	SOP	Primes	BDD/PTDD /WORK	Time	Cou.	ROTDD /WORK	Time	s_t/s_c
cordic	23/2	1206	914	1754	174/196/1982	0.7		169/15688	3.1	0.13
life	9/1	140	84	224	30/44/105	0.5	0.2	29/631	0	0.37
rd73	7/3	147	127	211	36/61/183	0.5	0.2	61/934	0	0.56
rd84	8/4	256	255	633	45/87/255	0.5		92/2077	0	0.40
ryy6	16/1	112	112	112	37/35/111	0.5		38/1392	0	0.29
sym10	10/1	837	210	3150	40/72/198	0.6	0.2	200/2074	0	0.99
t481	16/1	481	481	481	159/357/2639	0.6		73/4610	0	0.13
xor5	5/1	16	16	16	11/11/20	0.5		11/85	0	0.88

some cases, our method is faster, but in other cases Coudert-Madre's is faster. It should be noted that the Coudert-Madre successfully solved larger problems than we, because of memory size overflow. One reason for this is that our TDD uses binary vectors to represent multiple-output functions. The merit of our method is its simplicity. Each 1-path of the TDD corresponds to a prime implicant, and thus the analysis is relatively easy. Also, it is fast enough for ordinary applications.

12.7.2 Representation of SOP by ROTDD

For ESPRESSO benchmark functions, we converted PLAs into ROTDDs. The amounts of memory necessary for ROTDD and cube representations are given as follows: Cube representation (Sc): # of products × (# of bytes for each cube=n+16) ROTDD representation (St): # of nodes × (# of bytes for each node=28). Table 12.7.1 also shows the number of products (the 4th column), number of nodes (the 9th column), and (RATE=St/Sc, the last column). Cube representations require smaller amount of memory except for rd73. However, we found many benchmark functions whose ROTDD representations are smaller than cube representations. Table 12.7.3 shows such functions.

Table 12.7.4 Comparison of BDDs and TDDs.

	Uniqueness of Representation	# of nodes	Expression Represented by 1-paths	Application
QROBDD	Function	$O(2^n/n)$	Canonical SOP	Representation of a function
ROBDD	Function	$O(2^n/n)$	Disjoint SOP	Representation of a function
QROATDD	Function	$O(3^n/n)$	Complete sum of implicants	Generation of implicants
PTDD	Function	$O(3^n/n)$	Complete sum of prime implicants	Generation of prime implicants
ROTDD	SOP	$np + 2$	SOP	Representation of an SOP

12.8 CONCLUSION AND COMMENTS

In this chapter, we presented various TDDs to represent functions and SOPs. Table 12.7.4 compares the properties of BDDs and TDDs. In a PTDD, each 1-path corresponds to a PI. So, PTDDs are useful to generate the PIs of functions. PI generation using PTDDs is easier to understand than Coudert-Madre's method. We also proved that the number of nodes in ROATDDs and PTDD is $O(3^n/n)$. In the case of symmetric functions, ROBDDs require $O(n^2)$ nodes while QROATDDs require $O(n^3)$ nodes. When the number of the products is large, ROTDDs sometimes require smaller amount of memory than conventional cubical representation of SOPs. There exist a function whose RO-TDD (PTDD) requires only $2p^2 + 2$ nodes, while whose ROBDD requires 2^{p+1} nodes.

We can also consider a TDD where the third child represents $f_0 \oplus f_1$ instead of $f_0 f_1$. This TDD is called an EXOR TDD, and is useful for simplifying AND-EXOR expressions [23].

Acknowledgments

This work was supported in part by a Grant in Aid for Scientific Research of the Ministry of Education, Science and Culture of Japan. Mr. Takahisa Amada developed programs for BDDs and TDDs. Prof. J. T. Butler reviewed this chapter.

REFERENCES

[1] S. B. Akers, "Binary decision diagrams," *IEEE Trans. Comput.*, Vol. C-27. No. 6, June 1978, pp. 509-516.

[2] A. Blake, "Canonical expressions in Boolean algebra," Ph. D. Dissertation, Dept. of Mathematics, Univ. Chicago, 1937. Published by Univ. Chicago, Libraries, 1937.

[3] K. S. Brace, R. L. Rudell and R. E. Bryant, "Efficient implementation of a BDD package," *Proc. 27th Design Automation Conference*, June 1990, pp. 40-45.

[4] R. K. Brayton, G. D. Hachtel, C. T. McMullen, and A. L. Sangiovanni-Vincentelli, *Logic Minimization Algorithms for VLSI Synthesis*, Boston, MA. Kluwer, 1984.

[5] J. G. Bredeson and P. T. Hulina, "Generation of prime implicants by direct multiplications," *IEEE Trans. Comput.* Vol. C-20, April 1971, pp. 475-476.

[6] F. M. Brown, *Boolean Reasoning: The logic of Boolean equations,* Kluwer Academic Publishers, Boston, 1990.

[7] R. E. Bryant, "Graph-based algorithms for Boolean function manipulation," *IEEE Trans. Comput.* Vol. C-35, No. 8, Aug. 1986, pp. 677-691.

[8] O. Coudert and J. C. Madre, "A new graph based prime computation technique," in Sasao (ed.), *Logic Synthesis and Optimization*, Kluwer Academic Publisher, pp. 33-57, 1993.

[9] B. Dunham and R. Fridshal, "The problem of simplifying logical expressions," J. Symb. Logic. March 1959, pp. 17-19.

[10] S. R. Das, "A new algorithm for generating prime implicants," *IEEE Trans. Comput.* Vol. C-20, No. 12, 1971, pp. 1614-1615.

[11] M. Fujita, H. Fujisawa, and N. Kawato, "Evaluation and implementation of Boolean Comparison method base on binary decision diagrams," *ICCAD-88*, Nov. 1988, pp. 6-9.

[12] M. A. Harrison, *Introduction to Switching and Automata Theory*, McGraw-Hill, 1965.

[13] H-T. Liaw and C-S. Lin, "On the OBDD representation of generalized Boolean functions," *IEEE Transactions on Comput.* Vol. 41, No. 6, June 1992, pp. 661-664.

[14] C. L. Liu, Elements of Discrete Mathematics, (second edition), McGraw-Hill, New York, 1985.

[15] S. Minato, N. Ishiura, and S. Yajima, "Shared binary decision diagram with attributed edges for efficient Boolean function manipulation," *Proc. 27th ACM/IEEE Design Automation Conf.*, June 1990, pp. 52-57.

[16] E. Morreale, "Recursive operators for prime implicant and irredundant normal form determination," *IEEE Trans. Comput.* Vol. C-19, No. 6. June 1970, pp. 504-509.

[17] S. Muroga, *Logic Design and Switching Theory*, John Wiley & Sons, 1979.

[18] R. J. Nelson, "Simplest normal truth functions," *J. Symb. Logic* June 1954, pp. 105-108.

[19] W. V. Quine, "A way to simplify truth functions," *Amer. Math. Mon.*, Vol. 62, November 1955, pp. 627-631.

[20] B. Reusch, "Generation of prime implicant from subfunctions and a unifying approach to the covering problem," *IEEE Trans. Comput.* Vol. C-24, No. 9, Sept. 1975, pp. 924-930.

[21] R. L. Rudell, "Multiple-valued logic minimization for PLA synthesis," *Technical Report*, University of California, Electronics Research Laboratory, Berkeley, CA. June 1986.

[22] T. Sasao, "Input variable assignment and output phase optimization of PLA's," *IEEE Trans. on Comput.*, Vol. C-33, No. 10, pp. 879-894, Oct. 1984.

[23] T. Sasao, "AND-EXOR expressions and their optimization," in Sasao (ed.), *Logic Synthesis and Optimization*, Kluwer Academic Publisher, pp. 287-312, 1993.

[24] J. R. Slagle, C. L. Chang, and R. C. T. Lee, "A new algorithm for generating prime implicants," *IEEE Trans. Comput.* Vol. C-19, No. 4. April 1970, pp. 304-310.

[25] P. Tison, "Generalization of consensus theory and application to the minimization of Boolean functions," *IEEE Trans. Electron Comput.*, Vol. EC-16, August 1967, pp. 446-456.

[26] S. Yang, "Logic synthesis and optimization benchmark user guide, Version 3.0," MCNC, Jan. 1991.

13

OR-AND-OR THREE-LEVEL NETWORKS

Tsutomu SASAO

Department of Computer Science and Electronics
Kyushu Institute of Technology
Iizuka, Japan

Abstract— This chapter considers the number of gates to realize logic functions by OR-AND-OR three-level networks under the condition that both true and complemented variables are available, and each gate has no fan-in and fan-out constraints. We show that an arbitrary n-variable function can be realized by an OR-AND-OR three-level network with at most $2^{r+1}+1$ gates, where $n = 2r$ and r is an integer. We also prove that for sufficiently large n, regardless of the number of levels, we need at least $2^{r+1}(1-\xi)$ gates to realize almost all functions of n variables by an AND-OR multi-level network, where ξ is an arbitrarily small positive real number ($0 < \xi < 1$). We developed a heuristic algorithm to design OR-AND-OR three-level networks, realized various functions, and compared the number of gates for OR-AND-OR three-level networks with AND-OR two-level ones. For arithmetic functions of 8 variables, three-level networks require, on the average, 40% fewer gates than AND-OR two-level ones. For other benchmark functions of 9 to 128 variables, three-level networks required up to 91% fewer gates. For randomly generated functions of 10 variables, three-level networks required 50% fewer gates.

13.1 INTRODUCTION

Representing logic functions by using logic gates is a fundamental problem in logic design. Optimization methods for AND-OR two-level networks are well established [6].

Lawler [5] considered a design method for multi-level networks, such as OR-AND-OR, AND-OR-AND-OR, ... , etc. He showed a systematic way to find simple networks.

OR-AND-OR three-level networks are interesting research topic, since they require many fewer gates than AND-OR two-level ones, and are relatively easy to design.

Bounded-depth networks [1, 3] are desirable since the propagation delays of them are easy to estimate. Also, they can be used as initial networks for multi-level logic networks with fan-in limited gates.

In this chapter, we consider the number of gates to realize arbitrary functions by AND-OR multi-level networks under the condition that both true and complemented variables are available as inputs, and each gate has unlimited fan-in and fan-out.

An arbitrary logic function can be realized by an AND-OR two-level network. In this case, $2^{n-1}+1$ gates are necessary and sufficient to realize a parity function of n variables, which requires the largest number of gates in two-level realizations [13, 16]. In a similar way, an arbitrary function can be realized by multi-level networks such as OR-AND-OR three-level networks, AND-OR-AND-OR four-level networks, etc. Then, how many gates are necessary to realize an arbitrary function by a multi-level network? And, which function requires the largest number of gates in a multi-level realization? This chapter considers the number of gates to realize arbitrary functions and tries to find the most complex function in multi-level networks.

First, we show that an arbitrary function of n variables ($n = 2r$, where r is an integer) can be realized by an OR-AND-OR three-level network by using at most $2^{r+1}+1$ gates. Second, we prove that, regardless of the number of levels, we need at least $2^{r+1}(1-\xi)$ gates to realize almost all functions of n variables by multi-level AND-OR networks, where n is a sufficiently large integer, and ξ is an arbitrarily small real positive number ($0 < \xi < 1$). Thus, an AND-OR two-level network requires $2^{n-1}+1$ gates in the worst case, while an OR-AND-OR three-level network requires at most $2^{r+1}+1$ gates, where $r = n/2$. From these facts, we know that there is a distinct difference between AND-OR two-level networks and OR-AND-OR three-level ones, but not so much difference between three-level ones and multi-level ones with more than three levels.

This chapter is organized as follows: In Section 13.2, we will show that an arbitrary n-variable function ($n = 2r$) can be realized with an OR-AND-OR three-level network with at most $2^{r+1} + 1$ gates.

In Section 13.3, we introduce the concept of almost all functions and prove that we need at least $2^{r+1}(1 - \xi)$ gates to realize almost all functions of n variables by multi-level AND-OR networks.

In Section 13.4, we show the number of gates to realize various functions obtained by computer experiments.

13.2 UPPER BOUND ON THE NUMBER OF GATES

In this section, we derive an upper bound on the number of gates to realize an arbitrary n-variable function by OR-AND-OR three-level networks.

Theorem 13.2.1 *An arbitrary n-variable function ($n = 2r$, where r is an integer) can be realized by an OR-AND-OR three-level network with at most $2^{r+1} + 1$ gates.*

(Proof) An arbitrary n-variable function is represented by

$$f(x_1, x_2, \ldots, x_n) = \bigvee_{\boldsymbol{a}=(0,0,\ldots,0)}^{(1,1,\ldots,1)} g_{\boldsymbol{a}}(x_1, x_2, \ldots, x_k) x_{k+1}^{a_{k+1}} x_{k+2}^{a_{k+2}} \cdots x_n^{a_n}, \quad (13.2.1)$$

where $1 < k < n$, and $a_i = 0$ or 1 $(i = k+1, k+2, \ldots, n)$, $x_i^0 = \bar{x}$, $x_i^1 = x_i$, $g_{\boldsymbol{a}}(x_1, x_2, \ldots, x_k) = f(x_1, x_2, \ldots, x_k, a_{k+1}, a_{k+2}, \ldots, a_n)$, and $\boldsymbol{a} = (a_{k+1}, a_{k+2}, \ldots, a_n)$. Note that the symbol $\vee$ denotes the logical sum, and it is taken for 2^{n-k} different combinations. $g_{\boldsymbol{a}}(x_1, x_2, \ldots, x_k)$ can be represented by a canonical product-of-sums expression:

$$g_{\boldsymbol{a}}(x_1, x_2, \ldots, x_k) = \bigwedge_{\boldsymbol{b}=(0,0,\ldots,0)}^{(1,1,\ldots,1)} [g_{\boldsymbol{a}}(b_1, b_2, \ldots, b_k) \vee x_1^{\bar{b}_1} \vee x_2^{\bar{b}_2} \vee \cdots \vee x_k^{\bar{b}_k}],$$

where $g_{\boldsymbol{a}}(b_1, b_2, \ldots, b_k)$ is a binary constant, $\boldsymbol{a} \in B^{n-k}$, $\boldsymbol{b} = (b_1, b_2, \ldots, b_k)$, $b_j \in B (j = 1, 2, \ldots, k)$, and $B = \{0, 1\}$. The symbol $\wedge$ denotes logical product, and it is taken for 2^k different combinations. A network that generates all the maxterms of k variables is called a k-bit decoder. By using a k-bit decoder and 2^{n-k} AND gates, we can realize 2^{n-k} functions as follows: $g_{\boldsymbol{a}}(x_1, x_2, \ldots, x_k)x_{k+1}^{a_{k+1}} x_{k+2}^{a_{k+2}} \cdots x_n^{a_n}$, where $a_i = 0$ or 1 for $i = k+1, \ldots, n$. By summing these functions by an OR gate, we have a network for the function f. The designed network contains 2^k OR gates for the k-bit decoder, 2^{n-k} AND gates, and one OR gate. So, the total number of gates is $\alpha(n) = 2^k + 2^{n-k} + 1$. Let $k = n/2$, and we have $\alpha(n) = 2^{r+1} + 1$. Hence, we have the theorem. □

Note that in the above proof, the value of k was chosen so that $2^k + 2^{n-k} + 1$ takes its minimum.

Example 13.2.1 *The function shown in Fig. 13.2.1 can be represented as* $f = g_{(0,0)}\bar{x}_3\bar{x}_4 \vee g_{(0,1)}\bar{x}_3 x_4 \vee g_{(1,0)} x_3\bar{x}_4 \vee g_{(1,1)} x_3 x_4$, *where* $g_{(0,0)}$, $g_{(0,1)}$, $g_{(1,0)}$, *and* $g_{(1,1)}$ *are represented by canonical product-of-sums expression :*

$$
\begin{aligned}
g_{(0,0)} &= x_1x_2 \vee \bar{x}_1\bar{x}_2 &&= (\bar{x}_1 \vee x_2)(x_1 \vee \bar{x}_2),\\
g_{(0,1)} &= x_1 \vee \bar{x}_2 &&= (x_1 \vee \bar{x}_2),\\
g_{(1,0)} &= \bar{x}_1 &&= (\bar{x}_1 \vee x_2)(\bar{x}_1 \vee \bar{x}_2), \text{ and}\\
g_{(1,1)} &= x_1\bar{x}_2 \vee \bar{x}_1 x_2 &&= (\bar{x}_1 \vee \bar{x}_2)(x_1 \vee x_2).
\end{aligned}
$$

Therefore, the given function is represented by

$$
\begin{aligned}
f &= (\bar{x}_1 \vee x_2)(x_1 \vee \bar{x}_2)\bar{x}_3\bar{x}_4 \vee (x_1 \vee \bar{x}_2)\bar{x}_3 x_4\\
&\quad \vee (\bar{x}_1 \vee x_2)(\bar{x}_1 \vee \bar{x}_2)x_3\bar{x}_4 \vee (\bar{x}_1 \vee \bar{x}_2)(x_1 \vee x_2)x_3x_4.
\end{aligned}
$$

Fig. 13.2.2 shows the OR-AND-OR three-level network for this function. Note that this network uses $\alpha(n) = 2^k + 2^{n-k} + 1$ *gates, where* $n = 4$, *and* $k = 2$. ■

Theorem 13.2.2 *Let* $(n + \log_2 m)/2$, $n/2$ *and* $m^{1/2}$ *be integers. Then, an arbitrary n-input m-output function can be realized by an OR-AND-OR three-level network with at most* $m^{1/2} \cdot 2^{(n/2)+1} + m$ *gates.*

(Proof) Consider a set of m expressions of the form (13.2.1). Consider the use of k-bit decoders, as shown in Example 13.2.1. Then, the total number of AND gates is at most $m \cdot 2^{n-k}$. We need m OR gates for m outputs. Therefore, the total number of gates is

$$m \cdot 2^{n-k} + 2^k + m \tag{13.2.2}$$

Let $k = (n + \log_2 m)/2$, and (13.2.2) becomes $m^{1/2} \cdot 2^{(n/2)+1} + m$. Hence, the theorem. □

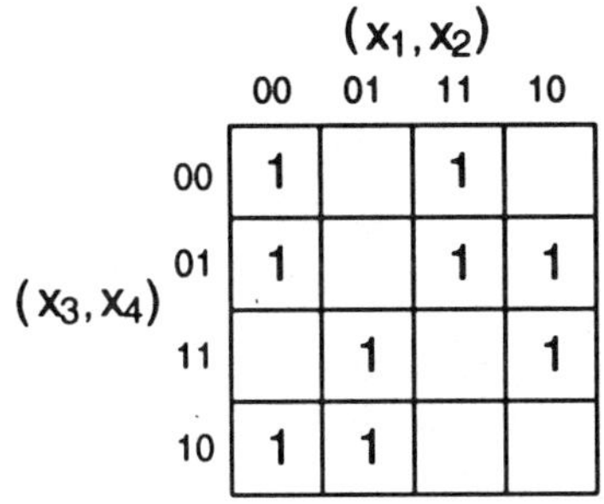

Figure 13.2.1 Example 13.2.1.

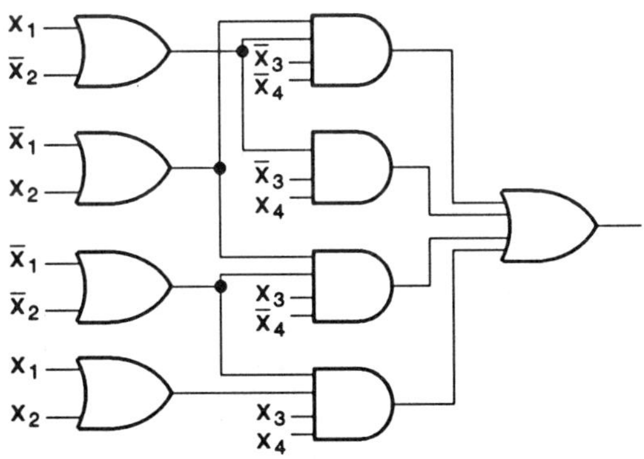

Figure 13.2.2 Example 13.2.1.

Again, in the above proof, the value of k was chosen so that $m2^{m-k} + 2^k + m$ takes its minimum.

13.3 LOWER BOUND ON THE NUMBER OF GATES

Among n-variable functions, it is well known that the parity function of n variables is the most complex ones to realize in an AND-OR two-level network. In fact, the parity functions of n variables require $2^{n-1}+1$ gates, and other functions require the same number of gates or less. So, upper bound on the number of gates to realize arbitrary function of n variables in an AND-OR two-level network is $2^{n-1}+1$. Then, the question arises how many gates are necessary to realize an arbitrary function of n variables by multi-level networks. To answer this question, we have to find an n-variable function requiring the largest number of gates in a multi-level realization. Unfortunately, it is extremely difficult to find such a function. This is partly due to the fact that the problem of realizing a particular function by a multi-level network using the least number of gates is extremely difficult. So, we use the concept of almost all functions, which was used to derive the complexity of series-parallel relay networks [8, 14].

Definition 13.3.1 *A property* $\boldsymbol{A}$ *is said to hold for* **almost all functions** *if the proportion of* n*-variable functions that do not satisfy* $\boldsymbol{A}$ *tends to zero as* $n \to \infty$*. In other words, let* $w(n)$ *be the number of* n*-variable functions that do not satisfy* $\boldsymbol{A}$*. Then* $w(n)/2^{2^n} \to 0$ *as* $n \to \infty$*.*

From here, we will show that we need at least $2^{r+1}(1-\xi)$ gates to realize almost all function of n variables, where $n = 2r$, and r is an integer, and $0 < \xi < 1$.

We will not show that any particular function requires $2^{r+1}(1-\xi)$ gates, but we will show that it is impossible for almost all functions to require less. In other words, there do not exist enough networks to represent almost all n-variable function with less than $2^{r+1}(1-\xi)$ gates. To enumerate the number of meaningful networks, we will define a class of AND-OR multi-level networks that do not contain a certain type of connections.

Definition 13.3.2 *A multi-level network consisting of AND and OR gates is said to be* **normal** *if no output of any AND (OR) gate is connected to the input of an other AND (OR) gates.*

Lemma 13.3.1 *An arbitrary AND-OR multi-level network can be converted into a normal one without increasing the number of gates.*

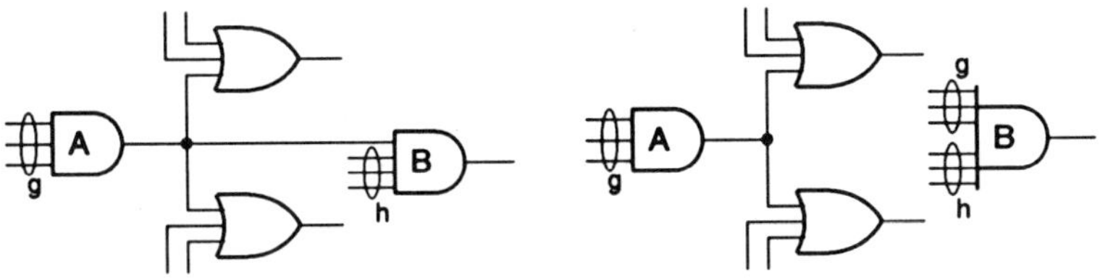

(a) (b)

Figure 13.3.1 Transformation to normal AND-OR network.

(Proof) If the output of an AND gate $\boldsymbol{A}$ is connected to an input of an other AND gate $\boldsymbol{B}$ (Fig. 13.3.1(a)), then delete the connection between $\boldsymbol{A}$ and $\boldsymbol{B}$, and connect all the input signals of $\boldsymbol{A}$ to the input of $\boldsymbol{B}$ in addition to the original signals (Fig. 13.3.1(b)). Apply this operation to all AND gates. Apply a similar operation to all OR gates. And, we have a normal AND-OR multi-level network. It is clear that these operations do not increase the number of gates. □

Lemma 13.3.2 *Let $\mu(n, m, N)$ be the number of different n-input m-output normal AND-OR multi-level networks with at most N gates. Then*

$$\mu(n, m, N) \leq (3^{nN}) \cdot (2^{N^2/4}).$$

(Proof) Consider normal AND-OR multi-level networks with p levels. Assume that all the output functions are obtained from OR gates. Suppose that p is an odd number. In this case, the first-level gates (those which are farthest from the outputs) are OR gates, the second-level ones are AND gates, third-level ones are OR gates, ... , the $(p-1)$-th ones are AND gates, and the p-th (output) ones are OR gates, as shown in Fig. 13.3.2. Let a_i be the number of i-th level gates $(i = 1, 2, \ldots, p)$. Because the total number of gates is at most N, we have $\sum_{i=1}^{p} a_i \leq N$.

First, consider the number of different patterns of connections in an OR gate in the first level. To each OR gate, either a true variable is connected, the complement is connected, or neither of them is connected, for each variable x_i. So, three possible cases exist for each variable x_i. Therefore, the number of different patterns of connections in an OR gate in the first level is 3^n.

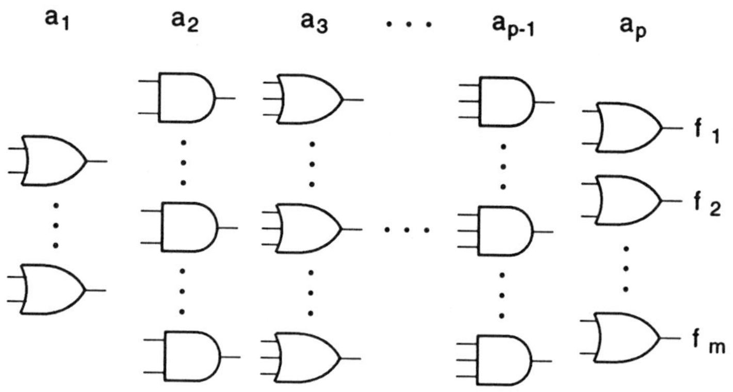

Figure 13.3.2 Normal AND-OR multi-level network.

Second, consider the number of different patterns of connections in an AND gate in the second level. To each AND gate, either x_i or $\bar{x}_i$ or none of them is connected for each x_i. Also, either an output of an OR gate in the first level is connected or unconnected. Therefore, the number of different patterns of connections in an AND gate in the second level is $3^n \cdot 2^{a_1}$.

Third, consider the number of different patterns of connections in an OR gate in the third level. To each OR gate, either x_i or $\bar{x}_i$ or none of them is connected for each x_i. Also, either an output of an AND gate in the second level is connected or unconnected. Therefore, the number of different patterns of connections in an OR gate in the third level is $3^n \cdot 2^{a_2}$.

Lastly, consider the number of different patterns of connections in an OR gate in the p-th level (output). To each output OR gate, either the outputs of the AND gates in the 2nd, 4-th, ... , and $(p-1)$-th level are connected or unconnected, in addition to the patterns of an OR gate in the first level. Therefore, the number of different patterns of connections in an output OR gate is $3^n \cdot 2^{a_2+a_4+\cdots+a_{p-1}}$.

So, the numbers of different patterns of connections in the gates are:

First level	3^n
Second level	$3^n \cdot 2^{a_1}$
Third level	$3^n \cdot 2^{a_2}$
Fourth level	$3^n \cdot 2^{a_1+a_3}$
Fifth level	$3^n \cdot 2^{a_2+a_4}$
Sixth level	$3^n \cdot 2^{a_1+a_3+a_5}$
$\vdots$	
$(p-1)$-th level	$3^n \cdot 2^{a_1+a_3+\cdots+a_{p-2}}$
p-th level	$3^n \cdot 2^{a_2+a_4+\cdots+a_{p-1}}$

Therefore, the total number of different connections in the network is

$$\begin{aligned}\mu(n,m,N) &= (3^n)^{a_1} \cdot (3^n \cdot 2^{a_1})^{a_2} \cdot (3^n \cdot 2^{a_2})^{a_3} \cdot (3^n \cdot 2^{a_1+a_3})^{a_4} \\ &\quad \cdot (3^n \cdot 2^{a_2+a_4})^{a_5} \cdots (3^n \cdot 2^{a_2+a_4+\cdots+a_{p-1}})^{a_p}. \\ &= 3^{n(a_1+a_2+\cdots+a_p)} \cdot 2^{\{a_1a_2+a_2a_3+(a_1+a_3)a_4+\cdots+(a_2+a_4+\cdots+a_{p-1})a_p\}}\end{aligned}$$

In the above, the expression inside of { } is simplified as follows:

$$\begin{aligned}& a_1a_2 + a_2a_3 + (a_1+a_3)a_4 + (a_2+a_4)a_5 + (a_1+a_3+a_5)a_6 \\ &\quad +(a_2+a_4+a_6)a_7 + \cdots + (a_1+a_3+a_5+\cdots+a_{p-2})a_{p-1} \\ &\quad +(a_2+a_4+\cdots+a_{p-1})a_p \\ &= (a_1+a_3+a_5+\cdots+a_p)\cdot(a_2+a_4+a_6+\cdots+a_{p-1}) \\ &= N_1 \cdot N_2, \text{ where}\end{aligned}$$

$N_1 = a_1 + a_3 + a_5 + \cdots + a_p$ and $N_2 = a_2 + a_4 + a_6 + \cdots + a_{p-1}$. Because $N_1 + N_2 = N$, we have $N_1 \cdot N_2 \le (N_1+N_2)^2/4 = N^2/4$. So, $\mu(n,m,N) \le 3^{nN} \cdot 2^{N^2/4}$. Thus, we have the lemma when p is an odd number. When p is an even number, we can prove the lemma in a similar way. Hence the lemma. □

Definition 13.3.3 *Let $\eta(n,m)$ be the tight upper bound on the number of gates to realize an n-input m-output function by a multi-level AND-OR network.*

Theorem 13.3.1 *For an arbitrary small positive real number ξ $(0 < \xi < 1)$, there exists a positive integer $n(\xi)$ satisfying the following condition: when $n \ge n(\xi)$, $\eta(n,m) > m^{1/2} \cdot 2^{(n/2)+1} \cdot (1-\xi)$.*

(Proof) The number of n-input m-output functions realized by AND-OR multi-level networks with at most N gates does not exceed $\mu(n,m,N)$, the number of normal AND-OR multi-level networks with at most N gates. There are 2^{m2^n} different n-input m-output functions. From here, we will show that if $N = (m)^{1/2} \cdot 2^{(n/2)+1} \cdot (1-\xi)$ then $\mu(n,m,N)/2^{m2^n} < 1$, for sufficiently large n. By taking the logarithm of the left hand side of the inequality, we have

$$\log_2[\mu(n,m,N)/2^{m2^n}] \leq nN(\log_2 3) + N^2/4 - m2^n,$$

where we used the result of Lemma 13.3.2. The right hand side of the above inequality is given by

$$\begin{aligned} & n\{(m)^{1/2} \cdot 2^{(n/2)+1} \cdot (1-\xi)\}(\log_2 3) + m \cdot 2^{n+2} \cdot (1-\xi)^2/4 - m \cdot 2^n \\ = \; & n\{(m)^{1/2} \cdot 2^{(n/2)+1} \cdot (1-\xi)\}(\log_2 3) + \{m \cdot 2^n(-2\xi + \xi^2)\} \\ = \; & m \cdot 2^n\{n \cdot (m)^{-1/2} \cdot 2^{-(n/2)+1} \cdot (\log_2 3)(1-\xi) - \xi(2-\xi)\} \\ = \; & m \cdot 2^n[2 \cdot A(1-\xi) - \xi(2-\xi)], \end{aligned} \tag{13.3.1}$$

where $A = \dfrac{n \cdot \log_2 3}{m^{1/2} \cdot 2^{n/2}}$.

The right-hand side of (13.3.1) becomes negative when

$$2 \cdot A < \frac{\xi(2-\xi)}{(1-\xi)} = \xi + \xi(1 + \xi^2 + \xi^3 + \cdots) = 2\xi + \xi^2 + \xi^3 + \cdots. \tag{13.3.2}$$

(13.3.2) is satisfied if $A < \xi$. In this case, the logarithm still remains negative. In other words, there exists $n(\xi)$, and if $n > n(\xi)$, then not all the functions can be realized by using at most $N = m^{1/2} \cdot 2^{(n/2)+1} \cdot (1-\xi)$ gates. Hence the theorem. □

Theorem 13.3.2 *In order to realize almost all functions of n variables by an AND-OR multi-level network, $N = (m)^{1/2} \cdot 2^{(n/2)+1} \cdot (1-\xi)$ gates are necessary, where $n = 2r$ and ξ is an arbitrary small positive real number* $(0 < \xi < 1)$.

(Proof) Let ξ be a constant such that $0 < \xi < 1$, then the left hand side of (13.3.1) in Theorem 13.3.1 tends to $-\infty$ as $n \to \infty$. Therefore, $\log[\mu(n,m,N)/2^{m2^n}] \to -\infty$ as $n \to \infty$. In other words, $\mu(n,m,N)/2^{m2^n} \to 0$ as $n \to \infty$. Thus, the fraction of functions realized by N or fewer gates tends to zero as $n \to \infty$. This implies we need at least N gates to realize almost all functions. □

Let $m = 1$ in Theorem 13.3.2, we have the following:

Corollary 13.3.1 *Regardless of the number of levels, AND-OR multi-level networks require at least $2^{r+1}(1-\xi)$ gates to realize almost all functions of n variables, where $n = 2r$ and ξ is an arbitrary small positive real number.*

Theorem 13.2.2 shows that an upper bound on the number of gates to realize a function of $2r$ variables is $2^{r+1}+1$. On the other hand, Corollary 13.3.1 shows that a lower bound on the number of gates is $2^{r+1}(1-\xi)$. Thus, the ratio of the upper and the lower bounds on the number of gates is

$$\text{RATIO} = \frac{\text{lower bound}}{\text{upper bound}} = \frac{2^{r+1}(1-\xi)}{2^{r+1}+1} \simeq 1-\xi.$$

Thus, we can make RATIO $\rightarrow 1$, when n is sufficiently large. Hence, we have the following:

Corollary 13.3.2 *For almost all functions, there is no design method for AND-OR multi-level networks that requires many fewer gates than the one in Theorem 13.2.1.*

Thus, for almost all functions, it is sufficient to consider AND-OR networks with no more than three levels. Hence, three-level OR-AND-OR networks (and their duals) are very important multi-level networks.

Note that there exists some class of functions whose minimum AND-OR multi-level realizations require many fewer gates than the ones obtained by the method in Theorem 13.2.1. Such a class of functions includes parity functions, partially symmetric functions and adders. Of course, the fraction of such functions tends to zero as $n \rightarrow \infty$. Table 13.3.1 compares the numbers of gates to realize various classes of functions [12].

From here, we will show the number of gates to realize almost all functions by AND-OR two-level networks.

Lemma 13.3.3 *The number of different single-output n-variable AND-OR two-level networks with N or fewer gates is at most 3^{nN}.*

(Proof) Consider the number of different patterns of connections in an AND gate. To each AND gate, either a true variable is connected, the complement is connected, or neither of them is connected for each variable. So, three possible

Table 13.3.1 Number of gates to realize various functions.

Function Class	AND-OR	OR-AND-OR
Arbitrary function of n-variables	$2^{n-1}+1$	$2^{r+1}+1$ (UB) $2^{r+1}(1-\xi)$ (LB)
Parity function of n-variables	$2^{n-1}+1$	$k\cdot 2^t+2^{k-1}+1$ $(n=k\cdot t)$ (UB)
Symmetric function of n-variables	$2^{n-1}+1$	$k\cdot 2^t+(t+1)^{k-1}+1$ $(n=k\cdot t)$ (UB)
n-bit Adder	$6\cdot 2^n-3n-4$	n^2+5n+2 (UB)

UB: Upper Bound
LB: Lower Bound

cases exist for each variable. Therefore, the number of different patterns of connections in an AND gate is 3^n. Hence, the total number of different patterns of connections is 3^{nN}.

Theorem 13.3.3 *For sufficiently large n, almost all functions of n variables require at least* $(\log_3 2)\cdot(1-\xi)2^n/n$ *gates in AND-OR two-level networks, where ξ is an arbitrarily small positive real number, and* $0<\xi<1$.

(Proof) The number of n-variable functions realized by AND-OR two-level networks with at most N gates does not exceed the number of different n-variable AND-OR networks with N gates. From here, we will show that if $N=(\log_3 2)\cdot(1-\xi)2^n/n$, then the fraction of functions requiring at most N gates approaches to zero as $n\to\infty$. i.e., $\gamma=3^{nN}/(2^{2^n})\to 0$. By taking the logarithm of γ, we have

$$\log_2\gamma = nN(\log_2 3)-2^n=(\log_3 2)(\log_2 3)(1-\xi)2^n-2^n=-\xi 2^n.$$

Therefore, $\gamma\to 0$ as $n\to\infty$. Hence, we have the theorem. □

From Theorems 13.2.1 and 13.3.3, we see that OR-AND-OR networks require many fewer gates than AND-OR networks to realize almost all functions.

Table 13.4.1 Number of gates to realize arithmetic circuits.

Input Data			# of gates		B/A	CPU
Name	IN	OUT	two-level (A)	three-level (B)	(%)	SEC
adr4	8	5	80	38	48	5.5
inc8	8	9	46	41	89	3.7
log8	8	8	137	85	62	42.3
mlp4	8	8	134	84	63	25.2
nrm4	8	5	125	77	62	30.4
rdm8	8	8	84	54	64	7.1
rot8	8	5	62	48	77	9.3
sqr8	8	16	196	123	63	60.2
wgt8	8	4	259	49	19	20.5

IN: Number of inputs
OUT: Number of outputs
CPU: SUN SPARC Server 1000

13.4 EXPERIMENTAL RESULTS

We developed an algorithm to obtain OR-AND-OR networks with near minimum number of gates [11]. We coded it in FORTRAN and implemented it on a Sun SPARC Server 1000. We designed OR-AND-OR three-level networks, and compared the number of gates with AND-OR two-level ones. Table 13.4.1 shows the number of gates for 9 arithmetic networks. This results shows that the OR-AND-OR networks require, on the average, 40% fewer gates than AND-OR networks. For example, to realize wgt8 (which is also called as rd84), 259 gates are necessary for a two-level network, but only 49 gates for a three-level network.

We also designed other benchmark functions by OR-AND-OR networks. Table 13.4.2 shows the selected results, where three-level networks require many fewer gates than two-level ones. Especially for cordic, ryy6 and t481, three-level OR-AND-OR networks require only 9 percents of the gates for two-level AND-OR networks.

We also realized randomly generated functions of 10-variables. Table 13.4.3 shows average numbers of gates to realize functions with various numbers of

Table 13.4.2 Number of gates to realize various functions.

Function Name	IN	OUT	# of gates two-level (A)	# of gates three-level (B)	B/A (%)	CPU (sec)
apex5	117	88	1176	611	52	7647
bc0	26	11	188	149	79	129
chkn	29	7	148	105	71	144
cordic	23	2	916	81	9	3152
ex4	128	28	307	208	68	3178
ex7	15	5	124	67	54	23
ibm	48	17	190	159	84	133
in2	19	10	145	111	77	62
intb	15	7	638	239	37	614
misex3	14	14	710	204	29	1465
misg	56	23	92	76	83	57
ryy6	16	1	113	10	9	21
seq	41	35	385	229	59	314
soar	83	94	451	360	80	708
sym9	9	1	86	35	41	35
t481	16	1	482	42	9	276
tial	14	8	595	255	43	840
vg2	25	8	118	81	69	36
x1dn	27	6	116	74	64	43
x7dn	66	15	553	300	54	398
x9dn	27	7	127	81	64	44

IN: Number of Inputs
OUT: Number of Outputs
CPU: SUN SPARC Server 1000

minterms. Except for the cases when the number of the minterms is 128, all the three realizations required $2^{r+1} + 1 = 65$ gates, where $r = 5$.

Prof. Muroga's group obtained all the optimum AND-OR multi-level networks for functions of four or fewer variables [2]. Their optimality means minimization of the number of gates as the primary objective and the number of connections as the secondary objective, regardless of the number of levels. They obtained optimum networks for 222 representative functions of NPN-equivalence classes of four or fewer variables by the branch and bound method. Their result is very interesting: Of all the 222 representative functions, only two representative functions have optimal networks exclusively of four levels. No representative

Table 13.4.3 Number of gates to realize randomly generated functions of 10 variables.

# of minterms	# of gates: two-level (A)	# of gates: three-level (B)	B/A (%)
128	87.1	64.0	73
256	134.8	65.0	48
384	155.9	65.0	42
512	159.6	65.0	41
640	153.5	65.0	42
768	130.9	65.0	50
896	92.4	65.0	70

Average of 10 functions.

functions have optimum networks of five or more levels. In other words, in the case of functions with four or fewer variables, most optimum networks are three or fewer levels. In the case of three-variable functions, all the optimum networks are at most three levels.

As for the number of connections, we need $(n+1)2^{n-1}$ connections in the AND-OR two-level networks, while need only $2^n + n \cdot 2^{n/2}$ connections in the three-level networks [12].

13.5 CONCLUSION AND COMMENTS

In this chapter, we considered the number of gates to realize arbitrary function by OR-AND-OR three-level networks under the condition that fan-in and fan-out of the gates are unlimited. We showed that there exists a distinct difference between multi-level realizations and AND-OR two-level realizations. In two-level realization, we need $2^{n-1}+1$ gates to realize arbitrary n-variable functions, while in multi-level realization, we need $2^{(n/2)+1}(1-\xi)$ gates. We have the following results for almost all functions:

1. OR-AND-OR networks require many fewer gates than AND-OR two-level networks. However, we cannot expect drastic reduction of gates by using four or more level AND-OR networks.

2. No algorithm produces networks with many fewer gates than one in this chapter.

We developed a heuristic algorithm for OR-AND-OR three-level networks, designed many arithmetic and control networks, and compared the number of gates for two level networks with three-level ones. In the case of the arithmetic networks, three-level realizations required 40% fewer gates than two-level ones. In the case of other benchmark functions, three-level ones required up to 91% fewer gates.

In this chapter, we assume that each gate has unlimited fan-in. As an example of networks where this assumption holds is a NAND array shown in Fig. 13.5.1. This is a programmable logic device (PLD) commercially available [15]. To realize a given function, first, design a three-level OR-AND-OR network and then transform it into a NAND three-level network. The NAND three-level network can be obtained by properly programming the NAND array. In this case, the fan-in of each gate is sufficiently large.

However, we cannot apply the present method to the networks where each gate has fan-in limitation, such as ones in gate array LSIs. Most multi-level synthesis algorithms assume that each gate has fan-in limitation [4]. In such a case, the number of gates necessary to realize an n-variable function is proportional to $2^n/n$ [7].

Karpovsky [4] considered a design method for AND-AND-OR-OR multi-level networks. He assumed that the cost of an r-input gate is $r-1$, and tried to reduced the cost of the networks. He showed that multi-level implementations are on the average, 60 percents of the cost of AND-OR two-level implementations. Note that if the gates have no fan-in limitations, then an AND-AND-OR-OR network can be converted into an AND-OR one without increasing the number of gates (this was be shown in Lemma 13.3.1). Thus, if the costs of all the gates are the same and do not have fan-in limitations, then the cost of the AND-AND-OR-OR network cannot be smaller than AND-OR two-level networks.

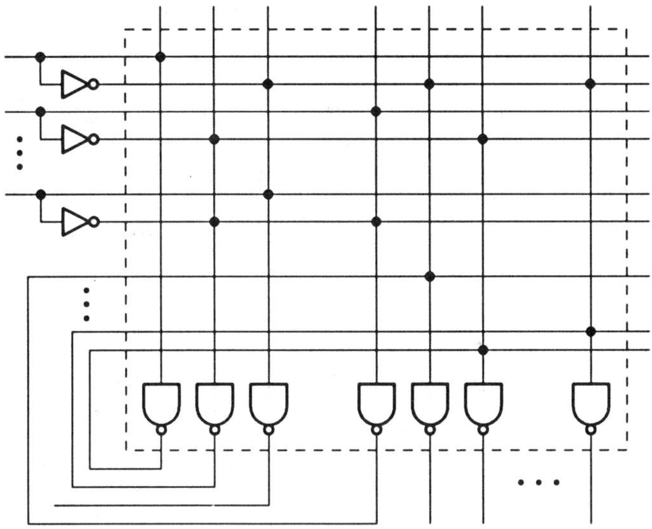

Figure 13.5.1 A PLA consisting of a NAND array.

Acknowledgments

Prof. J. T. Butler and Prof. I. Wegener carefully reviewed this chapter. The author thanks to M. Higashida, who developed the minimization program. [1] and [3] were mentioned by I. Stojmenovic and O. Watanabe, respectively.

REFERENCES

[1] B. Brustmann and I. Wegener, "The complexity of symmetric functions in bounded-depth circuits," *Information Processing Letters 25*, pp. 217-219, 1987.

[2] J. N. Culliney, M. H. Young, T. Nakagawa, and S. Muroga, "Results of the synthesis of optimal networks of AND and OR gates for four-variable switching functions," *IEEE Trans. on Comput.*, Vol. C-27, No. 1, pp. 76-85, Jan. 1979.

[3] J. T. Hastad, *Computational Limitations for Small-Depth Circuits*, The MIT Press, Cambridge, 1989.

[4] M. Karpovsky, "Multilevel logical networks," *IEEE Trans. on Comput.*, Vol. C-36, No. 2, pp. 215-226, Feb. 1987.

[5] E. L. Lawler, "An approach to multilevel Boolean minimization," *Journal of ACM*, Vol. 11, No. 3, pp. 283-295, July 1964.

[6] E. J. McCluskey, *Introduction to the Theory of Switching Circuits*, McGraw-Hill, 1965.

[7] D. E. Muller, "Complexity in electronic switching circuits," *IRE Trans. Elec. Computers*, EC-5, 1, pp. 15-19, 1956.

[8] C. A. Papachristou, "Characteristic measure of switching functions," *Information Sciences*, 13, pp. 51-75, 1977.

[9] T. Sasao, "Multiple-valued decomposition of generalized Boolean functions and the complexity of programmable logic arrays," *IEEE Trans. Comput.*, Vol. C-30, pp. 635-643, Sept. 1981.

[10] T. Sasao, "Input variable assignment and output phase optimization of PLA's," *IEEE Trans. Comput.*, Vol. C-33, No. 10, pp. 879-894, Oct. 1984.

[11] T. Sasao and M. Higashida, "A design method for three-level logic circuits," (in Japanese), *The Technical Papers of IEICE Japan*, VLD88-84, Dec. 1988.

[12] T. Sasao, "On the complexity of networks realized by fiber-optic logic elements," *SPIE's Technical Symposium on High Power Lasers and Optical Computing, Conference 1215*, Los Angeles, Jan. 17, 1990 (invited talk).

[13] J. E. Savage, *The Complexity of Computing*, John Wiley & Sons, 1976.

[14] C. E. Shannon, "The synthesis of two-terminal switching circuits," *Bell Syst. Tech. J.* 28, 1, pp. 59-98, 1949.

[15] Signetics, *Designing with Programmable Macro Logic*, 1987.

[16] I. Wegener, *The Complexity of Boolean Functions*, John Wiley & Sons, Stuttgart, 1987.

EXERCISE

(E) means that the exercise is easy.
(M) means that the exercise is of medium difficulty.
(D) means that the exercise is difficult.

Chapter 1

1.1. (E) Let $H(u,v)$ be an arbitrary switching function of two variables. Show that $H(f,g) = \bar{x}H(f_0, g_0) \vee xH(f_1, g_1)$, where $f = \bar{x}f_0 \vee xf_1$, and $g = \bar{x}g_0 \vee xg_1$.

1.2. (E) Assume that the ordering of the variables is $x_1, x_2, \ldots, x_{2n}$. Show that the number of nodes in the ROBDD for $f = x_1x_{2n} \vee x_2x_{2n-1} \vee \cdots \vee x_nx_{n+1}$ is 2^{n+1}.

1.3. (E) Draw the ROBDDs for $f = xy \oplus xz \oplus xw \oplus yz \oplus yw \oplus zw$ and $g = x \oplus y \oplus z \oplus w$.

1.4. (E) The full adder is a three-input two-output network generating two functions: $s = x \oplus y \oplus z$ and $c = xy \oplus yz \oplus zx$. Represent the full adder by an MTBDD and a shared BDD (BDD vectors).

1.5. (E) Represent the function $f = \bar{x}_1\bar{x}_2\bar{x}_3$ by an ROBDD and a ZBDD. Derive the sizes of DDs for $f = \bar{x}_1\bar{x}_2 \ldots \bar{x}_n$

1.6. (E) Let $f = x_1x_2x_3 \vee (\bar{x}_1 \vee \bar{x}_2 \vee \bar{x}_3)$. Represent f by an ROBDD and an ROBDD with complemented edges.

1.7. (E) Let $f = \bar{x}\bar{y}a \vee \bar{x}yb \vee x\bar{y}c \vee xyd$. Represent f by an ROBDD. Find the ordering that minimizes the size of the BDD. Find the ordering that maximizes the size of the BDD.

1.8. (E) Consider the three-valued input two-valued output function shown in Table E.1. Represent f by an MDD.

Table E.1:

x	y	f
0	0	0
0	1	1
0	2	1
1	0	1
1	1	1
1	2	0
2	0	1
2	1	0
2	2	1

Chapter 2

2.1. (E) Let $f = x_1x_2x_3 \vee \bar{x}_1\bar{x}_2\bar{x}_3$. Draw the Shannon tree and the positive Davio tree.

2.2. (E) Let $n = 2r$, and $f = \bar{x}_1\bar{x}_2 \ldots \bar{x}_n \vee x_1x_2 \cdots x_n$. Prove that the minimum FPRM requires $2^{r+1} - 2$ products. Prove that the minimum PSDRM requires n products.

2.3. (E) Let $f = x_1x_2 \vee x_3x_4 \vee \cdots \vee x_{n-1}x_n$, where $n = 2r$. Show that the PPRM requires $2^r - 1$ products.

2.4. (M) Show that an arbitrary symmetric function of $n = 2r$ variables can be represented by an ESOP with at most $2 \cdot 3^{r-1}$ products.

2.5. (M) Find the following switching functions f and g: 1) The size of the BDD for f is smaller than that of the FDD. 2) The size of the BDD for g is greater than that of the FDD.

2.6. (M) Find a switching function f such that the size of the minimum KDD for f is smaller than that of the BDD and the FDD.

2.7. (M) Find a switching function f such that the size of the minimum PKDD for f is smaller than that of the minimum KDD.

2.8. (D) Let $f = x_1y_1 \vee x_2y_2 \vee \cdots \vee x_ny_n$. Show that any ESOP for f requires at least $2^n - 1$ products to represent f.

Chapter 3

3.1. (E) Let $f = \bar{x}_1\bar{x}_2\bar{x}_3$. Obtain the Reed-Muller spectrum, the algebraic spectrum, and the Walsh spectrum.

3.2. (E) Suppose that the positive Davio expansion is used for x_1, the negative Davio expansion is used for x_2, and the Shannon expansion is used for x_3. Derive the transform matrix $Q(3)$.

3.3. (E) Let $f = \bar{x}_1\bar{x}_2\bar{x}_3$. Represent f by an arithmetic expression.

3.4. (E) Represent the full adder by an ACDD (a BMD).

3.5. (E) Find a switching function whose WDD is smaller than the BDD.

3.6. (E) Let f be an arbitrary switching function. Show that the FDD is not larger than the ACDD (BMD).

3.7. (E) Find a function whose ACDD is smaller than the corresponding BMD.

3.8. (E) Find a function whose ACDD is larger than the corresponding BMD.

3.9. (M) Prove the followings: In the $(1, -1)$ coding, the Walsh spectral coefficients for an n-variable function (scaled by 2^n) are even numbers in the range -2^n to 2^n, and the sum of the coefficients is 2^n.

3.10. (M) Is there any switching function where the algebraic expression requires fewer products than the SOP? If yes, show an example.

3.11. (M) Show the following: To represent an n-bit adder, an MTBDD requires $O(2^n)$ nodes, while a BMD (ACDD) requires $O(n)$ nodes.

3.12. (M) Suppose that the BDD reduction rules are used to simplify BDDs and FDDs. Show that the average sizes of the BDDs and the FDDs to represent switching functions are the same.

3.13. (M) Prove that the size of the WDD to represent an n-bit adder is $O(n)$.

3.14. (M) Prove that the size of the WDD to represent an n-bit multiplier is $O(n^2)$.

3.15. (M) Prove that the size of the WDD to represent X^k is $O(n^k)$, where X is an n-bit number.

3.16. (D) Enumerate numbers of distinct non-zero terminal nodes in MTBDDs for the n-bit multipliers up to $n = 8$.

Chapter 4

4.1. (E) Find a switching function f whose HDD is smaller than the BDD.

4.2. (M) Discuss the merit of BMDs over BDDs.

4.3. (M) How many linearly independent (2×2) matrices are there? Consider the following three cases:

1) Elements are 0 or 1.
2) Elements are 1 or -1.
3) Elements are -1, 0 or 1.

4.4. (D) Enumerate the number of matrices that satisfy the following conditions:

1) 4×4,
2) elements are 0 or 1, and
3) linearly independent.

Chapter 5

5.1. (E) Represent a full adder by an EVBDD.

5.2. (M) Suppose that we have to represent integer valued functions. Discuss the merit of EVBDDs over MTBDDs and BDD vectors.

5.3. (M) Show a class of functions whose EVBDDs are much greater than the corresponding BMDs.

5.4. (D) Is there any class of functions whose EVBDDs are much smaller than the corresponding BMDs?

5.5. (D) Is there any class of functions whose EVBDDs are much smaller than the corresponding BDD vectors?

Chapter 6

6.1. (E) Prove Theorem 6.2.3.

6.2. (M) Find a switching function f whose free BDD is smaller than the RO-BDD.

6.3. (M) Discuss the merit of A-transform.

Chapter 7

7.1. (M) Find the following expressions F and G: F is not an FPRM but an RKRO, and G is not a KRO but a GKRO.

7.2. (M) Discuss the merit of K*BMDs over HDDs.

7.3. (M) Find a class of functions whose BMDs are much greater than BDDs.

7.4. (D) Find a class of functions such that the sizes of the *BMDs are much greater than that of BDDs.

7.5. (M) Explain the difference between K*BMD and FEVBDDs.

Chapter 8

8.1. (E) Derive the EXOR ternary decision tree and the EXOR ternary decision diagram for $f = x_1\bar{x}_2 \vee x_3$.

8.2. (E) Derive the weight vector for $f = x_1\bar{x}_2 \vee x_3$.

8.3. (M) Prove the following: The number of nodes to represent an arbitrary function of n variables by a reduced ETDD is $O(3^n/n)$.

8.4. (M) Prove the following: The number of nodes to represent an arbitrary symmetric function of n variables by a reduced ETDD is $O(n^3)$.

Chapter 9

9.1. (M) Discuss the merit of EDDs over BDDs and FDDs.

9.2. (D) Estimate the number of n-variable functions that are partially symmetric.

9.3. (D) Estimate the number of n-variable functions that are monotone increasing.

9.4. (D) Estimate the number of n-variable functions that have disjoint decompositions.

Chapter 10

10.1. (M) Show an algorithm for counting the satisfying assignments in an OBDD with running time $O(|G_f|)$, where $|G_f|$ denotes the size of the given OBDD.

Chapter 11

11.1. (M) Find a function, where the replacement of variables by constants increases the size of the OFDD.

Chapter 12

12.1. (E) Derive the PTDD for $f = x(wy \vee w\bar{z} \vee y\bar{z}) \vee \bar{x}(yz \vee \bar{x}w \vee \bar{z}\bar{w})$.

12.2. (E) Show that for any function f, there exists an ROTDD for f whose size is not greater than the ROBDD.

12.3. (E) Find a sum-of-products expression F, where the ROTDD for F is smaller than that of the ROBDD for the function.

12.4. (E) Discuss applications of QROATDDs.

12.5. (M) Prove the followings: The number of nodes to represent an arbitrary n-variable function by an ROBDD is $O(2^n/n)$.

12.6. (D) Suppose that the order of the input variables is fixed. Is there any class of functions whose TDDs are of polynomial size, while whose BDDs are of exponential size?

Chapter 13

13.1. (E) Derive the OR-AND-OR three-level network for $f = x(wy \vee w\bar{z} \vee y\bar{z}) \vee \bar{x}(yz \vee \bar{x}w \vee \bar{z}\bar{w})$ by using the method shown in Theorem 13.2.1.

13.2. (E) Show that an arbitrary function of n variables is realized by an OR-AND-OR three-level network, where the number of connection is at most $2^n + n2^{n/2}$.

13.3. (M) Show that an arbitrary symmetric function of $n = kt$ variables is realized by an OR-AND-OR three-level network with at most $k2^t + (t+1)^{k-1} + 1$ gates.

13.4. (M) Show that a parity function of $n = kt$ variables is realized by an OR-AND-OR three-level network with at most $k2^t + 2^{k-1} + 1$ gates.

13.5. (M) Show that an n-bit adder is realized by an OR-AND-OR three-level network with at most $n^2 + 5n + 2$ gates.

APPENDIX

A

About the Authors

Bernd Becker studied mathematics and computer science at University of Saarland, Germany, from 1973 to 1982. He received the Dipl.-Math., Dr.rer.nat. and the Dr. habil. degrees in 1979, 1982 and 1988, respectively. Between 1979 to 1988, he was with Sonderforschungsbereich "Electronic Speech Recognition" (79-81), Institute for Computer Science and Applied Mathematics" (81-83) and Sonderforschungsbereich "VLSI Design Methods and Parallelism" (84-88), all at University of Saarland. He was an Associate Professor for "Complexity Theory and Efficient Algorithms" at the J. W. Goethe-University in Frankfurt am Main during 1989-1995. Presently, he is with the University of Freiburg im Breisgau as a Full Professor. His research interests include data structures and efficient algorithms (for circuit design), design, test and verification of VLSI-circuits.

Beate Bollig received the Master degree in Computer Science from the University of Dortmund, Germany, in 1992. Presently, she is a Ph.D. student of Prof. Dr. Ingo Wegener in Dortmund.

Jörg Bullmann received his diploma degree in Computer Science from the University of Karlsruhe, Germany in 1995. He is currently working as research assistant at the University of Tübingen, Germany.

Edmund M. Clarke received a B.A. degree in mathematics from the University of Virginia, Charlottesville, VA, in 1967, an M.A. degree in Mathematics from Duke University, Durham NC, in 1968, and a Ph.D. degree in Computer Science from Cornell University, Ithaca NY, in 1976. After receiving his Ph.D., he taught in the Department of Computer Science, Duke University, for two years. In 1978 he moved to Harvard University, Cambridge, MA, where he was an Assistant Professor of Computer Science in the Division of Applied Sciences. He left Harvard in 1982 to join the faculty in the Computer Science Department at Carnegie-Mellon University, Pittsburgh, PA. He was appointed Full Professor in 1989. In 1995 he became the first recipient of the FORE Systems Professorship, an endowed chair in the School of Computer Science. His interests include software and hardware verification and automatic theorem proving. Dr. Clarke has served on the editorial boards of Distributed Computing and Logic and Computation and is currently an editor-in-chief of Formal Methods in Systems Design. He is on the steering committees of two international conferences, Logic in Computer Science and Computer-Aided Verification. He is a member of the Association for Computing Machinery, IEEE Computer Society, Sigma Xi, and Phi Beta Kappa.

Rolf Drechsler received his diploma and Ph.D. degree from the J. W. Goethe-University in Frankfurt am Main, Germany, in 1992 and 1995, respectively. He is currently working at the Institute of Computer Science at the Albert-Ludwigs-University of Freiburg im Breisgau, Germany. His research interests include logic synthesis, testing, and genetic algorithms.

Masahiro Fujita received all of his academic degrees from the University of Tokyo. He received a B.S. in Electrical Engineering in 1980, an M.S. in Information Engineering in 1982, and a Ph.D. in Information Engineering in 1985. His Ph.D. thesis advisor was Prof. Tohru Moto-oka, chairman of the Japanese Fifth Generation Computer Project. From 1985 to 1993 he was employed as a research scientist by Fujitsu Laboratories in Kawasaki, Japan. Currently, he is manager of VLSI CAD research at Fujitsu Laboratories of America in Santa Clara, CA. He received the Best Research Award from the Information Processing Society of Japan in 1992 and the Sakai Award from the Information Processing Society of Japan in 1994. He has been on program committees for many conferences dealing with digital design. He is an Associate Editor of the IEEE Transactions on Computer-Aided Design and Formal Methods on Systems Design. His primary interest is in Computer-Aided Design for digital systems, especially logic synthesis and verification.

Thilo Harich studies computer science at the University of Frankfurt.

Fumitaka Izuhara received the B.E. degree in Electronic Engineering from Kyushu Institute of Technology, Iizuka, Japan, in 1995. Currently, he is a an M.E. student of Kyushu Institute of Technology, Iizuka, Japan.

Jawahar Jain received his B.E. in Electronics and Communication from the University of Bhopal, India in 1986; M.S. and Ph.D. from the University of Texas at Austin in Electrical and Computer Engineering in May 1989 and January 1993, respectively. He worked jointly as a post-doctoral fellow at the University of Texas at Austin and Texas A&M University in 1992 and 1993, and joined Fujitsu Laboratories of America in 1994 where he is a member of research staff in VLSI CAD Division. His research interests are in areas of analysis of Boolean functions especially in problems of verification, function representation, synthesis, testing etc.

Udo Kebschull received his diploma degree in Computer Science from the University of Karlsruhe, Germany in 1989. From 1989 to 1994, he worked as research assistant at the Forschungszentrum Informatik, Karlsruhe and at the University of Tübingen, Germany. In 1994, he received his Ph.D. degree in Computer Science from the University of Tübingen. Since 1995, he is Professor at the Department of Computer Science at the Fachhochschule Karlsruhe.

Yung-Te Lai received the B.S. degree in Urban Planning from National Cheng Kung University, Taiwan, in 1977, the M.S. degree in Computer Science from San Diego State University in 1982, and the Ph.D. degree in Computer Engineering from University of Southern California in 1993. He is currently a senior CAD engineer of Hitachi Micro Systems, San Jose. His research interests focuses on formal verification and logic synthesis.

Martin Löbbing received the Master degree in Computer Science from the University of Dortmund, Germany, in 1994. Currently, he is a Ph.D.-student of Prof. Dr. Ingo Wegener in Dortmund.

Shin-ichi Minato was born in Japan, on August 30, 1965. He received the B.E., M.E., and D.E. degrees in information science from Kyoto University, Japan, in 1988, 1990, and 1995, respectively. In 1990, he joined NTT (Nippon Telegraph and Telephone) corporation. He is currently a research engineer at NTT LSI Laboratories, Kanagawa, Japan, working on research and development of logic design systems. His current interest includes the representation and manipulation of Boolean functions for logic synthesis and verification. He has published "Binary Decision Diagrams and Applications for VLSI CAD", Kluwer Academic Publishers, 1995. He is a member of IEEE.

Claudio Moraga received his B.Sc. from the Catholic University of Valparaiso, Chile in 1961; his M.Sc. from the Massachusetts Institute of Technology, USA in 1962 and the Ph.D. degree from the Technical University "Federico Santa Maria" of Valparaiso, Chile in 1972. He was with the Catholic University, and the Technical University "Federico Santa Maria" in Chile. 1974-76 he was an Alexander von Humboldt Research Fellow at the University of Dortmund. Later he joined the University of Bremen in Germany. Presently he is Associate Professor at the University of Dortmund, Germany. He has published over 100 papers covering aspects of Multiple-valued Switching Theory, Spectral Techniques, CAD for Micro-electronics and Computational Intelligence. He has been Program Chairman for the IEEE International Symposium on Multiple-valued Logic several times and he is a permanent Member of the Program Committee of the International Conference on Signal Processing (China). He has organized several International Workshops on the Theory and Applications of Spectral Techniques. He is Member of the Editorial Board of "Mathware and Soft Computing" and of "Multiple-Valued Logic".

Massoud Pedram is an Assistant Professor of Electrical Engineering - Systems at the University of Southern California. He received his B.S. degree in Electrical Engineering from the California Institute of Technology in 1986 and his M.S. and Ph.D. degrees in Electrical Engineering and Computer Sciences from the University of California, Berkeley in 1989 and 1991, respectively. Dr. Pedram is a recipient of the National Science Foundation's Research Initiation Award (1992) and the Young Investigator Award (1994). His research has received a number of awards including one ICCD Best Paper Award and a Distinguished Paper Citation from ICCAD. He has served on the technical program committee of a number of conferences and workshops, including the Design Automation Conference. Dr. Pedram is the co-founder and General Chair of the International Symposium on Low Power Design (1995) and the Technical Co-chair of the International Symposium on Low Power Electronics and Design (1996). He has given several tutorials on low power design at major CAD conferences and forums including, ICCAD and DAC. His research interests span many aspects of design and synthesis of VLSI circuits, with particular emphasis on layout-driven synthesis, theory and applications of function graphs, and design for low power. Dr. Pedram is a member of IEEE and ACM.

Tsutomu Sasao received the B.E., M.E., and Ph.D. degrees in Electronic Engineering from Osaka University, Osaka Japan, in 1972, 1974, and 1977, respectively. He was with Osaka University, IBM T. J. Watson Research Center and Naval Postgraduate School in Monterey, California. Currently, he is a Professor of Kyushu Institute of Technology, Iizuka, Japan. He has published six books on switching theory and logical design, including "Logic Synthesis and Optimization," Kluwer Academic Publisher, 1993. He has served Program Chairman for the IEEE International Symposium on Multiple-Valued Logic many times. He organized the International Symposium on Logic Synthesis and Microprocessor Architecture, Iizuka, Japan in 1992, and the IFIP WG 10.5 Workshop on Applications of the Reed-Muller Expansions in Circuit Design (Reed-Muller '95), Makuhari, Japan in 1995. He received the NIWA Memorial Award in 1979, and a Distinctive Contribution Award from IEEE Computer Society MVL-TC in 1987. He is a Fellow of IEEE.

Martin Sauerhoff received the Master degree in Computer Science from the University of Dortmund, Germany, in 1994. Presently, he is a Ph.D. student of Prof. Dr. Ingo Wegener in Dortmund.

Radomir Stanković received the B.E. degree in Electronic Engineering from Faculty of Electronics, University of Nis and M.E., and Ph.D. degrees in Applied Mathematics from Faculty of Electrical Engineering, University of Belgrade. He was with High School of Electotechnics, Nis from 1976 to 1987. Form 1987 to date he is with Faculty of Electronics, Nis. Currently, he is an Associate Professor teaching logic design.

Sarma B. K. Vrudhula (a.k.a. Sarma Sastry) is an Associate Professor of Electrical and Computer Engineering at the University of Arizona in Tucson, AZ. He received the B.Math (Honors) degree from the University of Waterloo, Ont., Canada, in 1976 and his M.S. and Ph.D. degrees in electrical engineering from the University of Southern California in 1980 and 1985, respectively. From 1978 to 1981 he was research assistant at the Information Sciences Institute, in Marina Del Rey, CA, where he worked on microprogram verification. During 1985-1992 he was a member of the faculty at the University of Southern California in the EE-Systems Department. His research interests span a number of areas in design automation and testing of digital systems. These include various optimization problems that arise in VLSI chip layout, logic synthesis, low power design, testability analysis and design for testability, and automatic synthesis of asynchronous designs. His teaches undergraduate and graduate courses in digital systems design, VLSI design, computer architecture, testing and diagnosis and discrete mathematics.

Ingo Wegener received the Diplom, Ph.D. and Habilitation degrees in Mathematics from the Bielefeld University, in 1976, 1978 and 1981, respectively. He was with the Bielefeld University and the Johann Wolfgang Goethe- University at Frankfurt. Presently, he is a Full Professor of the Computer Science Department of the Dortmund University, Germany. He has published five books on complexity theory and algorithms, including "The Complexity of Boolean Functions", Wiley-Teubner, 1987.

Ralph Werchner received his diploma and doctoral degree from the Computer Science Department at the University of Frankfurt in 1991 and 1994, respectively. In his doctoral thesis he investigated the complexity of routing and related problems on the hyper-cube. After a one year postdoctoral stay at the International Computer Science Institute in Berkeley, California, he is now at the Mathematics Department of the University of Frankfurt. His current research interests include theory of parallel computing, combinatorial optimization, and data structures for Boolean functions.

Xudong Zhao received his B.S. and M.S. degrees in Computer science in University of Science and Technology of China in 1985 and 1988, respectively. He has been working on Ph.D. in Computer science in Carnegie Mellon under Professor Edmund M. Clarke since 1991. His main research include verification of hardware systems, theorem proving and efficient representation and manipulation of large matrices and functions.

B

IFIP WG 10.5 Workshop on Applications of the Reed-Muller Expansion in Circuit Design (Reed-Muller '95).
August 27-29, 1995,
Makuhari, Chiba, Japan.

Program Committee

Bernd Becker, University of Freiburg, Germany
Jon T. Butler, Naval Postgraduate School, U.S.A.
Randal E. Bryant, Carnegie Mellon University, U.S.A.
Edmund M. Clarke, Carnegie Mellon University, U.S.A.
Olivier Coudert, Synopsys Inc., U.S.A.
Masahiro Fujita, Fujitsu Laboratory America, U.S.A.
Kiyoshi Furuya, Chuo University, Japan.
Kiyoharu Hamaguchi, Kyoto University, Japan.
Udo Kebschull, Forschungszentrum Informatik (FZI), Germany
Christoph Meinel, University Trier, Germany
Shin-ichi Minato, NTT LSI Lab., Japan
Marek A. Perkowski, Portland State University, U.S.A.
Wolfgang Rosenstiel, University of Tübingen, Germany
Tsutomu Sasao, Kyushu Institute of Technology, Japan
Jonathan M. Saul, Oxford University, U.K.
Fabio Somenzi, University of Colorado, U.S.A.

Local Arrangement Chair

Kazuhiko Iwasaki, Chiba University (Currently, Tokyo Metropolitan Universit

Reviewers other than Program Committee Members:

P. Pal Chaudhuri	Yusuke Matsunaga
Debatosh Debnath	D. Michael Miller
Rolf Drechsler	Claudio Moraga
Bogdan J. Falkowski	Amar Mukherjee
Jordan Gergov	Jon C. Muzio
D. H. Green	Jeffrey Nowlin
Berthold Harking	Hiroyuki Ochi
Mark A. Heap	V. V. Bapeswara Rao
Se June Hong	Kewal Saluja
Stanley Hurst	Endric Schubert
Kazuhiko Iwasaki	Radomir S. Stanković
Mark Karpovsky	Yasuhiko Takenaga
Norio Koda	Anh Tran
Tomasz Kozlowski	Bapriaju Vinnakota
Michael Lightner	Ingo Wegener
Chen-Shang Lin	Koichi Yasuoka

INDEX